EXOTICIZING CONSUMPTION

EXOTICIZING CONSUMPTION

European Drug Cultures, 1670–1740

Edited by
E. C. SPARY
and
JUSTIN RIVEST

University of Pittsburgh Press

Published by the University of Pittsburgh Press, Pittsburgh, Pa., 15260
Copyright © 2025, University of Pittsburgh Press
All rights reserved
Manufactured in the United States of America
Printed on acid-free paper
10 9 8 7 6 5 4 3 2 1

Cataloging-in-Publication data is available from the Library of Congress

Cover art: detail from "Habit d'Apoticaire" (Paris, ca. 1700) by Nicolas de L'Armessin. Collection Michel Hennin. Estampes relatives à l'Histoire de France. Tome 74, Pièce 6569. Credit: Bibliothèque Nationale de France.

Cover design: Melissa Dias-Mandoly

Hardcover: 978-0-8229-4870-4
Paperback: 978-0-8229-6775-0

Publisher: University of Pittsburgh Press, 7500 Thomas Blvd., 4th floor, Pittsburgh, PA 15260, United States, www.upittpress.org
EU Authorized Representative: Easy Access System Europe, Mustamäe tee 50, 10621 Tallinn, Estonia, gpsr.requests@easproject.com

To our children, Oliver, Simon, and Helena

CONTENTS

Acknowledgments
xi

Introduction: The Troubling Exotic
3

E. C. Spary and Justin Rivest

PART I. MEANING AND VALUATION OF THE EXOTIC

1. The Unexotic World: Medical Drugs, Global Exchanges, and Moscow's View of the Foreign, ca. 1690
33

Clare Griffin

2. Galenic Bodies and Jesuit Beans: Consuming Drugs in Manila at the Turn of the Eighteenth Century
57

Sebestian Kroupa

3. Theriac as a Domestication Technology: The Indigenous and the Exotic in Galenic Pharmacy
85

Barbara Di Gennaro Splendore

PART II. MATERIA MEDICA: SUBSTANCES AND TEXTS

4. Competing Medical Substances during an Epidemic: Causes and Consequences of the Interference of Peruvian Bark and Cascarilla, 1720–1740
105

Wouter Klein

5. The Medical Reception of Sassafras in Early Modern English Print
131

Katrina Maydom

6. Accumulation of the Exotic in the *Palestra pharmaceutica* (Madrid, 1706) of Félix Palacios
153

Paula De Vos

PART III. NETWORKS OF COMMODIFICATION

7. A Science of Things: The Jesuits and the Introduction of American Materia Medica in Europe
183

Samir Boumediene

8. Masters of the Exotic? The Stocklists of Parisian Grocers and Apothecaries, 1650–1730
203

E. C. Spary

9. Consuming the Exotic in Eighteenth-Century Switzerland: Laurent Garcin's "Maduran Pills"
231

Alexandra Cook

10. Mercurius in the Storehouse: The Exotic in the Here and Now of Early Modern European Experience
255

Hjalmar Fors

Notes
271

Selected Bibliography of Secondary Sources
337

List of Contributors
347

Index
351

ACKNOWLEDGMENTS

This volume began as a workshop entitled "Consumers of the Exotic: European Commerce and the Consumption of *Materia Medica*, 1670–1730," held at the University of Cambridge April 5–6, 2017. In the first place, the editors would like to thank all the contributors and participants at that workshop. Besides the contributors to this volume, they also wish to thank Irene Fattaciu, Harun Küçük, and Victoria Pickering for their participation. The editors are grateful to *The Recipes Project: Food, Magic, Art, Science, and Medicine* (https://recipes.hypotheses.org/) for publishing a report on the workshop. The workshop itself was funded by Leverhulme Trust Research Grant 2014–289, "Selling Exotic Plant Products in Paris, 1670–1730," on which E. C. Spary was the principal investigator, with the collaboration of Samir Boumediene, Laia Portet, and Justin Rivest. The editors would like to thank Abby McAllister at the University of Pittsburgh Press for her interest in the project, and for all of her work in bringing it to fruition. They would also like to thank the anonymous peer reviewers for their thoughtful feedback and encouragement.

    E. C. Spary owes much to the contributions of the members of the Leverhulme research project—including her coeditor, whose patience and fortitude in persisting with the volume through pandemic, changes of circumstance, and other events has been legendary. She is grateful for the support and love of her sons, Oliver and Simon.

ACKNOWLEDGMENTS

Justin Rivest would like to thank his co-editor for bringing him onboard her Leverhulme project, for her mentoring and support, and for dozens of discursive journeys back in time to the seventeenth century that generally began, innocuously enough, at the King's Parade Caffè Nero. He is also forever grateful to Nathalie Demirdjian for her unfailing support in bringing this volume to completion.

EXOTICIZING CONSUMPTION

INTRODUCTION

THE TROUBLING EXOTIC

E. C. Spary and Justin Rivest

Today France is perhaps the best-known nation of coffee drinkers in the world, while tea is regarded as quintessentially English. What could be more French than the café, or more English than afternoon tea? Such national distinctions have even served as grounds for hypotheses linking the intrinsic properties of particular consumables with national characteristics.[1] Yet there is good reason to question such essentialism, historical sources suggest. For, in the mid-seventeenth century, tea was all the rage at the French court. It was well known that Pierre Séguier (1588–1672), the royal chancellor, particularly supported tea's health-bringing properties. But despite this, tea never entered either the pharmacopoeia or quotidian consumption in France. The craze for this drink lived and died to a great extent with Séguier himself. If the game of attributing particular essential properties to medicines and comestibles based on their location of origin was already commonplace in European medicine, thanks to Renaissance interest in the Galenic and Hippocratic corpora, the historical contingency that allowed tea to become an everyday item in England, but dwindle into a comparative rarity in France, was no consequence of national differences in taste. Rather, it reflects the varying paths along which such materials traveled as they entered European cultures. These routes of entry were dictated, as the chapters in this volume show, neither by national character nor by the intrinsic properties of the substances involved.

Unlike our present-day, more restricted usage of the term "drug," all such materials, serving simultaneously medicinal, pleasurable and dietary purposes, were classified as drugs in many early modern cultures. Through the eyes of the acerbic Paris faculty physician Guy Patin, we can follow tea's rise and fall in order see how a drug like this could mean very different things, not just in England and France, but even in Paris's court and its medical faculty. In 1648, the faculty dean approved the presentation of a doctoral thesis by Philibert Morisset, entitled *Ergo the Chinensium menti confert* ("Thus tea confers mind upon the Chinese"). Patin sneered at both the candidate and his own colleagues, in the first instance for their ignorance of the proper Latinization of "Chinese": it should be spelled *Sinensium* and not *Chinensium*, he moaned to his best friend, Charles Spon in Lyon. "This president only wrote that thesis on this herb, on tea," he went on, "in order to flatter the Chancellor, from whom the reputation of this drug derives, . . . and even those who praise it can't swear to its goodness, since they're unable to assign any good effects to it."[2]

While it might seem obvious to a reader today that tea is not a *medicinal* substance, we should still beware of taking Patin's judgment concerning the medical efficacy of tea as an "expert opinion." For, as his subsequent correspondence shows, his insistence that tea had no medicinal virtue stemmed far more from his general opposition to all non-European drugs, and antagonism towards the court as a space of medical practice, than from any experiments that might have proven or disproven that claim. In fact, Patin probably had difficulty actually getting hold of tea at all in 1640s Paris: its cost was high, supplies were irregular, and the ongoing civil war of the Fronde was disrupting supply routes. In keeping with the humanist principles to which Patin subscribed, his main source of knowledge of tea's purported virtues was not personal experience but books. And, at the time he wrote Spon, he was struggling to find anything at all about the drug in textual sources. This is clear from a letter he addressed to the same friend a fortnight later. Harping on about the correct Latinization of "China"—Dutch and Flemish medical authors were *all* misspelling the word!—Patin added, "I've seen nothing written on tea apart from Jacobus Bontius, we're making fun of it here."[3] Bontius was a medical practitioner working for the Dutch East India Company, a leading European tea advocate, and the author of a recently published book, *De Medicina Indorum libri quatuor* (Franciscus Hackius, 1642), which contains the earliest European image of the tea plant.

INTRODUCTION

FIGURE I.1. Charles Le Brun, *Le Chancelier Séguier* (1660). Musée du Louvre, Paris. In this portrayal of the chancellor's ceremonial reception of Louis XIV in Paris, Séguier's predilection for the "exotic" is evident in the inclusion of parasols, which, as Benjamin Schmidt has shown (*Inventing Exoticism: Geography, Globalism, and Europe's Early Modern World* [Philadelphia: University of Pennsylvania Press, 2015], 241–54), were becoming a generic signifier of the entire extra-European world in precisely these years. Online at https://commons.wikimedia.org/wiki/File:Charles_Le_Brun_-_Pierre_Séguier,_chancelier_de_France_(1655-1661).JPG. Public domain.

Over the next few years, discussion of tea appears to have died down within the Paris medical faculty. It would revive when a second doctoral candidate, Pierre II Cressé, defended the thesis *An Arthritidi Thee Sinensiam?* ("Is Chinese tea suitable for gout?") in 1657. Once again, this was linked to an attempt to capture the patronage of Chancellor Séguier, whose engraved portrait presided over the debate.[4] Patin's correspondence shows he had still not done his homework where this new drug was concerned. His initial response was to scoff to Spon that Cardinal Mazarin, the queen's favorite, was drinking tea to ward off the gout. Only three months *after* this did he finally capitulate, and write to a Dutch correspondent, Johannes Antonides Vander Linden, for help:

> I have another question to submit to you: please teach me what these leaves of a certain Indian plant called tea are. What is this plant, what are its faculties? Many ignorant people are recommending it here and they're abusing rather than using decoctions of it; but in fact, they don't assign any power or virtue to it which is certain or established; and yet it has a number of public criers who praise it at the tops of their voices over all other medicaments. Some say the herb is Indian, others that it's Chinese; some suspect fraud and adulteration, and think that it's not as exotic as it's said to be.[5]

In other words, Patin based his original judgment of tea on a position of near-zero knowledge: not even its region of origin was known to him. In the face of multiple claims by courtly consumers about its miraculous health effects, his only responses were either derision or bewilderment. Also clear is that, at this juncture in tea's historical trajectory in France, it remained a medicinal plant—neither courtiers nor doctors foresaw its eventual stabilization as a recreational beverage in Europe. Either way, Patin perceived tea as a threat, and he actively discouraged a magistrate's wife who was hoping to try it on her dropsy. His reasons for this attitude were, again, not based on anything resembling "clinical evidence" in today's sense, but rather on a general principle that all new drugs were to be avoided. "All my life," he informed Vander Linden, "I have avoided every new and unknown medicament with horror, and abstained from it." To do otherwise, in Patin's view, was to lower the standards of medical practice: "I would be placing myself on a par with the empirics and chymical imposters; may God spare me such a fate."[6] It is this aspect of Patin's response that might strike the modern reader as being most remote from current pharmacological norms of bioprospection coupled with clinical or laboratory research as ways to "develop" new drugs.[7] Yet it may well be that growing demand for tea among his elite clientele had led Patin to a realization that blanket condemnation, coupled with ignorance of this new drug, was not a position of strength.

The influx of foreign drug traders who brought such alien commodities to the Parisian medical marketplace clearly troubled Patin greatly. While shopkeepers who imported tea from the Dutch Republic trumpeted its marvelous qualities, the Paris faculty physician insisted, "I've yet to discover a single one."[8] Indeed, he speculated, it might even be that tea as they knew it in Paris was simply fake: "The whole rumor surrounding this Chinese herb that one calls tea has transmogrified into

INTRODUCTION

a fairytale; in fact, everything certain courtly good-for-nothings and medicasters have proclaimed about its powers is pure fable. . . . Our shopkeepers deny it, but the true tea is not available to us, we'd have to go and find it in China and [in reality] we are replacing it fraudulently with another herb. The glamour of novelty is deceiving the world."[9] So far from seeking out or experimenting upon new drugs, this Paris physician actively opposed them both because they were exotic and because they were new, two qualities he associated with charlatanry. Only thirteen years after his first encounter with tea did he finally cave and ask another physician, with closer ties to its trade routes, for information. The Dutch, thanks to their contacts with Far Eastern cultures that regularly used tea, were far more familiar with this drug than the French; over the period, several Dutch authors would publish on its virtues.

Patin's position as dean of one of Europe's leading medical faculties and a bourgeois citizen of one of its most cosmopolitan cities means that he cannot be dismissed as a lone eccentric. Indeed, he was by no means alone in suspecting tea. The vast distances over which this drug had to travel to reach European shops and homes presented significant difficulties, also mentioned, for example, by the Danish physician Simon Paulli.[10] These problems afflicted all attempts to pump-prime European interest in exotic drugs: what trial outcomes—and on whose bodies—counted as sufficient proof of a drug's medicinal efficacy? How could Europeans be sure they had secured a supply of the real thing? Was it charlatanical even to report on the efficacy of new drugs? As the case of tea shows, new drugs could sharply divide Europeans. Yet, over ensuing decades, numerous such substances would find a foothold in European consuming cultures.

This book is concerned with the process of procuring, introducing, using, trusting, and explaining extra-European drugs in Europe in the decades around 1700: with the people who specialized in accounting for their effects, the people who had access to them, the places and times they were used, and especially the connection between the fabulous world of the "exotic" they represented and the mundanity of actual commerce, dosage, and consumption. These were some of the people who, strand by strand, wove exotic drugs into European culture, crafting hybrid meanings from them that united knowledge from their places of origin with the new uses and meanings they accreted in the locations where they arrived.

In studying these debates over new drugs, historians cannot afford to treat the substances themselves in a deterministic way. The striking thing about Patin's vehement opposition to tea and to other novel and exotic drugs is that it extended to medicinal materials still accorded therapeutic efficacy today, such as Peruvian bark or cinchona, widely known as "Jesuits' bark" for its association with the eponymous missionary order, discussed by Samir Boumediene in this volume. Neither tea nor cinchona would be consigned to the dustbin of history; rather, both came to number among the most heavily imported drugs into Europe. Though tea went on to become associated with politeness, hospitality, and above all Englishness, it continues to be drunk in France as well.[11] Likewise, in modern biomedical science, Peruvian bark continues to be regarded as an effective therapy for fevers: recognized as the main source of the alkaloid quinine in the nineteenth century, it remained a frontline treatment for malaria well into the twentieth.[12] Tea and Peruvian bark thus traveled in different directions, the former ending up as a pleasurable drink, while the latter's medicinal efficacy endures even in modern pharmacology. Over the same time span, the grounds for reaching agreement about the effects of drugs on the body have radically transformed. We cannot afford to assume, therefore, either that physicians had a monopoly over the meaning and uses of exotic drugs, or that the drugs themselves had an intrinsic power to compel cultural consensus as to their effects. To write of "intoxicants" or "psychedelics" is to write with the advantage of hindsight: and the winner's viewpoint may efface what a drug originally meant to a culture upon arrival. Thereby opens up a need for historical explanations of both how and why new substances like this found their way into the pharmacopoeia, the dispensary, or the coffeehouse, and of how they failed to do so. Uptake cannot be treated as an inevitable consequence of availability.[13]

Drugs have proven to be remarkably fertile subjects when it comes to de-naturalizing and de-essentializing our assumptions, offering useful analytical models for the recent material turn in historiography. As Carla Nappi observes for the case of ginseng, "the trans-historical object does not exist. Even if there is a stable material entity that persists over time, its meaning, identity, and thing-ness change sometimes dramatically in different (historical, geographic, epistemic) contexts."[14] To borrow Renata Ago's metaphor, drugs "are solids and yet behave like fluids, taking a shape that is imposed from the outside."[15] This volume records efforts to delineate and understand the processes that imposed "shape" on these substances by treating them as distinctly cultural products. It is precisely the variability of cultural and historical contexts that should lead us to interrogate how and why some substances crossed between cultures, while others did not; why opposition existed, and how it was overcome in some—but only some—cases.[16] For Patin was far from alone in his suspicion of the exotic drugs like tea that were entering the European medical marketplace in the seventeenth and eighteenth centuries. European archives and libraries are full of documents attesting not only to early modern curiosity over new drugs, but also to deep unease about the untested, unverified, unusual, and foreign.[17] Patin's position was fairly common in an ongoing debate over the relative value of ancient and modern knowledge. Should scholars seek to improve European societies by relying upon established authority, or by embracing the new and unknown? Humanist scholars of the former cast, even doctors and naturalists, would reach for books before they conducted trials.[18] Far from being self-evidently valuable, or compelling agreement over their advantages for health and happiness, exotic drugs—from tea to opium—thus often seemed alien: their effects were deemed uncertain, unmeasurable, and even unsuitable for European bodies. To others, by contrast, these substances held promise as new and marvelous cures.

LOCALIZING THE "EXOTIC"

In his original Latin reference to the "exotic" drug tea, Patin used a term applied to characterize non-European animals and plants a half-century before these events, by the Flemish naturalist Charles de l'Écluse—better known as Carolus Clusius—in his *Exoticorum libri decem*. This book juxtaposed New World drugs like tobacco or jalap with familiar Eastern materia medica like ginger, nutmeg, or betel. The transfer of the term "exotic" from Latin into the vernacular accompanied early modern processes of transculturation of the drugs themselves from distant locations into European consumption and scholarship.[19] The chapters in this volume address the problem of exoticism within Europe, but they start with the fundamental recognition that Europe itself is local, and a European perspective is no more central, universal, or modern than any other global vantage point.[20] The spatial relativity of the term "exotic" is evident in its very etymology: the ancient Greek word ἐξωτικός refers to that which is "foreign," from the Proto-Indo-European particle "eghs," Latinized as "ex-," meaning "outside."[21] By definition, therefore, the exotic is that which is *out of place*, and its use in relation to drugs attaches them to specified locales. In European writing on drugs around 1700, the exotic is often simply conflated with the extra-European. As one dictionary of 1704 put it, "an exotic plant is a foreign plant, such as those brought from America and the East Indies, and which do not grow in Europe."[22] Yet the relationship between Europe and the rest of the world was not simple "Eurocentrism." European scholarly culture was built upon texts from the Mediterranean world, many of which attributed sacred power to foreign drugs from the East. Scripture itself invoked the efficacy of many such precious substances—cinnamon, spikenard, gold, frankincense, myrrh. European medical traditions, especially, depended upon Galenic pharmacy texts written at a time when the Roman Empire was conducting long-distance trade in drugs around the world.[23] Christianity attributed great importance to the region we now term the "Middle East," known to Europeans around 1700 as the "Levant": this was both the location of the biblical Garden of Eden, mapped by medieval cartographers, and also the site of long-standing geopolitical contestations over the Holy Land from the Crusades onward.[24] Medieval and early modern scholarship perpetuated views of the East and the "Spice Lands" as regions from which pleasure, fragrance, and healing flowed. For most of the early modern period, Europe itself remained a pharmacological periphery, an "outsider" to more fortunate lands to the East. Elite medicine and cuisine continued to depend, all through the medieval and early modern periods, upon these Eastern drugs with their fabled efficacy and beneficial properties, which adhered to them in consequence of their sacred place of origin, but also in response to perceptions of the Eastern empires—Mughal, Safavid, Ottoman—as cultures of greater luxury and politeness than Europe, whose affluence and cultivation antedated European preoccupations with "civilization" by centuries.[25] The rise of court culture formalized this association between exotic drugs and elite consumption: as their sacred value declined, they retained cultural prestige, spreading from rulers to nobles. The drugs sold in the shops of medieval grocers and apothecaries were commodities of high price and therefore luxury goods. It is hard for the modern reader to imagine a time when cinnamon was literally worth its weight in gold, but that realization gives us some sense of the awe and covetousness these materials evoked among early modern elites.

Thanks to these long associations with sanctity, high-status consumption and alien-yet-exciting lifestyles, exotic drugs and spices were a key driver of European commercial expansion from the sixteenth century. Asiatic consumer goods, including drugs, were as coveted in Europe as was access to their markets, but their trade routes were already populated by other, more experienced and more privileged merchant networks, such as the Armenian and Jewish traders who conveyed drugs to the farthest points of European consumption, or monopolized their trade in many port towns. European merchants, healers, and clients had little power within these Old World drug networks, where trade was controlled by a range of non-European actors. Venetian, Genoan, Portuguese, French, English, and other merchants went cap in hand to local Ottoman or Mughal rulers for trading privileges, courted Siamese kings and Chinese emperors to permit the construction of factories, and paid Armenian caravaneers for safe travel overland with their cargoes of luxury drugs and spices, ivory, gems, and silk.[26] Control of the Levant by these other trading cultures meant that drugs from this part of the world provoked conflicting responses within European societies. Levantine drugs were necessary and delightful; also potential causes of ill health, wasteful expenditure, intoxication, decadence, and immorality. All these things were associated with spatial and cultural "otherness." Early modern medical critiques of exoticism led to the invention of the exotic's converse: the "indigenous" or "domestic."[27] If an "exotic" drug was present in early modern European gardens or cabinets at all, it was often a great rarity, certainly not a product of one's own native polity. In its exotic status—neither local, familiar, nor wholly "known"—lay both its promise and its risk.

Considered from the vantage point of the ancient trade network of the Old World known as the "Silk Roads," Europe was thus merely a terminus, and a comparatively impoverished one to boot. All of this sheds a different light upon "Orientalism." In its early form, European Orientalism took the form of rivalry or emulation, rather than domination, of Eastern societies. Around 1700, travel narrators might flag European superiority over Eastern cultures, but many also expressed awe at Persian or Ottoman learning, cultivation, or military power, or even favorably contrasted the doctrinal unity of Islam with the European Reformation's confessional divides. The European "Occident" was certainly less cultivated, a newer, rawer periphery to older and much more sophisticated Eastern centers with privileged access to miraculous substances. Orientalism thus wore a more ambivalent guise for early moderns, distinct from its later form as famously characterized by

Edward Said.[28] Yet, if Europeans' global agency around 1700 often still boiled down to attempts to hack into existing trade networks, or broker favorable deals with port officials on the Malabar or Red Sea coasts, European expansion in the New World was gradually changing the stakes. This was true not only in terms of European access to new drugs, but also for European involvement in the global drugs trade.[29] In most places colonized by Europeans, from the Mascarenes to Manila, from Lima to Jamaica, drugs whose curative virtues were well-known to local healers became objects of appropriation and/or cultivation. These materials, and knowledge of how to use or grow them, were assimilated into European use and trade by colonizers, captains, missionaries, soldiers, merchants, and healers.[30] In households, marketplaces, universities, mission and trading posts, and military installations, in Europe's mainland as in its colonies, in metropoles as in provinces, these new drugs were subjected to experiment and empirical investigation. They were objects of knowledge, but it was their value as objects of exchange that encouraged enterprising individuals to endeavor to create tastes for new drugs, both in Europe and elsewhere in the world. Over these same centuries, as European colonial empires expanded, curiosity, commerce, and consumer demand came to reshape existing trade networks. European trade in drugs attests not merely to the state of affairs within the European marketplace, but also to the many interconnections between Europe and the rest of the world, the complex chains of social relations and material transactions that gave European consumers access to drugs *qua* global commodities.[31]

This spatial relativism meant that what counted as "exotic" was thus impermanent, subject to broader geopolitical and commercial transformations as well as responsive to changes within European consuming cultures. Over time, and especially during the period addressed by our volume, foreign drugs could also become *de*-exoticized. When drugs were transplanted, some came to be domesticated within European cultures, often through trade and consumption, as in the cases of coffee and tea. Others might become naturalized in a more literal sense, rather than as imported goods. The French druggist Pierre Pomet noted in 1694 that chili pepper was now so widely grown in Languedoc that there were "very few Gardens that lack it, and it even serves to adorn some shops." But, of the three varieties he knew, only one was grown in France: "The others are too acrid, which means that it is only the Savages who make use of these, and are great lovers of them."[32] The

INTRODUCTION

domestication process operated a transformation not only in space but also in character, fitting the chili pepper to what, in Pomet's view, were more refined French tastes. Thus, Eurocentrism had a more than merely spatial dimension: it was also a conquest of meaning, character, and virtues, as Marcy Norton suggests was the case for chocolate and tobacco.[33]

This ongoing transition—at the midpoint of which our volume falls—also marked a transformation in European relations with the "Orient." By the end of the eighteenth century, something much closer to Edward Said's Orientalism was apparent in the way European epistemic violence supported the use of physical force in attempts at Western hegemony within Asia.[34] Setting this volume at the crux of this shift from one Orientalism to another means we can also ask how the shift occurred. For the distinctive relations with Eastern drugs that are evident in medieval and early modern Europe should not be mistaken for a *general* European openness to the exotic, as we have seen. While Old World, "Eastern," drugs were seen as civilizing influences, those from the New World were often portrayed as needing to be "exorcized, sanitized, and civilized" by Europeans. Tobacco, for example, had to be shorn of its idolatrous and superstitious connections to Mesoamerican religious practices in order to be fitted to European tastes.[35] The case of tobacco also illustrates how drugs from the Americas began to offer Europeans something of value for dealing with Ottoman or Chinese rulers, merchants, and clients. These rare substances with new virtues often came from points westward to which Asiatic cultures and merchants had no access. The disruptions caused by such substances sometimes provoked controversy in the receiving culture—in both China and the Ottoman Empire, tobacco gave rise to anti-smoking campaigns—but they were also means by which Europeans could break into Eastern drug markets.[36] By the 1720s, Europeans were selling coffee grown in their American colonies back to the Ottoman Empire, and sage and Canadian ginseng to the Chinese. They re-exported cargoes of New World drugs like jalap or cinchona arriving at Cádiz or Seville, both within Europe and beyond. This was an enterprise that capitalized on the increasing frequency of Atlantic maritime traffic afforded by the slave trade—itself a development of ballooning reverse trade in one particular Levantine drug, cane sugar.[37] By the eighteenth century, many "exotic" drugs were thus becoming profitable products sold *by* as well as *to* the European empires now spreading around the globe.

THE EUROPEAN MEDICAL MARKETPLACE

Exoticizing Consumption addresses some of the ways in which these new drugs imported "from the Americas and the East Indies" disrupted the European medical marketplace. Their vendors and promoters often ran counter to established commercial interests or traditional medical doctrines; their efficacy and even safety were frequently challenged by licensed medical practitioners. Yet the wariness with which exotic drugs were often regarded receives less attention in the work of modern historians than the viewpoint of those who ardently embraced them. Scholars should beware of conferring a power of self-advancement upon the new substances which entered into European consumption in the period 1670–1740, in place of historical inquiries into their acceptance or rejection.[38]

One way of expressing the altered perspective of this book is to ask why early modern people were prepared to put their lives at risk by consuming these strange substances from distant lands, with unknown effects. In the decades around 1700, no central, national, or international system for policing drug production and consumption existed, not even an international botanical classification that might have permitted plant materials in trade to be cross-referenced to their original source. If today's consumers grapple with concerns about the fraudulent substitution of one drug for another, the problems confronting early modern consumers in knowing whom to trust concerning a drug's effects upon themselves were an order of magnitude greater. The uptake of non-European materia medica by European consuming publics did not amount to consensus over their virtues. If we accept that foreign drugs' value was historically contingent rather than biologically inherent, how then was their efficacy constructed over time? The chapters in *Exoticizing Consumption* show that "embedding" a new drug in European cultures was frequently a contested process. "Exoticism" could signify the mystery of distance, and hidden or lost knowledge; but it could also impede commodification, triggering suspicion in place of wonder. Novel substances raised concerns not just about efficacy (did they really work? how could one prove it?) but also about cost. New medicinal materials that were taken up by elites or widely publicized could trigger fashions. The high prices they then commanded prompted quests for cheaper, indigenous *succedanea*—substitute drugs—either to enable wider access to their healing properties, or just to piggyback on the market boom.

INTRODUCTION

The dual status of drugs as both commodities and objects of knowledge thus opens up many historical complexities. When were they seen as "green gold," eagerly consumed by a drug-hungry European populace? And when did they become those expensive, ineffective, "outlandish herbs" decried by a succession of European observers, from Paracelsus in the sixteenth century to Nicholas Culpeper or Guy Patin in the seventeenth? Our volume starts from the assumption that the exotic was an *acquired* taste—that demand for novel drugs was itself a cultural construct, emerging over time from the medieval spice trade up to the first transoceanic European empires in the sixteenth century. The implications of claiming that tastes were constructed by historical processes have generally been neglected in two genres of drug history in particular. Histories of pharmacy have tended to treat the appropriation of new drugs in the teleological manner mentioned above, ascribing their uptake to innate efficacy rather than cultural circumstances. Socioeconomic studies, often subscribing to similar scientific norms, attempt to account for drug consumption in terms of innate propensities for sweetness, or inherent curative or psychoactive attributes.[39] Especially in studies of drugs covering longer timespans, a fine-grained historicism all but falls away: the need to generalize and produce readable overviews necessarily obliterates the rich canvas of interactions and meanings that drugs had in European cultures, both individually and collectively—a phenomenon giving rise to an extensive printed corpus and iconography.

New and fruitful methodological approaches from related disciplines such as anthropology and literary theory have focused scholarly attention upon meaning as well as materiality. We can ask how drugs changed in meaning as they moved *between* cultures; but also, and forming a central theme of this volume, we can ask how their meaning was constructed during transaction rituals of use, experimentation, and exchange *within* a particular culture. Following drugs as they travel between places, peoples, and bodies can offer historians a tracer with which to construct broader social, cultural, and economic relations in the early modern world.[40] One reason why drugs are interesting to the historian is because writing their history necessitates attention to the history of embodiment. Historians such as Pablo F. Gómez and Ralph Bauer have highlighted the transformative effect of alchemical styles of knowledge upon the global drugs trade, and the centrality of new histories of the emotions, senses, and body for writing the history of material culture.⁴¹ Drugs offer "sampling devices" for cultural historians exploring conceptions of materiality and embodiment worldwide. Lastly, drugs generate a revised understanding of experiment and empiricism in the history of science.⁴² In all these senses, we can view drugs as productive substances from a scholarly standpoint, since they united knowledge-claims about health, taste and embodiment, political and commercial considerations, and material culture.

In this book, we do not propose a singular answer to the question of why people elected to take new drugs. Initially, a variety of different hypotheses might present themselves: some took drugs out of curiosity, as a novel means of exciting their senses or altering their embodied state; others sought new cures for various ailments; while still others turned to them only as a last resort to relieve suffering, a process that continues into the present day. Yet individualistic explanations like these do not sufficiently account for the *collective* nature of the societal shifts that characterize the period around 1700. Shifts in demand were prompted by the stabilization of court culture, the expansion of European empires, and the consolidation of print culture with the rise of the newspaper press. Were European consumers of the exotic following medical trendsetters at court, or were they swayed by clever marketing in print? When and why did they express doubts and suspicions, and when and how did they become convinced of a substance's efficacy and utility? All of these questions require answers which combine local and global perspectives and explanations.

The chapters collected in this book aim to challenge teleological assumptions about uptake by highlighting some of the myriad routes by which particular exotic drugs were described, explained, valued and used in Europe, how they reached European markets, who sold them and who consumed them. The diverse contributions to *Exoticizing Consumption* span Europe, from Spain in the south to Russia in the north, drawing upon extensive data culled from manuscript and printed sources—apothecary stock lists, recipe books, pharmacopoeias, newspapers, correspondence, and more—and analyzing these using techniques drawn from both traditional scholarship and digital humanities. The focal point between 1670 and 1740 affords a cross-section and comparative view of drug culture across Europe at a time of rapid change and development. Although the sixteenth century saw the introduction of many new substances to Europe, it was not until the later seventeenth century that some became objects of everyday consumption. Louise Hill

INTRODUCTION

Curth's work on almanacs, a key site of medical advertising, has shown a more than six-fold increase in English medical advertising—much of which concerns drugs—in just two decades, the 1670s and 1680s. Similarly, Patrick Wallis's comprehensive survey of port records shows that English drug imports increased faster between 1620 and 1690 than in the whole of the following century.[43] Other studies have shown similar trends for other areas of Europe, though not all. These phenomena suggest that the years around 1700 were a decisive phase in which the influx of "exotic" drugs into Europe grew from a trickle to a torrent, and when the substances themselves moved from being curiosities and rarities to becoming commodities, with a deepening penetration into different social groups of consumers, in part tracking the contemporaneous expansion of print culture.[44] These shifts in availability of, and information about, drugs were accompanied by changes in the ways they were viewed in governmental, institutional, and mercantile settings, accompanying the expansion of the first European colonial empires. Between 1670 and 1740, drugs moved from being arcane secrets possessed by individuals to becoming the focus of active governmental programmes of appropriation, cultivation, and marketing to an unprecedented degree. While a flurry of interest in the potential of natural resources to enrich the crown is evident at the end of the sixteenth century in Spanish and Portuguese imperial enterprises, it is well known that this process was limited in duration. By the end of the eighteenth century, the Spanish Empire was forced to play catch-up to other European powers engaged in consolidating their own rival forms of colonial productivity.[45] For instance, it was not until 1751, more than a century after the drug had entered widespread use in Europe, that the Spanish Crown established a royal reserve of cinchona trees. Even then, as Matthew Crawford has shown, these late-in-the-game efforts largely failed to generate a stable, high-quality, "imperial" supply of the bark.[46] But during the seventeenth and eighteenth centuries, the relationship between states, trade, and drugs underwent significant transformation, leading to the establishment of the first colonial botanical gardens, which often served as clearing-houses for the study of new exotic drugs and plant commodities, with a view to profit.[47]

What the history of drugs demonstrates particularly well are the problems of assuming a "largely unproblematic confluence of interests among state and non-state actors" for the period 1670–1740, in the words of Loïc Charles and Paul Cheney.[48] Historians have recognized that early modern states and their princes did not necessarily play a significant role in conveying exotic goods to consumers. Transnational trade networks might follow confessional lines instead, like the Protestant networks of Laurent Garcin in Alexandra Cook's contribution to this volume, or the Jesuits in Samir Boumediene's. Such disseminated trade networks, whether of missionaries or merchants, could play significant roles in shuttling drugs between different parts of the world. Early modern rulers' involvement in commerce, by comparison, was more tangential; these "connected" histories, to borrow Sanjay Subrahmanyam's term, were thus not always *imperial* histories.[49] Nor was "Europe" a homogeneous marketplace. Its markets varied extensively in almost all respects, from their degree of access to foreign trading ports to rulers' sumptuary legislation; from the taxation of imports to the extent and character of consumer demand. Commercial connections were often focused around very specific "paths of possibility," and trade was anything but freely practiced around the world. A drug did not necessarily reach a consumer in Moscow, for example, via the same networks and trade routes as might enable its arrival in Paris or Rome. Political and commercial differences of this kind meant that the landscapes of transformation of European tastes, and the patterns of availability of individual drugs, were also highly variable around 1700. And it was these differences that served to determine and diversify Europeans' encounter with "the exotic."[50]

SCOPE OF THE VOLUME

The nine contributions to the volume speak to three broad themes. The first theme interrogates in three case studies what it meant for a substance to be "exotic" in European eyes. The second addresses the complex relationship between exotic drugs as material and textual objects: how did texts structure the experience of the exotic? The third theme addresses commodification, posing the question of how specific drugs went from being objects of interest and curiosity within small groups, to becoming mass commodities in globalizing medical markets.

The Significance of the Exotic

The chapters in part I reckon with the meaning and valuation of the exotic in Europe. Each of them highlights, in one way or another, the role of terminology in shaping the categories not only of early modern actors, but also of modern-day historians. Opening up "the exotic" as a category for question necessitates decentering the privileged subject-position of Europe (and especially Western Europe), as well as questioning a series of modern critical binaries. How does our view of the trade in exotic goods change when we shift the cultural and geographical vantage point? Did Europeans always seek out substances to commodify in distant colonies? And does the flow of indigenous European drugs to distant locations throughout the globe provide a counterpoint to narratives of bioprospecting and extraction?

Clare Griffin's chapter shows how looking at the early modern world from a different vantagepoint—Russia, rather than the Atlantic-facing kingdoms of England, France, Spain, or Portugal—reveals some of the inherent subjectivity of categories like "the exotic." By looking outward from the eastern rather than western edge of Europe, the early modern world appears less maritime: in place of western imperial geographies, where goods increasingly transited between metropole and colony by sea, we find a vast Eurasian terrestrial continuum, "east of Delft but west of Edo," which depended more upon the land-bound Silk Roads than the transoceanic empires of Western Europe. Tellingly, Griffin points out, the Russian version of the term exotic, *ekzotichnii*, is a loanword adopted from the French only in the nineteenth century. In the early modern period, Russians thus had more immediate access to those peoples and objects Western Europeans might consider exotic, yet Western ideas of exoticism were reworked in Russian contexts. Griffin thus raises a second, fundamental question for our understanding of the larger processes at stake in the transformations of the global drugs trade between 1670 and 1740: how far the maritime trade which was allowing Europeans to bypass older land routes mattered to shifts in patterns of global trade. Under Peter the Great, rhubarb and other drugs traveled overland across an expanding Russian Empire from China and the Ottoman Empire, supplying the Russian Empire itself, while victory in the Great Northern War of 1700–1721 gave direct access to North Sea trade routes to European port cities.

Sebastian Kroupa's chapter moves even further afield, to Manila in the Philippines, to decenter Western European notions of "the exotic." He shows that, even in tropical colonies, Europeans often clung to traditional Galenic materia medica, on the grounds that climate, constitution, and broader policies of ethnic segregation demanded a distinct regime of bodily management and cure. In this sense, "exotic" drugs had first to be "processed" by receiving an imprimatur in European medical cultures, before they could enter European networks of trade in drugs. Kroupa's contribution shows that not all colonial encounters yielded new drugs for European exploitation. For, while European drugs flowed into Manila, very few medicinal materials flowed back. The principal exception was the St. Ignatius bean, successfully commodified for use in European medical encounters by the Jesuit order, thanks to their global communication networks. Kroupa demonstrates the need to reconsider and even reverse much of the received historiography of colonial "bioprospecting."

Theriac offers perhaps the best reminder of the inherent relativism of any "indigenous vs. exotic" binary. As Barbara Di Gennaro Splendore's chapter shows, the antidote known as theriac, a staple of Galenic pharmacy since the first century CE, was known to Western Europeans as a powerful compound drug from the eastern Mediterranean. But their ceaseless efforts to replicate its ancient recipe began to reverse the flow of theriac and its healing reputation between east and west. Western European cultures, initially consumers of the Venetian treacle (as it was known in English), became producers, and state-sponsored producers at that. By the eighteenth century, Venetian theriac was helping western Christian emissaries to curry favor at the Ottoman Porte. Even more tellingly, theriac, a complex product with over sixty ingredients, eventually escaped the East-West dyad entirely to become a multifarious and thoroughly *global* substance, mutating through adaptations and domestication in different environments, from Portuguese *Triaga brasilica* and Dutch/Sri Lankan Andromachus theriac, to the chymical *Theriaca coelestis* of Northern Europe. Theriac became, Di Gennaro Splendore argues, a "domestication technology," a means of absorbing new substances into the Galenic pharmacopoeia, and a manifestation of European versatility in adapting this commodity to the demands of different lands. As such, it neatly demonstrates how even an ancient drug could mutate over time to implant in new environments and contexts, and in so doing, blur the lines between exotic and indigenous, local and global.

Materia Medica: Substances and Texts

The term "materia medica" refers both to medicinal materials themselves, and also to the corpus of texts describing their nature and uses. Part II explores the interplay between texts and substances, and more particularly what texts, from popular advertisements to learned pharmacopoeias, can tell us about the presence and demand for exotic drugs. Recent work on the Atlantic world has highlighted the importance of textual genres in shaping the reception of material substances, and the fraught position they often occupy as mediators between cultures of secrecy and accessibility, inclusion and exclusion, in the circulation of healing knowledge.[51] Most of all, these studies have pointed to the unstable relationships between names and things. Short- and long-term case studies of individual drugs and their associated textual assemblages offer some of the most powerful means of revising current narratives. Our understandings of the significant degree of contingency in the European uptake of well-known drugs like cinchona bark have been substantively revised over the past few years, thanks to such work.[52]

Katrina Maydom's study of the penetration of the New World drug sassafras into English print and commercial cultures spans the century from 1577 to 1680. She demonstrates that sassafras became increasingly familiar and less exotic over time to English consumers, appearing in a wide range of printed materials. The assimilation process began slowly, and faced considerable resistance from different configurations of actors. For instance, chymists' critique of sassafras stemmed from the ease with which it could be assimilated to Galenic pharmacology, which the chymists opposed. Competing groups of healers thus played an active role in the fate and appropriation of a particular drug. Sassafras displaced guaiacum by the 1660s as the leading New World drug mentioned in English texts, but Maydom reminds us that successful acculturation of a drug was not an inevitable outcome. Her account of delays and obstacles to the drug's uptake also applies to other drugs: in the case of cinchona bark, for example, Klein and Pieters have documented a similarly high degree of contingency and accident surrounding its transition to and uptake in European cultures.[53]

By contrast, in his chapter, Wouter Klein adopts a much shorter timespan in order to compare the changing reputation of two febrifuges, Peruvian bark and cascarilla, during a 1727–1728 fever epidemic in the Dutch Republic. Drawing upon a rich store of newspaper and ephemeral printed sources, he shows how print functioned as a tool for promoting the sale and documenting the efficacy of new drugs. Together with Maydom's chapter, his study demonstrates the ways in which digital history techniques can open up entirely new research avenues and questions for the historian of the European drugs trade. Through his novel juxtaposition of epidemiology and printed advertising, Klein demonstrates how an epidemic could affect the market for a drug, driving up demand for novel curative media, and thus contributing to the acculturation of exotic substances. This is also a case that allows us to see how two different drugs might be considered by contemporaries to serve the same or similar purposes; thus it permits consideration of how a particular historical culture negotiated and reached agreement about efficacy and identity.

Where Maydom's and Klein's chapters each depend upon the analysis of a large corpus of texts, Paula De Vos's chapter offers an in-depth quantitative analysis of all of the drugs mentioned in a single text, Félix Palacios's 1706 *Palestra pharmaceutica*. Through her excavation of the different strata of drugs in Palacios's work, De Vos shows to what extent the entire textual tradition which European medicine had inherited was itself "exotic," a corpus of knowledge borrowed from a different time and culture, with a different geographic center of gravity which she terms the "Indo-Mediterranean world." It is easy to forget that the Dioscoridean-Galenic tradition itself, though so central to medical knowledge in western and northern Europe, was itself a foreign one for many parts of Europe: it relied almost entirely upon a combination of materia medica local to Mediterranean and Middle Eastern regions, and others that traveled along the Silk Roads from further east (the Indian subcontinent, China) or from Africa. Most of these plants did not grow in most of Europe. Even when rare plants from Indo-Mediterranean regions did become available in western and northern Europe, they could not always be cultivated there. Renaissance botanical gardens were the scene of a long process of appropriation and assimilation of both the plants and their curative powers, mediated through the rise of medical and botanical teaching in universities, in particular.[54] This "domesticated" Mediterranean tradition, which De Vos explores from the vantagepoint of the Spanish Empire, constituted the core of Western pharmacy by the mid-seventeenth century; over time, newer layers or strata accreted around it as Europeans forged new commercial and cultural contacts with both East and West. At the same time, De Vos's work urges caution in assessments of the extent of transformation that changing geopolitical, trading, and colonial relations actually made possible.

Networks of Commodification

How did exotic plants go from being mere curiosities to commodities with growing European markets? The chapters in this part of the volume approach the question at different scales of analysis, from highly localized case studies via trading networks and companies to global models, which tend to privilege the state, macroeconomics, and international politics.[55] On the interpersonal scale, a rich literature on recipes, a central source of knowledge about exotic drugs, has been informed by studies of household, family, and kinship networks, with a particular emphasis upon gender.[56] At the other end of the scale, historians of empire have emphasized how plants traveled through quasi-global networks. The chapters in this part take up Stephen Harris's recommendation to attend to "long-distance corporations," particularly missionary orders and trading companies, as a "meso-level" between households on the one hand, and empires on the other. Missionaries like the Jesuits (Boumediene), urban guilds like the Parisian grocers (Spary), or trading companies like the Dutch VOC (Vereenigde Oostindische Compagnie, or "United East Indies Company") all offer windows on the contribution of highly networked groups and organizations to the commodification of exotic drugs in Europe.

Catholic missionary orders, and particularly the Society of Jesus, are unavoidable in any account of the European appropriation of exotic drugs. Samir Boumediene's chapter explores the Jesuits' role in the commodification of exotic remedies as a handmaid to broader Jesuit goals of education and conversion. Learning the virtues of locally available plant substances could be critical for the survival of a mission in a hostile natural environment far from European trade routes. But it also offered the means to interact with local healers. The Society's mission allowed for knowledge transmission between non-European cultures and Europeans in the process of conversion and salvation: a Jesuit missionary might learn of useful drugs while hearing the confession of an Amerindian woman, or might watch a shaman at work. These spiritual goals also shaped the Jesuits' transmission of drugs within European publics. Back in Europe, exotic drugs were not only commodified in the open marketplace, but also entered parallel economies of gift-giving and charity.

Once drugs arrived in Europe as objects of consumption, they might become embedded within existing corporate and regulatory structures whose histories have been central to reappraisals of the European medical marketplace in recent decades.[57] E. C. Spary's contribution explores a specific urban network of commodification through the notarized stock lists that were compiled on the death of Parisian apothecaries and grocers (and sometimes of their wives) as part of probate inventories. Whereas grocers largely dealt in unprocessed plant substances, apothecaries concentrated on producing compounds, extracts, and chymical preparations. Beyond this repartition of substances and techniques, Spary's analysis reveals a complex drug market with a variety of distinct niches and areas of specialization—from the learned merchant-*curioso* who trafficked in a wide variety of rare substances, via the bulk suppliers who probably served as wholesalers to other merchants in the drug trade, down to the sellers of *grabot* or "garble": the dregs of exotic drugs like senna or coffee, which may have afforded access to such substances even to less affluent Parisians. Her study uses stock lists to trace uptake, consistency of availability, and changes in market value in the commodification of many exotic drugs, as a possible model for comparative studies of other urban contexts. Spary fleshes out a highly stratified drug world, extending from urban suppliers incorporated into guilds, whose trade linked them to distant parts of the globe, via the herbalists of Les Halles, who relied on herb women to forage for local plants growing outside the city, all the way up to privileged court merchants who used their royal connections to monopolize given exotics across the entire kingdom.

Alexandra Cook's chapter complements this transnational and local approach to networks to interrogate the micro-level through the career of a single practitioner—Laurent Garcin—and how he commercialized a single proprietary drug, the so-called Maduran pills. Garcin served as a surgeon for the Dutch East India Company (VOC), a globally active "long-distance corporation," as iconic among trading companies as the Society of Jesus is among missionary orders. Cook shows that Garcin's position in the VOC not only enabled him to collect information on the plants of the Indies, it also afforded him the opportunity to learn a medical secret from a Maduran Brahmin in Ceylon. Decades later, he successfully marketed this secret from his base in Neuchâtel as a proprietary drug. Against any simplistic narrative of appreciation of the exotic among European publics, Cook's chapter argues that the very foreignness of Garcin's drug may have impeded its commodification. She also demonstrates how the making and marketing of exotic drugs facilitated the transmission of non-European practices into European medicine, in this case, Tamil Siddha medical traditions. As recent work by scholars such as Kevin Siena on guaiacum or Anna Winterbottom on the China root has demonstrated, the introduction of new drugs was a source of transformation and innovation in European medical practice.[58]

INTRODUCTION

Collectively, these chapters not only invite us to take stock of the impact of exotic drugs in Europe, but also pose questions about perspective. If "exotic" is always a subjective, relational, and contextual term, as the chapters in this volume suggest, we might pose the question of whether it is helpful to view the exotic as something tied to provenance from specific places. Might "the exotic" rather reside in responses of wonder, confusion, or even the sensuous material attributes of a substance? Must a substance remain *elusive* in order to be truly *exotic*? This is the subject of the closing chapter by Hjalmar Fors. He argues that seventeenth-century Europeans had a very different conception of the exotic, one more closely tied to the unusual properties of the objects that reached them in Europe than to the associations such objects possessed with particular places or peoples. For early modern people, "exotic" expressed the confusion and general instability of knowledge surrounding the geographic origin of substances. European attraction to the strange and wondrous disappeared in the same measure as substances were disambiguated, classified, and "known," making them less exotic as they became more familiar.[59] Commodification and classification stand at the vanishing point where exoticism and commerce meet. In making this point, Fors cautions historians against importing modern categories of exoticism—a geographical definition of Europe, a model of center and periphery, of colonizers and subaltern peoples—into our analysis of the past. This does not mean that such categories did not exist. Rather, it is the historian's task to explain how such relations came into being; or, to put it another way, to ask how European interest in extracting "exotic" (valuable, rare, efficacious) materials metamorphosed into economies of exploitation and cultural hegemony. The question we need to pose, according to Fors, is "how an older culture of trade and wonder is replaced with a new culture, preoccupied with colonial-territorial domination."

THE POLEMICS OF THE EXOTIC

If Guy Patin had had his way when it came to drugs in the seventeenth century, there would have been no further exoticization of European consumption at all. The future of European medicine would have reflected a therapeutic minimalism, staunchly hostile to novelty, and faithfully adherent to a revived (and ideally pristine) Galenism. Patin's conservative impulse, it bears note, was rooted in more than a slavish reverence for classical medicine, inculcated by the values of a fading late

Renaissance humanism. Rather, there was what we might call, somewhat anachronistically, a political economy to his critique of exotic drugs. According to Patin, "the great abuse in medicine comes from the plurality of useless remedies. . . . An apothecary who has a grand boutique for his gilded jars shouldn't need more than a sideboard or a cupboard to store five or six containers."[60]

It is telling here that Patin focused on apothecaries' ostentatious practices for displaying their drugs. The apothecary shop was the original of all shops: the terms *boutique* in French and *botica* in Spanish derive from it. Its walls were lined with costly sets of matching jars, known as *albarelli*, filled with a profusion of drugs. This impressive visual exhibition of trade was very different from the self-presentation of faculty doctors like Patin himself. The sheer variety of remedies on the medical marketplace in the mid-seventeenth century French capital was one feature that drew Patin's ire; but we also know from other sources that the "plurality of useless remedies" he had in mind was typified by a trio of expensive compounds: bezoar, theriac, and antimony. The first two, ironically, symbolized a more eclectic Greco-Arabic polypharmacy, inherited by Europe from the classical Mediterranean world. Their use, although itself ancient, had been revived in the early modern period alongside the very Galenism Patin prized. Bezoar stones, described by Cook in her chapter, were concretions from the digestive tracts of animals, often porcupines, in high demand as antidotes to various forms of poison. Bezoar was notoriously difficult to authenticate, leading to a trade in artificial and sometimes fraudulent stones. Theriac, as mentioned above, was a widely-applied and versatile antidote, composed of over sixty ingredients (depending upon the recipe), prominent among which was the flesh of vipers.[61] The third expensive drug peddled by apothecaries, perhaps the most loathed by Patin, was the potentially toxic mineral emetic antimony, available in a variety of preparations, widely used by non-Paracelsians, but often anathematized as a product of chymical medicine.[62] Patin called antimony "the devil's medicine" (*diabolum medicamentum*), and recorded a "martyrology" of patients he believed had lost their lives to the drug. Between 1641 and 1671, he accounted for eighty-nine.[63]

The physicians from all over Europe who corresponded with Patin often shared his alarm over the growing popularity of exotic drugs and the commercialization of medicine. One widespread response was to insist upon physicians' corporate right to regulate and inspect drug-sellers' shops, taking merchants who infringed the law to court. But litigation was both costly and of uncertain success. Like other physicians, Patin also responded to the proliferation of drugs enriching apothecaries and itinerant healers by spreading knowledge of better, simpler remedies, both among the upper classes in private practice, and among the poor through charitable healing.[64] The model for this kind of activity came from *Le médecin charitable* (1623), a popular medical handbook written by Patin's own teacher, Philibert Guybert, which ran to some thirty editions.[65] But, like his mentor Guybert, Patin failed to acknowledge that the purified Galenism he sought to revive was, ironically, itself a product of an earlier moment of "globalization" in the ancient Indo-Mediterranean world: a Greco-Roman fascination with exotica, facilitated and enabled by the empires of the Hellenistic era and the Pax Romana. The gentle purgative senna, one of Patin's favorite drugs, grew nowhere near Paris. It was a product of the Mediterranean coast; the most highly valued variety, *Senna alexandrina*, was an import from Egypt. Even Guy Patin, in other words, was unable to escape the pull of the exotic, which provided conservative doctors like him with indispensable and time-sanctioned therapeutic instruments. Patin's loathing for exotic drugs may have been articulated in terms of a Eurocentric resistance to the exotic, but at bottom it responded to far more local disputes, ones in which he was enmeshed throughout his working life, from concerns over medical innovation to clannish faculty opposition to market encroachment by apothecaries, operators, and chymical physicians in the French metropolis. These last were peddlers of the exotic whom his urban medical clientele might easily encounter. In short, Patin and the old guard of the Paris faculty he represented were upset over the influx of new drugs primarily because they were controlled by rival healers. When their autonomy and economic model was challenged, the faculty doctors chose to emphasize what Harold Cook has called "good advice"—defined as judicious, learned counsel on how to maintain health—over the "strong medicine" touted by apothecaries and operators.[66] The local character of this conflict also underscores why it is essential not to reify "European" and "indigenous" as polar opposites in studying debates over exotic drugs. For, as we have seen, Europeans themselves did not constitute a homogeneous "center," but rather a diverse conglomerate of cultures, medical, political, social, linguistic, each standing in a specific relation to these "outside" drugs. Disagreement rather than consensus was the norm, not the exception, in the uptake of new drugs.

FIGURE I.2. Photograph of the Apothicairerie at the Hôpital Saint-Jacques, Besançon, France, showing the multiplicity of vases or *albarelli* considered appropriate at the time this dispensary was constructed in 1686. Theriac holds pride of place. Online at https://commons.wiki media.org/wiki/File:Apothicairerie_Besan%C3%A7on_0023.JPG. Credit: Arnaud 25, CC BY-SA 3.0 <https://creativecommons.org/licenses/by-sa/3.0>, via Wikimedia Commons.

In the event, Patin's ideal, minimalistic world of drug consumption was never to rematerialize. Medieval apothecaries' shops may have housed small cupboards with but a modest display of jars, but their early eighteenth-century counterparts might stock hundreds of different kinds of drugs. Patin ended life a curmudgeon, demoralized by the victory of the chymists, empirics, and apothecaries he had fought so hard to contain, some of whom he even lived to see penetrate his own faculty's inner sanctum.[67] Nonetheless, his critique reminds us that exotic drugs were not necessarily greeted with open arms (or rather mouths) in Europe. Their victory in new markets of European consumption was no foregone conclusion, dictated by essential features of the chemical substances they contained. In fact, we face a classic problem of historical hindsight in writing the history of drugs: we only get to read about the success stories, not the obstacles exotic drugs faced, nor the many cases of failed transfer.[68]

If we eliminate essentialism, and with it the presumption of inevitable dissemination, where does that then leave explanations of how and why substances do (or do not) transfer between cultures? In place of factors intrinsic to drugs, our volume proposes modes of transmission arising from complex interactions between cultural circumstances, ranging from accidental "discovery" via colonial appropriation to patronage from princes. Successful transmission to a new culture should be taken as the circumstance most acutely requiring historical explanation, as recent literature shows.[69] As Sarah Easterby-Smith suggests, the vast majority of transfers likely occurred tacitly, and without generating any textual record. Failure to transfer was perhaps the norm, rather than the exception to be explained away.[70] In his study of the failed transmission of psychedelic drugs to Europe, Benjamin Breen invokes the concept of "assemblages" as a way to express how items of materia medica do not travel alone, but rather in tandem with a whole set of associated knowledge, practices, technologies, and norms. Differences in the uptake of given substances across different spaces and cultures can be explained by appealing to the extent to which such "assemblages" can successfully be accommodated in diverse cultural contexts, and the duration of their "career" there.[71] The globalization of drugs in the early modern world is best explored through scrutiny of locally prevailing actors' categories, power relations, and knowledge-claims. Thus, the answers to both how and why exotic drugs found (or indeed failed to find) favor among European consumers often lie in Europe itself. At the start of this introduction, we mentioned one such case, where the success of tea was coterminous with the lifespan of a single patron. In today's world, we might point to the erosion of legal and societal approval of tobacco, after centuries of implantation into European society; or perhaps to the growing legal and societal approval of cannabis characterizing the second decade of the twenty-first century. The results of applying such approaches are the chapters collected in this volume. They remind us that historical phenomena—both at the meta-level, with globalization, and at the micro-level, in quotidian decisions to ingest or refuse individual substances—are, and always have been, highly contingent.

PART I
MEANING AND VALUATION OF THE EXOTIC

1

THE UNEXOTIC WORLD

Medical Drugs, Global Exchanges, and Moscow's View of the Foreign, ca. 1690

Clare Griffin

For scholars working on Western Europe's interactions with the rest of the world, Edward Said's *Orientalism* has long been the canonical text, making a substantial impact on a variety of subtopics, including the study of early modern Western European perceptions of the rest of the non-European world.[1] This is already an expansion of Said's original mission, as he himself was particularly concerned with nineteenth-century British and French views of the Middle East. Since *Orientalism* was first published in 1978, such applications of Said's ideas to a set of circumstances that were beyond his original framing of his work have been common, and have demonstrated both the value and the difficulties of such an approach.

Said's work has been applied to European conceptions of the Americas, where the European gaze is focused not eastward, but westward; the geography is rearranged, but the European/non-European dichotomy remains stable.[2] Inventively, scholars working on East Asia have used Said's work to create the concept of Occidentalism, which Xiaomei Chen defines as "a discursive practice that, by constructing its Western Other, has allowed the Orient to participate actively and with indigenous creativity in the process of self-appropriation, even after being appropriated and constructed by Western others."[3] Benjamin Schmidt has alternatively proposed the existence of an exoticizing gaze that the Dutch of the late seventeenth and early eighteenth centuries turned toward the rest of the early modern world; critiquing Said for overemphasizing the importance of differences between Europe and not-Europe, Schmidt argues that difference should be understood as existing alongside a flattening-out of global differences. As Europe and not-Europe were seen to be different to each other, the not-Europe of the Americas and the not-Europe of East Asia were presented as markedly similar.[4]

As a Russianist, I then have to ask: Where could Russia's place lie in the discourses of Saidian Orientalism, Chenian Occidentalism, and Schmidtian Exoticism? By the 1690s, the well-connected denizens of Moscow were aware of places, peoples, and goods that originated from both the East and the West, and for that matter the North and the South as well. Taking that as our subject, we can then leave aside Occidentalism, because it deals more with how views of the indigenous self have been shaped. Yet that leaves many other questions open. How did early modern Russians think about the world around them, and its various foreign places, peoples, and goods? Did they apply an orientalizing or exoticizing gaze? In which direction? And if Muscovites did so, from their location east of Delft but west of Edo, would they be constructing an Orientalist discourse?

For the purposes of the present volume, we are concerned specifically with medical drugs, one small yet vital part of the complicated interactions of the early modern world. Turning to the Russian documents themselves, we see a flurry of words that tell us about things and about places. The main repository of medical records for this period is the Apothecary Chancery, the palace medical department which existed from the 1570s until the 1710s. This official department tells us at least something about unofficial practices, as during this period the department was—somewhat cynically—both investigating various unofficial practices, but also using unofficial medicine salesmen as a source of ingredients for the department. Happily for us, this means we have records of the most local sales of the great empire: we can see what was being sold in Moscow, here focusing on the 1690s. One such document is a seventy-eight-page list of items the Apothecary Chancery bought from the various Moscow markets in 1694. On this list we find: lavender and rosemary, perhaps collected in the fields surrounding Moscow; cinnamon, brought in from China; nutmeg and mace, from the Dutch-controlled Spice Islands; spermaceti, from sperm whales hunted in the North Sea by the British; French wine; and sassafras, from all the way across the Atlantic in European colonies along the East Coast of North America, notably Florida.[5] In the 1690s, customers of the central Moscow markets could buy the products of the early modern global world.

These medical commodities did not travel the global world alone. Those objects were bound up with places, peoples, and ideas, all of which had to be dealt with and unpacked. Moreover, medical commodities interacted with both people and with places: the climate from which a commodity originated was thought to influence its nature, and the commodity could pass that nature on to someone consuming it. This linking of bodies, consumables, and places was so strong that white Europeans feared becoming more like their colonial subjects through consuming colonial products.[6] To see if Muscovites had something akin to the Western European exoticizing or orientalizing gaze, it is then necessary to think more broadly than medical drugs. We should ask, did Muscovites view foreign materia medica as exotic? And we should also ask, how did Muscovite ideas about foreign peoples and foreign places affect and intersect with their views on foreign materia medica? Here, then, thinking of foreign things, peoples, and places will help us understand whether Floridian tree bark would have been seen as exotic by Muscovites in the 1690s.

GRAMMARS OF SAMENESS, GRAMMARS OF DIFFERENCE

From Said to Schmidt, central to understanding how Western Europeans viewed the rest of the world has been examining issues of Self and Other, what is the same and what is different. Schmidt has proposed the existence of a "grammar of sameness" that Dutch geographers used to flatten out the variety of the non-European world; in this he plays with Ann Stoler and Frederick Cooper's Saidian concept of a "grammar of difference," the ways the colonial Other was constructed as different and distant from the European Self.[7] This is hugely important to Said and the Saidists, but certainly not unique to them; indeed, Russian-language folkloric scholarship has long had an analogous concept, *svoi/chuzhoi*, which can be translated as own/other, and the Moscow-Tartu School of Soviet Semiotics, unofficially led by Iurii Lotman and Boris Uspenskii, also made extensive use of that concept, and similar binaries.[8]

What is own and what is other can also be a useful concept when discussing historiography and scholarship. Work on exotic drugs in Western Europe belongs squarely within the discipline of English-language history of science and medicine; work on exotic drugs in Russia, and on

Russian science and medicine more generally, does not fall within that discipline so easily. The overwhelming view of Russian science within English-language scholarship is that this is a post-1700 issue, or even post-1725, when the Russian Academy of Sciences was established; for example, Loren Graham's classic survey of Russian science begins with the assumption that there was no science in Russia before 1700, and that Tsar Peter the Great imported science from Western Europe.[9]

Within Russian-language history of science, however, such a chronology makes little sense. For decades the Moscow-based scholar R. A. Simonov has written numerous works on pre-1700 Russian science, in particular mathematics and astronomy, and has been a key figure in a circle of other scholars working on similar topics.[10] There are exceptions to this divide: Eve Levin's work on medicine, and Valerie Kivelson's on cartography, are about pre-1700 Russian science but written in English; Rachel Koroloff's work on Russian botany sometimes crosses the Rubicon of 1700.[11] But those scholars are much better known, and more commonly cited, in scholarship on Russian history than on early modern science. Scholarship on pre-1700 Russian science is "own" to scholars working in Russian, but distinctly "other" to those interested in similar topics but who work in English.

Indeed, I am somewhat rehearsing a version of this dichotomy here. English-language scholarship on non-Western Europe is often guilty of taking sources from non-Western Europe, information from publications by non-Western scholars, and then explaining that material through the theoretical lens of English-language scholars, often white men. That I began a chapter on Russian history with Benjamin Schmidt and Edward Said—both non-Russians—is problematic, then. But Schmidt and Said are familiar figures to English-language scholars, and can lead us to equally vital Russian-language scholarship.

As well as historians of science like Simonov, an important scholar for the topic of exotic drugs in early modern Russia is K. A. Bogdanov, and in particular his book *On Crocodiles in Russia*.[12] Bogdanov takes his title from a famous wood-cut image of the Russian fairytale witch Baba Yaga in conversation with a crocodile (as fairytale witches are wont to be), in which the crocodile looks rather less crocodile-like, and rather more like a large squirrel with a beard. It may not have the artistic finesse of Dürer's *Rhinoceros*, but it is a similar attempt to recreate visually an unseen animal.[13] This starting point indicates Bogdanov's chronological focus: the image in question is conventionally dated to the reign of Peter the Great (1698–1725), but the most famous extant examples are from the 1760s: Bogdanov is primarily interested in the Russian exoticizing view of the eighteenth century, not the seventeenth.[14]

Nevertheless, Bogdanov is hugely important to understanding the 1690s, as there are continuous concerns between the seventeenth and eighteenth century. When Western Europeans wrote about exotica in the 1690s, they had both a grammar and a vocabulary of difference to mark those items out, including the term "exotica."[15] But the Russian equivalent term, *ekzotichnii*, is a loan word taken from French in the nineteenth century. This is a point of as great importance to Bogdanov's eighteenth-century texts as it is to our seventeenth-century documents; indeed, Bogdanov spends time on the various similar terms we might need to consider—such as *chuzhdyi*, meaning "other," *inozemnyi*, meaning "from another land"—and how they differ from *ekzotichnii*.[16] In whatever way Russians in the 1690s were thinking about and writing about foreign medical drugs, foreign places, and foreign peoples, they were certainly not thinking of them as *ekzotichnii*, because that word did not yet exist. Following Bogdanov, we need to contemplate the complex mappings of own and other that were common to late seventeenth-century Russia, and avoid forcing foreign or modern concepts onto Muscovite texts.

The push-pull between the West and Russia that we see in the scholarship was also an issue for the Apothecary Chancery. From the beginning, that department was a combination of different kinds of servitor of different backgrounds. The head of the department was always a high-ranking member of the Moscow court with close family ties to the tsar, and who was officially responsible for reporting on the department's activities to the tsar and his counselors; that elite Russian leader was supported by a number of Russian secretaries, whose duties were to keep the records and correspond with the numerous other departments (around seventy) that ran every aspect of life in the Russian state. As the Apothecary Chancery was responsible not only for treating the royal family and courtiers, but also providing medical services to the army, conducting investigations into potential outbreaks of the plague, assisting in witchcraft trials and—increasingly toward the end of the seventeenth century—also policing medical practices outside of the Kremlin, the work of these secretaries was vital. They and their heads of department were Russian, and so on the *svoi* side of the Self/Other divide.

On the other hand, there were workers in the Apothecary Chancery who were distinctly Other to Muscovy: medical practitioners in that department were almost always foreign, primarily being recruited from Russia's key trading and diplomatic partners, the German Lands, the Netherlands, and Britain. These university-trained physicians and guild-trained apothecaries and surgeons relied upon Western European ideas and practices, communicated between themselves in German or Latin, and knew little to no Russian. They were the expats of the early modern world, pampered foreign talent who had no incentive to and little interest in becoming part of the Russian world. Mediating between the *svoi* of the Russian bureaucrats and the *chuzhoi* of the Western European medical experts were the translators. Before 1685, when the Slavo-Greko-Latin Academy opened in Moscow and began formally training translators (among their other educational pursuits), these translators were informal, situational translators, often taken from Polish communities.[17] The staff of the Apothecary Chancery constantly dealt with the divide between Russian and not-Russian.

The Apothecary Chancery relied upon Western European medical practitioners; it also relied upon Western European medical texts. The department maintained a library full of Latin- and German-language medical texts on a range of topics, including a number of pharmacopeia, such as Johann Daniel Horst's 1651 *Pharmacopoiea Galeno-Chemica*, and natural histories like Ole Worm's 1655 *Museum Wormianum*.[18] The department and its employees also created a number of reports on medical issues, and medical books, both in Russian, which heavily drew upon Western European sources. The Apothecary Chancery was then a hybrid institution, bringing together the Other of Western European experts and expert texts—with their ideas about exotica—with the Self of Muscovite bureaucrats who had a different set of references.[19] Considering Russian ideas of exotica in this period means also considering Western European ideas.

The issue of inter-linguistic and cross-cultural interactions between Russians and Western Europeans takes us back to Bogdanov. Bogdanov notes that the long history of cultural and linguistic borrowings in Russia and elsewhere is "first of all the history of the rethinking (*pereosmyshleniia*) of borrowed values."[20] Otherwise put, when words and concepts were taken from Western European texts, contexts, and languages, and introduced into contemporary Russian, they were reformed in that new situation. Take for example the word *travnik*. In many Western European languages, there are herbals (books of medico-botanical knowledge), herb-collectors or botanists (people with knowledge of plants), and herbaria (dried collections of plants). As Koroloff has discussed, in early modern Russia one term—"travnik"—served across the sixteenth to eighteenth centuries to mean one or all of these things.[21] "Travnik" was a Russian translation for the Western European herbal, but it was also a rethinking of that concept, as Bogdanov has it, that more strongly linked one form of plant knowledge with related forms in Russian than in, say, English. Late seventeenth-century ideas on exotica in part came from Russian ideas but were fundamentally formed in the interlinguistic encounter between Western European concepts and Muscovite culture.

SPACES AND PLACES

Exotica is related to a division of space, the idea of the presence of the foreign in the local reality. Space and commodities were closely linked in this period: in early modern Europe the Maluku islands in present-day Indonesia were known as the Spice Islands, conflating the region with the valuable commodities found there. We can then begin by thinking about space, and more specifically about how Russians around 1690 depicted and wrote about foreign spaces and places. The words "space" and "place" are significant. Schmidt's idea about the Dutch flattening out the global world's distinctions speaks to the Dutch seeing the rest of the world as a space, a unified region without distinct characteristics.[22] Alternatively, both Orientalism and Occidentalism deal with places, distinct and different regions held in a tight, mutually reflective binary of West vs. East.[23] By the 1690s, Russians were producing both textual and visual geographic works that depicted their empire and the rest of the early modern global world. Can we then helpfully understand Muscovite geographic thought through the lens of Schmidtian Exoticism, Saidian Orientalism, or Chenian Occidentalism?

A helpful example of a foreign place is China. China was often the object of the Western European exoticizing gaze in the late seventeenth century; several of the works Schmidt focuses on deal in part with China. The interactions of the Dutch with the Chinese were very different from the relationship of the Russians with the Chinese. This relationship can be exemplified on the Russian side by the much-discussed Godunov ethnographic map, a work depicting Siberia and its environs and peoples and commissioned by the governor of Siberia, Petr Ivanovich

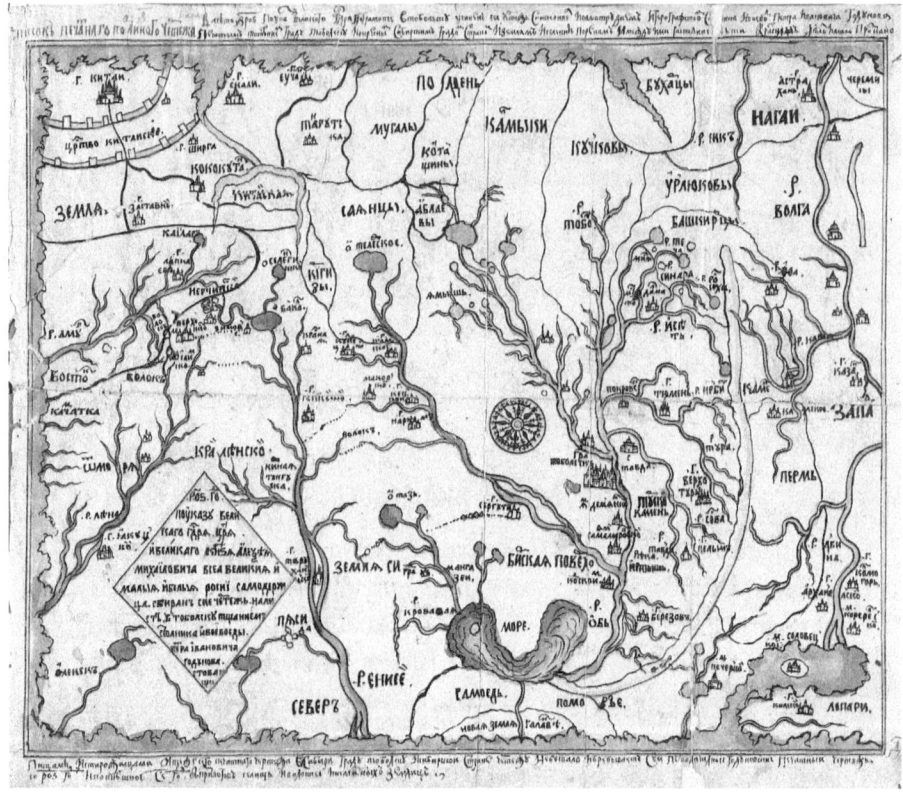

FIGURE 1.1. S. U. Remezov's copy of the Godunov map of 1666–67. MS Russ 72 (1). Houghton Library, Harvard University. Online at https://iiif.lib.harvard.edu/manifests/view/ids:23066694. Public domain.

Godunov, in the Siberian capital of Tobolsk in 1666–67. This map, as was typical of pre-1700 Russian maps, is oriented with the north at the bottom of the page; in the top left-hand corner, China appears as a small region behind a wall. The inclusion of China in maps of the easternmost territories, and the delimitation of China in those maps by representing a wall between the territories, became typical of late-seventeenth century Russian productions.[24] The Dutch representations of China discussed by Schmidt do not put Dutch territories alongside Chinese lands in this way, for obvious reasons. Moscow itself was far from Qing territory, but Tobolsk was much nearer, and the Godunov map presents a Tobolsk-centric gaze on China, one which was sufficiently accepted elsewhere in the empire for the Godunov map to be repeatedly copied and emulated. In the Russian case, the emphasis is not on difference and distance, but on proximity. Muscovy's China was not a distant, exotic land; the China of the Godunov map is, if anything, too close for comfort.

The Godunov map visualizes the political, diplomatic, territorial, and commercial interactions of the Russian and Chinese realms of the late seventeenth century. According to Gregory Afinogenov, the Russians did not gather a lot of information about China at the beginning of the seventeenth century, but "in the second half of the seventeenth century, the production of both foreign-authored and Russian works about Siberia began to create a new and distinctive frontier literature about Russo-Qing relations."[25] In part, this interest was about Russo-Qing competition over the loyalties of the peoples between their empires; one notable incident involved the Tungus chieftain Gantimur, who fled Siberia—and Russian demands for tribute—across the border into Qing territory and service in 1653, only to desert the Qing over a decade later, in 1666–67, remaining in the Russian Empire until his death.[26] As well as unofficial border-crossers like Gantimur, there were official crossers like Nicolae Milescu Spathary, who led a Russian embassy to Peking in 1675. Yet "border" is not quite the right word, as the boundaries of the Russian and Chinese Empires in this region were contested at least until the Treaty of Nerchinsk in 1689, which officially resolved the question, as well as setting up official trading policies between the empires.

The Treaty of Nerchinsk attempted to regularize a much older and unofficial system of commodity exchange between Russia and China. The most famous example of this would be the rhubarb trade, in which Russia obtained valuable Chinese rhubarb to sell on to Europe, where it was prized as a medicament. This trade had been going on since at least the sixteenth century.[27] Other medicaments were also crossing the border: in 1690, the Apothecary Chancery requested that the Siberian towns send them "Chinese cinnamon."[28] Such a request was based on the understanding that Siberian settlements had regular access to substantial quantities of cinnamon being moved into Russia from the Qing Empire, suggesting a more everyday interaction of Russians with China, in particular Russian subjects in Siberia, than the political machinations of Gantimur and Spathary. All this is very different from the Dutch experience of China. Simply put, with all the vital interactions between Russia and China over their shared borders and mutually contested imperial spaces, the experience of and ideas about China in Moscow and Tobolsk were very different from those in Amsterdam and Delft.

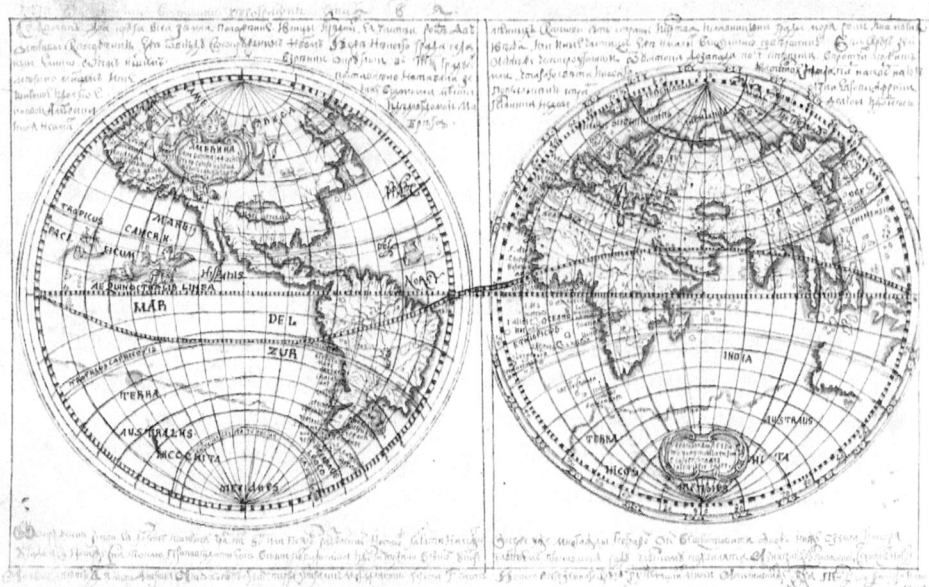

FIGURE 1.2. The double-hemisphere map in S. U. Remezov, "Chorographic Sketchbook," 1697–1711. MS Russ 72 (6). Houghton Library, Harvard University. Online at https://iiif.lib.harvard.edu/manifests/view/drs:18273155$6i. Public domain.

These various binding ties of Russia and China in the late seventeenth century had been visualized in the 1660s by the Godunov map; that map continued to be copied across the late seventeenth century, including by the most prolific Muscovite cartographer, Semyon Remezov (ca. 1642–ca. 1720). Remezov, who produced maps for the Russian state, both created his own maps and copied and adapted the works of various other cartographers besides Godunov. His work helps uphold Schmidt's claim that the Dutch provided much cartographic work for the rest of Europe, as many of his copies are from Dutch work, including a Dutch two-hemisphere world map he included in his "Chronographic Sketchbook," produced between 1697 and 1711.[29] Notably, this world map includes the Americas. This is important as it expanded the visual geographical repertoire of Muscovite cartography: the Americas had been written about within Russia since at least the 1530s, but this may be the first map of that continent produced within the Russian Empire.[30]

Remezov's "Chronographic Sketchbook" is also significant because of the way he positions Siberia in geographical relationship to the

Americas. In his conclusion to this work, Remezov calls his work an atlas "of the entire interior of Siberia, with the reigning city of Tobolsk and the cities, settlements, forts and parishes under its jurisdiction, especially between the countries [sic] of Asia, Europe and America."[31] The mention of Tobolsk is a typical Remezov move: Tobolsk was his hometown, and he inserts its presence into all of his world maps. Yet it also plays a bigger role than merely hometown pride. As with the representation of China in the Godunov map, in this depiction the rest of the early modern global world is shown to be part of a geographic contiguity with the Russian Empire.

Valerie Kivelson, writing of Remezov's atlas, has argued for "betweenness," a focus on connections of geographic space, being a key and unique feature of official Muscovite geography, claiming that official geography in this period presented the empire as standing between other major world regions.[32] By 1690, the Russian Empire stretched from the Caspian Sea in the south to the Barents Sea in the north, from Okhotsk on the Pacific coast of Asia in the east to Smolensk in the west, and was multi-ethnic and multi-religious. The geographical works Kivelson discusses show only one view from that sprawling and heterogenous empire, but it was an influential one. Kivelson's argument about official Russian geography mirrors those by other scholars who have looked at the interconnectivities of the early modern world, such as Kapil Raj's insistence upon the co-construction of knowledge texts between Europe and South Asia, and Sanjay Subrahmanyam's concept of the "connected histories" of Eurasia.[33] Early modern Russian geography's "betweenness" had unique Muscovite features, but this focus on connections also linked it to other knowledge practices of the early modern world. The Americas might have been farther away from Tobolsk than China, but in late seventeenth-century Russian officialdom both places were depicted, and thought about, as a part of a continuous whole. This "betweenness" is a different kind of geographic thought from that examined by Said and Schmidt: official Muscovite world maps seek to bring together all geographic space, not to create and maintain a division between the European and non-European worlds. The official geographies of Muscovy did not make space for the exotic.

PEOPLES AND CIVILIZATIONS

At least from the seventh century, when Western Europeans were producing the T-O maps that depicted which tribes of Israel were supposed to have settled in which region of Afro-Eurasia, geographic thought went hand-in-hand with anthropological thought; this depiction of peoples and places together continued to be a key element in the Western European Renaissance.[34] The interest in different groups of humans, like that expressed in these ethnographic maps, have led a number of scholars to argue for the existence of premodern ideas of race, and to create the field of premodern critical race studies.[35] That field provides vital insight into ideas of the foreign and the exotic via its focus on the racialization of different groups. The supposed geographic arrangement of humanity—what Geraldine Heng has referred to as cartographic race—substantially intersected with ideas of exotic lands. Indeed, the idea of humanity is rather tricky here: there was a common medieval idea that semi-human monsters lived on the margins of the civilized world.[36] In western European texts like the fourteenth-century *Travels of Sir John Mandeville*, those monsters were first ascribed to Asia. After 1492, the same ideas informed European perceptions of Native Americans; Christopher Columbus took a volume of *Mandeville* with him on his voyages, and his writings on Native Americans show that text having a substantial influence.[37]

A major part of the intersection of anthropological and geographical thought was the idea that the nature of the human body was fundamentally shaped by both the environment in which it lived and the commodities it consumed. Surekha Davies has discussed how ancient humoralism saw the nature of the human body as shaped by the climate it was in, again linking bodies to places.[38] The climate and commodities of a location were thought to have the potential to change bodies, posing a danger to colonists that they might "acclimatize" to the colonial world either through presence in the environment or through continuous consumption of the commodities of that environment. This idea was both long-held and widespread, with Rebecca Earle discussing its presence in early New Spain, and Matthew P. Romaniello, in the eighteenth-century Russian Empire.[39] Commodities coded as "exotic" were thought to impact the people who consumed them; conversely, as we will see, some commodities were linked to specific racialized groups in an attempt to denigrate the commodity by association. This complex relationship between premodern ideas of human difference, place, and commodities means that premodern critical race studies has much to tell us about how to understand ideas of the exotic, and thus the early modern global drug trade.

Attempts to deal with human difference were also important in early modern Russia. By the late seventeenth century, the Russian Empire included not only Slavic-speaking Orthodox Christians of the Western Steppe, but also include Tatar-speaking Muslims and Finnic-speaking shamans, among many others. By the 1690s the Russian Empire had reached the regions which Western Europeans had once claimed to be populated by monsters. Like their Western European contemporaries, Muscovite cartographers produced maps—in particular maps of Siberia—that detailed the human populations to be found in each region.[40] In contrast to *Mandeville*, Muscovite depictions of the region's inhabitants saw them as humans, not monsters. Russians did have a contemporary concept of semi-human monsters—Peter the Great in fact promulgated a decree in 1717 that such creatures were to be sent to him and not to be killed—but those monsters were not associated with a specific geography.[41] Peter's monsters were not exotic. It is vital to keep in mind these differences: as numerous scholars have emphasized, we can find ideas of race in the premodern world, but those ideas differ from modern conceptions of race, and also differ across regions. The ideas of race Heng has found in medieval Western Europe are not the same as those Bruce S. Hall has found in early modern Muslim West Africa.[42] To further explore the Muscovite concept of the exotic, we should now turn to ideas of human difference and their expression at the intersection of anthropological and geographical thought, and how these played out in early modern Russia in particular.

One figure of particular interest here is the man known in Russian as Avram Petrovich Gannibal. Although popularly associated with the Assyrian Empire and so Ethiopia, Gannibal was actually born further to the west, in the region of present-day Cameroon, in 1696; he was kidnapped as a child, sold into slavery in the Ottoman lands, and given as a present to Peter the Great in 1704.[43] In this, he was not alone. A document by the Russian servitor Franz Lefort speaks to the casual and regular importation of enslaved people by the Russian court by the 1690s. In 1698, Lefort wrote to Peter the Great, who was then on his Grand Embassy to Western Europe, in some fairly questionable if comprehensible Russian: "poujalest nie zabouvat coupit arapi" (please do not forget to buy Arabs).[44] Most of the enslaved "Arabs" purchased in this way ended up as ceremonial page boys; Gannibal's story was different. Peter the Great took a liking to him, made him his godson, sent him to be educated in Paris, and made him an advisor and a regional governor. Gannibal

FIGURE 1.3. Portrait, ca. 1721, attributed to Jean-Baptiste van Loo (1684–1745). The identity of the individual in the picture is disputed, but one possibility is that this is a young Avram Gannibal, perhaps painted when he was a student in France. Online at https://commons.wikimedia.org/wiki/File:Petit_Hanibal.jpg. Public domain.

died in St. Petersburg in 1781, a member of the Russian elite, but not before fathering several children, a line that can be traced down to the famous Russian poet Alexander Pushkin.

But was Gannibal, and were those enslaved pageboys, viewed as exotic? Contemporary depictions are significant. There are a number of images of men who could be Gannibal in Russian artworks, or artworks about Russia, but these are stereotyped and generalized depictions of dark-skinned men in European or Turkish garb, rather than representations of specific and identifiable individuals. This trend means that singling out images of Gannibal himself has often proved tricky.[45] Both Hall and Heng have noted that skin color played a role in marking human difference, even as other elements were also important.[46] The fact that depictions of Gannibal heavily rely on skin color to represent him shows the longevity of this, as Heng has it, "politics of the epidermis."[47] Those politics were long-lived, but the period in which images of Gannibal were created was one in which skin color as a marker of human difference was on the rise. Indeed, art historian Anne Lafont has written about how skin color became the major marker of race in European artwork in the eighteenth century.[48] In paintings, the specific individual Gannibal was made into a generic dark-skinned man, following the specific "politics of the epidermis" of eighteenth-century European art.

It is also interesting that when Lefort so offhandedly referred to the violent enslavement of fellow humans he referred to them specifically as Arabs. Depictions of enslaved persons at the Russian court show dark-skinned individuals commonly dressed in the Turkish style. What is the ethnic group being depicted here? Can we associate them even with the extremely broad linguistic-regional-ethnic groups of Turks, Arabs, or Africans? Can we—did the artists—meaningfully distinguish between the inhabitants of Anatolia, the Arabian Peninsula, and Gannibal's West Africa? Both in texts and in images, even dividing these depictions into such broad categories does not seem possible. Despite the presence of one notable West African, in this time and place individuals such as Gannibal were treated as part of an undifferentiated group of dark-skinned southerners.

Other documents speak to this conflation of the Arabian Peninsula with the African continent. In 1664, the British physician and Apothecary Chancery employee Samuel Collins provided the department with a description of that vital early modern eastern African beverage, coffee. Collins, who had studied at Cambridge and Padua and had been practicing medicine in London, was recruited to the department by a British merchant under commission by the Russian court in 1659, and then spent nearly seven years as a top physician at the Russian court, entrusted with prescribing to the tsar and compiling reports for the department head. Collins's report begins with a geographic overview of coffee. He states that it is in use by the Persians, Turks, and English, and that London now has two hundred coffee houses. He also notes the best methods of preparing coffee (with nutmeg and sugar), and the medical benefits of the drink including, of course, *"mnogosonie otgoniaet"* ([it] drives away much sleep). The origin of these assertions is unclear, as he gives no specific attribution for these claims.[49] Significantly, Collins had visited London two years before composing this text. As well as perhaps sampling some coffee at one of the coffeehouses he mentioned, he may also have read about it. In the same year, the Royal Society, with which Collins had links, deposited *Discourses about Cyder and Coffee* in their archive.[50] That text certainly would have been of use to Collins when he wrote his report shortly after his return to Moscow. Collins's Latin draft of the coffee report was certainly shaped by Western European ideas about the beverage, and was perhaps even based on a specific Royal Society document.

As was common practice, the Russian translation of the coffee report differs somewhat from the Latin draft. Notably, the translation inserts a sentence on the geographical origins of coffee, noting that "coffee is the berry of certain bushes that grow in Arabia."[51] However, coffee does not originate in the region commonly referred to in English as Arabia, but rather the horn of Africa, to the south of the historical Arabic-speaking lands. Why identify an African plant as an Arabian one? By the 1660s, coffee and coffeehouses were popular in the Ottoman Empire, although attendance at the coffeehouse had a complicated relationship to contemporary socioeconomic, gender, and political norms.[52] Much coffee was also moved through the Arabian Peninsula and the Ottoman Empire into Europe. There were reasons for Europeans to think of coffee as Arabian.

However, presenting coffee as Arabian may have less to do with contemporary commodity consumption and circulation trends and more to do with Islamophobia. The Russian Empire had a complicated relationship with Islam and Muslims. Russia maintained diplomatic and trade links with Muslim polities, including Safavid Persia and the Ottoman Empire. By the 1690s the Russian Empire already had Muslim subjects, such as the Tatars of Kazan, whose Khanate the Russians conquered in

1552. The Russian Empire also regularly fought with Muslim armies in the seventeenth and eighteenth centuries. In 1696, the Russian Empire defeated the Ottomans at Azov, in the far south of the Russian Empire, and Peter the Great commissioned an etching entitled *An Allegory of the Victory of Christianity over Islam* to mark that event, presenting a geopolitical win as a triumph of faith.[53] Russian responses to Islam were never monolithic, but there were Islamophobic tendencies in late seventeenth-century Muscovy.[54]

There is a notable history of links between Islamophobia and coffee, in which the Russians may well have taken part. Nabil Matar has discussed how late seventeenth-century British writers—contemporaries of Collins and his employers—commonly sought to demonize the beverage through associating it with Muslims, and so to imply that to drink coffee was to lose both one's Christianity and one's Britishness. Those same texts commonly referred to coffee as the "Arabian berry," just as the Russian report does.[55] In Britain, such Islamophobia seems to have had little effect on the popularity of coffee; the same discourse may have had more of an effect in Russia. Although Collins recommends the beverage, coffee was in little use in late seventeenth-century Russia. According to Bogdanov, there was a certain amount of recreational coffee consumption in St. Petersburg in the 1720s, but Audra Yoder has found only limited evidence of coffee use in Russia before the twentieth century.[56] The Arabian berry was not popular in Muscovy, perhaps because of its Muslim connotations.

When early modern British and Russian writers used terms like "the Arabian berry," they were conflating Arabia and Arabs, Turks and the Ottoman Empire, Islam, and the various peoples, religions, and polities of Africa. There was a shared view of southern, heathen lands that flattened out regional, linguistic, ethnic, botanical, and religious differences to produce the Southern Muslim Other. This takes us back to Said, who notes of a later period that "European culture gained in strength and identity by setting itself off against the Orient as a sort of surrogate and even underground self."[57] The idea of superiority through contrast would seem to apply as much to Matar's seventeenth-century British authors' views of coffee as polluting the morally superior white Christian British as it does to Said's later writers. Although the Russian texts do not express themselves as directly on this point as Matar's British sources, the 1664 coffee report is generally indicative of a Muscovite orientalizing gaze toward an undifferentiated Southern Muslim Other.

Was this orientalizing gaze directed only towards southerners? Or was it part of a broader understanding of non-Russians? On this point we can look at Russian law. As Russians increasingly employed foreigners, they included them in their law codes. T. A. Oparina has noted that in the seventeenth century the word "foreigner" (*inozemets*) did not mean a person from a foreign land, but a person not belonging to the Russian Orthodox Church. Indeed, a longer version of this phrase was "nekreshchenye inozemtsy," unbaptized foreigners. Children of foreigners who were born in Russia did not automatically become Russians, but were known as "old foreigners," "foreigners of earlier immigration," or "Moscow foreigners" (*starye inozemtsy, inozemtsy starogo vyezda, Moskovskie inozemtsy*). Similarly, the non-Russian-speaking, non-Orthodox Christian peoples of the Russian Empire were also considered foreigners, albeit "internal" foreigners.[58] These legal categories show an interesting approach to the issue of own and other: foreignness is not an issue of borders; it is possible to be internal to the empire and still be foreign. This category of internal foreigner based on a religious distinction takes us again back to the work of Heng, who has argued for Jewish populations in medieval Western Europe being treated as a "religious race" by their Christian neighbors; a group can be resident in a region without being accepted as a part of that region.[59] Especially in official documents, there was a Muscovite Self: a Russian-speaking Orthodox Christian born to Russian-speaking Orthodox Christians within the Empire. To speak of a unified non-Muscovite Other would be to go too far, but peoples who would, for various reasons, not be seen as that Muscovite Self were considered part of a generalized category of "foreigner." Geographically, the official Muscovite view saw Russia as connected to the rest of the early modern global world. Anthropologically, Russian officials created a greater distance, one that was understood primarily in terms of religion.

DRUGS AND THINGS

And so we can now return to the point with which we began: things as exotic (or not). We earlier noted that the Moscow markets of the 1690s stocked any number of foreign medicaments, like cinnamon, nutmeg, mace, spermaceti, French wine, and sassafras.[60] Those same medicaments were prescribed by the Apothecary Chancery, and written about in contemporary Russian-language medical books. This issue of foreign medicaments in early modern Russia has been an understudied topic, as scholars working on Russian medicine have thus far focused on medical institutions and practitioners, and those concerned with Russian botany have focused on local plants and plant-collecting practices.[61] So, do any of those sources treat foreign materia medica as exotic? And how did those ideas about foreign peoples and foreign places that we just examined intersect with and impact ideas about foreign materia medica?

One source in particular demonstrates that peoples, places, and medicaments were considered together. Discussing the variety of medical treatments from around the early modern global world in 1664, Samuel Collins wrote:

> There are in creation peoples whose lives, customs, and thoughts are only of medical drugs; not one nation appears so unthoughtful as to—whether by accident or out of certain unavoidable necessities—fail to acquire and put into use drugs whose virtues are hidden from others. Brazilian tribes in America, naked and illiterate, nevertheless have their own, not unworthy medicines, having brought into use the tree sassafras, guaiacum, jalap, and many other plants of unusual powers. Miracles are told of the Chinese doctors, by whose art serious diseases are driven away without venesection or blood-letting, and with only the use of herbal simples of their own creation. Indians drive out illnesses with the steam from boiling herbs with special properties, not with the oils that slaves apply and gently scrape from nobles by hand, to which habit the nobles are so accustomed that they cannot sleep without it. Every year the great Mogul [sic] emperor of the Indians would elevate one of his doctors, and that doctor would hold forth his judgements on the excess or deficits of the emperor's body.... Arabs, Chaldeans, Greeks, Latins, allow any medicines (judged in their eyes to be healthy).[62]

As with his coffee report, here Collins lists botanical substances and healing practices by linking them both to a specific geographical space, and in this case to particular groups of people. This exemplifies the interconnectedness of ideas of the foreign. Just as anthropological thought was commonly joined to geographical, botanical and medical thought could also be linked to both. Ideas about materia medica were formed and reformed in interaction with the geography and peoples associated with those objects.

What are Collins's ideas about these clusters of drugs, places, and peoples? Let us begin where he does, with the Brazilians. The statement that they are "naked and illiterate" closely follows a common Western

European stereotype of Native Americans as in a state of nature.[63] Moreover, Collins's botanical geography is similarly the product of a particular Western European view on the Americas. All of the plants he mentions grow substantially to the north of Brazil—already understood to occupy the northeastern part of South America by the 1660s— and so were used by other Native American groups than those living in Brazil. Notably, both guaiacum and jalap were used by the Mexica, often referred to in the English-speaking world as the Aztecs.[64] The early modern Mexica may have been less heavily dressed than the elaborately attired Europeans, but they were hardly naked. More importantly, the Mexica possessed both a tradition of sophisticated thought, and archives of official documents.[65] The Mexica were neither naked, nor were they illiterate. Similar to the Southern Other, Collins here is using a concept of a Native American Other, where a singular, stereotyped "Brazilian" stands in for all the peoples of the Americas.

The other peoples described by Collins are also of interest. Schmidt has argued that Dutch geographical works of this period flatten out global differences by comparing non-European to non-European, rather than non-European to European.[66] Although Collins makes a brief reference to the Greeks and the Latins, he focuses on comparing Brazilians to Chinese to Indians to Persians, in other words, to comparing non-Europeans to non-Europeans. Schmidt sees this collapsing of global differences in his Dutch sources as leading to the creation of a Disneyfied global world: "Instead of a hotly contested space of exploding imperial antagonisms, the non-European world created by the Dutch abounded with curiosities, diversions and delight."[67] Collins's topic is sickness and healing, which certainly has its violent and gruesome sides. Yet—aside from the mention of bloodletting—his text similarly presents a pleasant view of productive and intriguing global healing practices, creating an image rather similar to that delightful non-European world Schmidt sees in his contemporary Dutch sources.

Collins—the British physician who put so many of his thoughts into the Apothecary Chancery documents—tended to adopt a Schmidtian exoticizing view of the non-European world, flattening out differences between broad geographical spaces. Other works circulating in Russia in this period similarly group objects from a broad geographic range into a single category. Across the seventeenth and eighteenth centuries, the most commonly circulating medical text was the *Garden of Health*, a herbal first translated from the 1485 low German *Gaerde der Sundheit* into the Slavonic *Blagoprokhladnyi vertograd* in 1534, and then copied and revised many times.[68] Among the many extant versions of this work was a new arrangement created in 1672, and given the slightly modified title of *Prokhladnyi vertograd*, a text that was still circulating at the end of the seventeenth century and into the eighteenth.[69] In that text, herbs are described as either Russian, or *zamorskie*, foreign, literally "overseas."[70] Other than being an interesting category for a land empire, the category "overseas" generally serves to create a boundary: foreign things come in from across the sea. Certainly this was often the case: a substantial quantity of foreign goods entered Russia via the great northern port of Arkhangelsk, a Baltic port like Narva, or (after its founding in 1703) the port of St. Petersburg. Conversely, a substantial quantity of goods from Asia entered Russia through long-established central Asian land routes in this period, being imported by key middlemen merchants, the Bukharans, the Armenians, and the Indians.[71]

But does "overseas" mean exotic? This term seems to mirror the English Nicholas Culpeper's use of the word "outlandish" to describe certain medicinal herbs in 1652.[72] According to Alix Cooper, Culpeper's use of this term was a part of his conceptual oppositions, where indigenous herbs were cheap, good, and healthy, and "outlandish"—or foreign—herbs were expensive, questionable, and potentially dangerous.[73] In the same way as the Muscovite category of "foreigner" elided differences between various groups, "overseas" puts all non-Russian plants into a large, non-distinct group of Other plants. In contrast to Culpeper, the Slavonic *Garden of Health* does not dwell on the properties of these "overseas" herbs, and so parsing the meaning of the category remains tricky.

Moreover, "overseas" was not the only category which Russians applied to foreign plants. There was a contemporary term closely analogous to the English word "spice": *priannost*, alternatively *priannoe zelie*. This is a common category from this period of Russian sources, such as in a trade document from 1662 about the department sourcing cloves and other unspecified spices (*priannye zel'i*) from the annual fair at Arkhangelsk.[74] When the category is used next to a list of objects described by it, we can see that it contains similar items to the modern English term "spice": East Asian plant commodities like nutmeg. These commodities had been available in the eastern Slavic lands for centuries, being moved along the same trade routes used to sell Siberian furs across Eurasia.[75] Like the term "overseas," here we have a specific category of foreign materia medica; unlike the term "overseas," "spices" acknowledges at least some differences in the non-Russian botanical world.

Following Cooper, we should consider not only about what Russians thought about foreign materia medica, but also what they thought about local herbs. A. B. Ippolitova has extensively examined early modern Russian ethnobotanical ideas, concepts relating to local plants, and in particular those plants in Russian texts that somehow earned fabulous names. Some plants were given royal names, like "The Tsar's Eyes" (*tsarevye ochi*), or Marian and Holy names, like "The Weeping Plant" (*plakun'*), which plant, according to Russian folklore, first sprung from the earth dampened by the falling tears of Mary as she watched the Crucifixion.[76] Ippolitova has shown that plants given these names typically distinguish themselves in some way, such as by being brightly colored or highly fragrant.[77] Royal, Marian, and Holy plants show a Muscovite trend toward categorizing the natural world, but these categorizations are based on the characteristics of the plant itself, not on botanical geography.

Alongside actors' explicit categories, we can sometimes also discern unvoiced groupings. In 1694 the Apothecary Chancery bought sassafras, the bark of a tree that grew in Florida and neighboring areas along the Eastern Seaboard of North America. It was then a very foreign drug. Sassafras had been available in Russia since at least 1602, and was one of several American herbal medicines regularly imported into Russia across the seventeenth century, alongside sarsaparilla, guaiacum, and cinchona.[78] In 1664, Collins wrote about several of those commodities together, suggesting that the Apothecary Chancery may have seen American drugs as a coherent category within the natural world.

In Collins's 1664 report, American drugs are treated as a discrete group.[79] When those American drugs were written about or prescribed by his colleagues, however, they were dealt with variously. The Apothecary Chancery's *Pharmacopoeia*, first created from Latin sources in 1676 and extant in a 1700 manuscript, the *House and Field Pharmacy* text presented to Peter the Great in the 1690s, and both the 1738 and 1760 editions of *Florin's Economy*, are all Russian-language medical books including recipes that call for American plants.[80] In these cases, the American items are prescribed alongside materia medica imported from elsewhere in the early modern global world. Similarly, in a collection of Apothecary Chancery prescriptions from 1698, the American drugs sassafras and sarsaparilla are prescribed alongside anise, cinnamon, and senna.[81] Extant recipes and prescriptions from this period jumble together American drugs, other foreign drugs, spices, animal parts, chemicals, and any number of locally sourced items. In the final usage, even things as foreign as Floridian tree bark were not separate.

What is "exotic," anyway? And what might Siberian cartography and portraits of African Russians have to do with understanding the exotic status of foreign drugs in late seventeenth-century Russia? To justify my openly troublemaking title for an essay in a volume on exotica, I should now try and answer that question. We already know that Muscovites would not have used the term "ekzotichnii," but they had any number of other terms for foreign places, foreign peoples, and foreign things. Moreover, the major medical institution of early modern Russia, the Apothecary Chancery, relied upon Western European medical experts and Western European medical texts; other Western European works—notably on geography—also circulated in the Empire. Russia had access to the very people and texts by whom and in which Schmidt sees the exotic as having been invented. To think about Russia's experience of the foreign is to deal with their rethinking—as Bogdanov has put it—of Western European ideas of the exotic.

Those Western European ideas, and Russian ideas reformed in the presence of Western European views on the exotic, can be found in late seventeenth-century and early eighteenth-century Russia. Collins's work on coffee, and the attitudes toward enslaved foreigners, seem both to indicate an orientalizing view of the southern lands as undifferentiated, and of the denizens of that region as an Arabic-speaking, coffee-drinking, dark-skinned Southern Muslim Other. The prolific Samuel Collins provided the Apothecary Chancery with other texts that cast an exoticizing gaze on the rest of the early modern global world; in these texts, he ignores differences between the various peoples of the Americas and collapses distance between his stereotyped Brazilians and various other healers from around the world. Muscovy, linked to Western Europe in any number of ways, also shared some Western European ideas about the non-European world.

Yet those ideas really were, as Bogdanov insists, rethought in the Russian context. The Muscovite attitude to foreign places, foreign peoples, and foreign things was complex and multifaceted, and what extant documents most clearly show us is the official view of the foreign. According to that view, foreigners could exist within the Empire; Africans could govern parts of it; commodities from around the world could be—quite literally—thrown together to make medicines. This should all be seen in the light of Kivelson's idea of the concept of "betweenness" as the driving force in official Muscovite geographic thought, how early modern Russian officials created a view of the Empire as a part of a geographically contiguous space. Remezov and other Russian officials saw themselves living in a continuous geographical space that linked them to the other major cities and empires of the early modern global world. As the Americas could exist in the same geographic spectrum as Tobolsk, putting sassafras, spices, and local goods together implies that each was a part of a greater whole, a more fundamental unity. Muscovite officials were influenced by the idea of the exotic they heard from their Western European contemporaries, but for them the Russian and non-Russian worlds had more that connected them than that set them apart. For Muscovites, the world was global, but also distinctly unexotic.

2

GALENIC BODIES AND JESUIT BEANS

Consuming Drugs in Manila at the Turn of the Eighteenth Century

Sebestian Kroupa

In 1730, while waiting in Seville for his ship to New Spain, the Swiss-German Jesuit missionary Philipp Segesser sent home "twenty-five St Ignatius beans" which "are greatly desired both in Rome and in Vienna, and one has to consider it a blessing to get one or two."[1] The precious item hiding under this name was a powerful drug native to the Philippine Islands. As the Jesuit missionary Pedro Chirino wrote in 1604, "this very appetite for drugs caused the Spaniards, or Castilians, to discover and settle the Philippines, as is well known."[2] After establishing a firm foothold in the archipelago in 1571, the Spanish crown had high hopes that its new dominion would become another set of Spice Islands. However, despite early Spanish efforts to bioprospect for valuable spices and other medicinals, by the late seventeenth century these projects had still failed to yield the desired outcome.[3] In eighteenth-century European works of materia medica, only one Philippine import features regularly: the St. Ignatius bean.[4]

Rather than flowing outward to lucrative markets in Europe, drugs tended to travel the other way in the early modern Philippines. Spanish imperial visions notwithstanding, the crown expended considerable sums to import to Manila vast quantities of proven Galenic remedies, associated with the humoral tradition.[5] These developments contrast with previous studies of the consumption of drugs in colonial worlds, which have tended to highlight the insufficiency of European medicines in such spaces and the colonists' hunger for "green gold," useful plants that could be turned into profitable commodities.[6] The principal exception to this reverse trajectory was the aforementioned St. Ignatius bean. As the name suggests, the bean's distribution was closely associated with the activities of the Society of Jesus, whose engagements in the drug trade are explored in chapter 7 of this volume by Samir Boumediene.[7] This chapter considers the reasons behind the continuous efforts of the Spanish crown to supply its colonies with Galenic remedies, alongside the circumstances under which an existing and long-standing pharmaceutical tradition could be disrupted by a novel drug like the St. Ignatius bean. By shifting the geographical and cultural vantage point to patterns of drug consumption in Manila, my aim is to question and decenter European notions of the "exotic."

To explore the kinds of drugs available in Manila around the year 1700, I draw on registers of drugs stocked by the Hospital Real de Españoles (Spanish Royal Hospital), and the works of Georg Joseph Kamel (1661–1706), a Jesuit lay brother pharmacist who ran the Society's local apothecary shop.[8] Kamel also wrote a treatise on the St. Ignatius bean that provides insights into the drug's appropriation by European colonizers, which is explored in the second half of this chapter. The records from the Hospital Real and Kamel's pharmacy point to the skepticism of colonial medical authorities toward foreign substances and their adherence to established European pharmaceutical traditions. While the crown supplied the Hospital Real largely with Galenic medicines brought at high cost across the Pacific, Kamel combed Philippine nature for plants that could serve as substitutes for Galenic remedies that he knew from Europe but were not available to him in Manila. Although Kamel adopted native plants, while the Hospital Real relied on imported medicines, I argue that both their understandings of what constituted a cure were underpinned by identical Galenic concerns with the body. Any differences in the remedies employed may largely be ascribed to disparities in funding.

This shared commitment to Galenic remedies conceals deeper anxieties of Spanish colonial authorities regarding foreign materia medica, as well as entanglements of drug consumption with political and economic matters. Rebecca Earle, Linda Newson, and Paula De Vos have called attention to Spanish suspicion of substances encountered in colonial spaces.[9] In her study of pharmacists' practices in colonial Peru, Newson has attributed Spanish preoccupation with Galenic drugs to deeply entrenched humoral doctrine and religious suspicions, reinforced by the institutions of the Protomedicato and the Inquisition.[10] While institutional frameworks contributed to the continued supply of Galenic remedies to Manila, here I examine the underlying colonial logics and the intertwined social and natural orders that underpinned Spanish adherence to Galenic drugs. This commitment was shaped by the political, religious, and economic concerns of Spanish colonial governments. But more than just a top-down imposition of bureaucratic authority, Galenism prevailed because it resonated with prevailing European ideas about climates, cures, and bodily constitutions: assumptions that made sense not only to colonial officials but also to consumers.

The humoral system posited an intricate link between bodies, substances, and environments: through the four qualities of hot, cold, wet, and dry, the human body was tied to the wider macrocosm. Substances and climates were characterized in terms of the four qualities, as well as featuring among the six Galenic non-naturals which exerted influence on the humoral equilibrium of bodies. Moreover, different environments bred distinct kinds of humoral balance, or constitutions. The Spanish, native to dry and hot Iberia, were considered fierce and choleric, whereas Indigenous bodies in Spanish colonial spaces were often conceptualized as full of phlegmatic humors, which made them prone to apathy and conquest. Variations in the stars, airs, and the substances consumed therefore offered coherent explanations for perceived differences in the bodies and characters among ethnic groups, including the colonizers and the colonized.[11]

The Spanish used ideas rooted in Galenic humoralism to validate their superiority and rule over subjugated populations. However, through exposure to inappropriate climates or substances, Spanish bodies risked humoral deterioration, which could occasion disease and even death. As Earle has argued, to protect their superior humoral constitution, the Spanish supplied their colonies with their own domestic foods. Due to the perceived links between bodies, diet, and ethnic hierarchies, the distinctions between European and American foods also developed into markers of the superiority of European culture.[12] Here, I extend this argument to the consumption of drugs. Given the blurred boundary between foods and drugs in humoralism, it is more productive to treat histories of foods and drugs together as histories of consumption. The crown, alongside colonial practitioners like Kamel, favored proven Galenic medicines because they were certified by centuries of tradition and considered better-suited for protecting the humoral constitutions of the colonists' bodies. As part of a sanctioned medical system deployed by the Spanish Empire to assert authority over colonial spaces and bodies, the drugs stood for the cultural and epistemological superiority of European medicine.

The case of the St. Ignatius bean reveals that the humoral theory not only underpinned Spanish commitment to Galenic remedies, but also provided a compelling framework for engaging with colonial natures. To appropriate foreign materia medica from colonial contexts into recognized European frameworks, Jesuit missionaries engaged in the "Galenization" of non-European substances, or their systematic translation into the Galenic humoral system.[13] Kamel identified the Philippine drug with the *Nux vomica*, or "vomic nut," of Serapion the Younger: a twelfth-century Christian physician writing in Arabic, whose work had become assimilated into the Galenic corpus.[14] Yet, in line with Galenic theory, the question of the bean's compatibility with European bodies remained an obstacle to its adoption. For European consumers, ingesting foreign or unknown substances was a matter of trust that posed its dangers. As a disciple of Galenism, Kamel therefore tested the suitability of Philippine drugs for European humoral constitutions—or for European medical clients, both in Manila and beyond. In these trials, he used the bodies of colonial, Indigenous, migrant, and enslaved individuals as instruments. As such, Kamel embedded the Philippine substance within medical humoralism, endowed its virtues with clear theoretical foundations comprehensible to European practitioners and customers, and paved the way for its deployment on both local and global scales and markets.

The continued Spanish provision of proven Galenic remedies to Manila's Hospital Real, along with the fortunes of the St. Ignatius bean, provide insights into the increasing globalization of European pharmacopoeias. Together, these case studies offer a lens through which to observe epistemic, bodily, and geographical translations of medical knowledge. To explore these dynamics further, the final section of this chapter touches upon the introduction of chemical and American medicines in Manila and their integration into Spanish pharmacopoeias. The Philippines thus emerge as an important node in the increasingly global networks of the drug trade, where medical knowledge, practices, and substances from different corners of the world intersected.

THE HOSPITAL REAL: SPANISH DRUGS FOR SPANISH BODIES

In 1571, fifty years after Ferdinand Magellan had claimed the Philippine archipelago for the Spanish crown, the conquistador Miguel López de

Legazpi established the Spanish city of Manila at the heart of the recently conquered Indigenous state of Maynila. Into the eighteenth century, the city supported a fragile Spanish presence which struggled to impose control over the broader archipelago due to a shortage of financial, human, and other resources.[15] Although the Philippine colony was formally governed from peninsular Spain, in practice much of the administration was managed by the Viceroyalty of New Spain for logistical and economic reasons. Under the Treaty of Tordesillas (1494), the route around Africa fell within the Portuguese sphere of influence and was later dominated by Dutch and English rivals of Spain. Moreover, by the late sixteenth century, Ming China's adoption of silver-based economy generated immense demand for the precious metal, extensively mined in Spanish America.[16] Every year, enormous quantities of silver were carried by the Manila galleon dispatched by the Viceroyalty of New Spain, alongside shipments of material, fiscal, and military aid. The impatiently anticipated *socorro* (aid) and *situado* (salary) were essential to sustaining Spain's frail colonial presence.

In addition to the disconnection from peninsular Spain, colonists in the Philippines faced dangers of unfamiliar environments, diseases, and diets that threatened their health. Lamentations about unhealthful climate and anxieties about sickness abound in colonial records. These concerns with health have attracted growing scholarly interest in histories of medicine in the early colonial Philippines.[17] The healing landscape in Manila was marked by persistent tensions between the visions of Spanish Catholic imperialism and the realities of the colonial situation. Due to the close ties between healing, Christian charity, and missionary work, as well as the scarcity of licensed medical practitioners, religious missionaries represented the main face of the colonial medical establishment. They founded and managed institutions, including hospitals and pharmacies, to cater for the physical and spiritual needs of the local communities of European, Indigenous (*Indios*), and Chinese (*Sangleyes*) ancestry, alongside various enslaved and migrant individuals. Healing practices that did not conform to the sanctioned religious and medical systems of Catholicism and Galenism were subject to demonization and policing. Yet recent studies of *hechicería* (witchcraft) Inquisition trials have revealed Manila as a site of vibrant medical pluralism, populated by practitioners from different cultural and ethnic backgrounds who drew upon European, Indigenous, Chinese, and other medical and spiritual practices.[18] The decrease in the number of *hechicería* trials after 1650 has been attributed to the success of the Spanish evangelization project.[19] However, rather than an actual decline in the incidence of *hechicería*, this change may reflect a shift in Spanish institutional priorities, or constitute an artefact of archival preservation; references to witchcraft persist in missionary accounts well into the eighteenth century.[20] These documents, alongside the developments discussed here, reflect the enduring anxieties of the colonial establishment over the regulation of remedies and the control over who could produce, prescribe, and consume them.

Spanish colonial medical authority in Manila found its manifestation in the establishment of the Hospital Real de Españoles. It was to protect the colonists, and due to "the much greater expense to the King, our sovereign, of transporting one soldier here than of supporting two," that the hospital was erected at the foundation of the city.[21] Similar royal hospitals were founded in every major settlement throughout Spanish colonial territories.[22] Since their institution was tied to military activity and efforts to control local populations, scholars have conceptualized royal hospitals as tools of colonization and conquest.[23] We gain insights into the workings of Manila's Hospital Real around the year 1700 from several official reports. In 1690, the new governor-general of the Philippines, Fausto Cruzat y Góngora, informed the crown that the hospital employed "a physician, a pharmacist, a blood-letter, and a surgeon's assistant."[24] The building held "four infirmaries with fifty to sixty patients," and "a pharmacy from which remedies necessary for its patients are dispensed [and] sold to different individuals."[25] In 1711, the newly appointed hospital chaplain Joaquín Ramírez described how, "every month, twelve servants come to work in the hospital . . . : two for the medical room, two for the chapel, two for the surgery, two for the porter's lodge and the laundry room, and two for the kitchen; and the two who remain are to apply themselves to the rooms with the sickest."[26]

Admission to Manila's Hospital Real was governed by colonial logics of social order rooted in Spanish policies of ethnic hierarchization, which fed into the development of the so-called *casta* system. The hospital was destined for the treatment of Spaniards in royal service, namely military personnel, and state and religious officials."[27] Spanish women received care separately, alongside enslaved and mixed-ancestry individuals, in the Hospital de San Juan de Dios. Additional institutions catered for lepers (the Hospital de San Lázaro), *Sangleyes* (the Hospital de San Gabriel), and *Indios* (the Hospital de Naturales).[28] Despite these policies of segregation, ethnic identities in Spanish colonial worlds were negotiable and contested. In the Philippines, individuals of non-European descent were commonly recruited into the Spanish army.[29] Yet access to Manila's royal hospital formally remained a Spanish male privilege.

Throughout the Spanish colonial period, the administration of the Hospital Real was a cause of rows between different religious orders, the secular clergy, the crown government, and other stakeholders.[30] Royal hospitals operated under the crown's patronage with their expenses covered by the royal treasury. The two most important positions—*mayordomo* (administrator) and *capellán* (chaplain)—were royal servants appointed by crown officials. The funds and the personnel required caused Philippine authorities constant headaches. In his fourteen-page complaint written in 1711, the chaplain Ramírez portrayed the institution as a crumbling ruin and a nest of vice, run by incompetents. Allegedly, the *mayordomo* commonly abused hospital servants for his personal needs; the physician came in when he wanted rather than when required; his assistant was a former delinquent who boozed all day; the patients fornicated with the servants and indulged in gambling; and the pharmacy lacked necessary supplies, many of which had been sold off to external buyers for profit.[31] Perhaps in response to this complaint, the Bourbon ascent to the Spanish throne in 1714 led to reforms of the hospital.[32]

The remedies supplied to the hospital lay at the heart of these reforms. Ramírez explained that medical provisions were shipped annually across the Pacific from New Spain as part of the *socorro*, a process that he found protracted, expensive, and unreliable, with many medicines arriving spoiled or damaged. The shipments were organized by colonial officials, and the cost was covered by the treasury. Every year, the hospital's *mayordomo* submitted a list of required goods to the governor-general of the Philippines, who forwarded it to New Spain. There, the requested provisions were purchased from contracted pharmacists, and added to the next *socorro*.[33]

These ventures took place at no small expense to the crown. In 1699, Cruzat y Góngora confirmed the receipt of goods for the hospital valued at nearly 21,500 pesos, equivalent in value to over half a ton of silver.[34] By comparison, the annual salary of the hospital's physician was 300 pesos.[35] These shipments included a wide assortment of supplies, from blankets and surgical instruments to books and wine for celebrating masses in the hospital chapel. Yet the most costly and prominent

TABLE 2.1. MEDICAL PROVISIONS BROUGHT FOR THE HOSPITAL REAL DE ESPAÑOLES IN MANILA WITH THE 1642 *SOCORRO*

Xaraves (syrups)	Combined weight: 16 *arrobas* (≈ 184 kg)	
Spanish term	English equivalent	Quantity
miel rosada espesada	thickened rose honey	4 *arrobas*
xarave de nueve ynfuciones	syrup of damask roses of nine infusions	1 *arroba*
arope de matlaliste especado*	thickened matlalitztic arrope*	2 *arrobas*
arope de agras	agraz arrope	2 *arrobas*
garave de cantueso	French lavender syrup	1.5 *arrobas*
oximiel sechilitico	squill oxymel	1 *arroba*
garave de artemisa	mugwort syrup	1.5 *arrobas*
xarave de menta simple	mint syrup	2 *arrobas*
xarave de ajenjos	wormwood syrup	1 *arroba*

Aceytes (oils)	Combined weight: 21.5 *arrobas* (≈ 247 kg)	
aceyte de almendras dulces	sweet almond oil	1 *arroba*
aceyte de mancanilla	chamomile oil	3 *arrobas*
aceyte rosado completo	oil of roses complete	3 *arrobas*
aceyte de menbrillos	quince oil	2 *arrobas*
aceyte de espique	spikenard oil	1 *arroba*
aceyte de ajenjos	wormwood oil	1 *arroba*
aceyte de aparicio	Aparicio de Zubia's oil	2 *arrobas*
aceyte de lenti[s]co	lentisk oil	1 *arroba*
aceyte de laurel	laurel oil	0.5 *arroba*
aceyte de catagueia[?]	??? oil	6 *arrobas*
aceyte comun	ordinary olive oil	1 *arroba*

Unguentos (ointments)	Combined weight: 9 *arrobas* (≈ 103.5 kg)	
ungunto rosado	rose ointment	2 *arrobas*
ungunto desopilativo de cumos	deobstruent juice ointment	1 *arroba*
ungunto apostolorum	Apostles' ointment	1 *arroba*
ungunto confortativo	comfortative ointment	1 *arroba*
trementina comun buena	ordinary good turpentine	2 *arrobas*
trementina de abeto	fir tree turpentine	2 *arrobas*

Letuarios y confeciones (electuaries and preserves)
Combined weight: 8 *arrobas*, 3 *libras*, 16 *onzas* (≈ 94 kg)

letuario diacatalicon	diacatholicon	2 *arrobas*
letuario diaphinicion	diaphoenicon	1 *arroba*
xerapliega	hiera picra	1 *arroba*
atriaca magna	theriac	2 *libras*
pildoras de fumaria	fumitory pills	4 *onzas*
pildoras agregativas	aggregative pills	4 *onzas*
pildoras coquias	cochiae pills	4 *onzas*
pildoras de hiera	pills of hiera	4 *onzas*
escamonea	scammony	1 *libra*
acucar rosado	rose sugar	4 *arrobas*

Emplastos (plasters): 7 *arrobas* (≈ 80.5 kg)

emplasto estomaticon	stomach plaster	1 *arroba*
emplasto diaquilon menor	simple diachylon plaster	1 *arroba*
emplasto diaquilon mayor	great diachylon plaster	0.5 *arroba*
emplasto diapalma	diapalma plaster	2 *arrobas*
emplasto geminis	geminis plaster	2 *arrobas*
emplasto meliloto	melilot plaster	0.5 *arroba*

Powders and simples **Combined weight:** 46 *arrobas*, 50 *libras* (≈ 554 kg)

polvos de sueldas	comfrey powder	0.5 *arroba*
polvos de almasiga	powdered mastic gum	4 *libras*
polvos de juanes	red precipitate (mercuric oxide)	6 *libras*
atutia preparada	tutty (cadmia)	2 *libras*
flor de mancanilla en manojos	chamomile flowers in bundles	2 *arrobas*
matlalistic*	matlalitztic*	2 *arrobas*
caqualtipan*	zacualtipán (a purgative sourced in Zacualtipán)*	2 *arrobas*
oja de sen	senna leaves	1 *arroba*
pinaza	pine needles	2 *arrobas*
alholvas	fenugreek	2 *arrobas*
sublimates	sublimates	4 *arrobas*
mirra	myrrh resin	2 *libras*

canfora	camphor	2 *libras*
almasiga blanca	white mastic gum	0.5 *arroba*
albayalde	ceruse	3 *arrobas*
todas las rayces diuireticas [= *esparraguera, hinojo, grama, perejil, rusco*]	the five diuretic roots (asparagus, fennel, dog's-grass, parsley, butcher's-broom)	1 *arroba*
simiente de adormideras blancas y negras	white and black poppy seeds	4 *libras*
rosa colorada y bu[en]a	red roses	2 *arrobas*
*sarsa de mechoacan**	sarsaparilla from Michoacán*	4 *arrobas*
pes griega	colophony	6 *arrobas*
agua rosada en frascos	rose-water in flasks	4 *arrobas*
polvos de rosa	rose-petal powder	4 *libras*
romero	rosemary	1 *arroba*
origano	oregano	0.5 *arroba*
*quannenepile**	cohuanenepili*	0.5 *arroba*
nueces de sipres	cypress nuts	0.5 *arroba*
las tres harinas [= probably *habas, cebada, yervo*]	the three flours (probably fava bean, barley, ervil)	3 *arrobas*
cardenillo	verdigris	1 *arroba*
arope comun	ordinary arrope	0.5 *arroba*
asarcon castellano	Castilian red lead	1 *arroba*
polipodio	polypody	0.5 *arroba*
polvos de polipodio	polypody powder	2 *libras*
anis	anise	0.5 *arroba*
semilla de ynojo	fennel seeds	8 *libras*
*polvos de sarsa**	sarsaparilla powder*	6 *libras*
polvos reales	royal powder (*pulvis basilicus*)	1 *arroba*
alquitira	tragacanth gum	6 *libras*
hermodatiles	hermodactyl	4 *libras*
*tecamehaca**	tacamahaca*	12 bottles

◀ *Table 2.1 Note:* Substances of American origin are marked with an asterisk. One *arroba* is the equivalent of ca. 11.5 kg; one *libra* is approximately equivalent to its modern English counterpart; one *onza* is about 290 g. See Juan Villasana Haggard, *Handbook for Translators of Spanish Historical Documents* (University of Texas, 1941), 72, 79, 81. The identities of the drugs and their English equivalents were determined using a combination of Spanish pharmacopoeias from the late seventeenth century, dictionaries and contemporary works in the history of medicine; see especially Luis de Oviedo, *Methodo de la coleccion y reposicion de las medicinas simples* (Madrid: Melchor Alvarez, 1692); Jerónimo de la Fuente Pierola, *Tyrocinio pharmacopeo* (Zaragoza: Manuel Roman, 1698); Henry Neuman and Giuseppe Marco Antonio Baretti, *Neuman and Baretti's Dictionary of the Spanish and English Languages*, 2 vols. (Boston: Hilliard, Gray, Little, and Wilkins, 1831); José Luis Fresquet Febrer, "El uso de productos del reino mineral en la terapéutica del siglo XVI. El libro de los 'Medicamentos simples' de Juan Fragoso (1581) y el 'Antidotario' de Juan Calvo (1580)," *Asclepio* 51, no. 1 (1999): 55–92; Charles Davis and María Luz López Terrada, "Protomedicato y farmacia en Castilla a finales del siglo XVI: Edición crítica del 'Catálogo de las cosas que los boticarios han de tener en sus boticas,' de Andrés Zamudio de Alfaro, Protomédico General (1592–1599)," Asclepio 62, no. 2 (2010): 579–626; Juhani Norri, *Dictionary of Medical Vocabulary in English, 1375–1550* (Routledge, 2016).

item was drugs for the hospital pharmacy. In 1719, Philippine officials estimated that "the yearly expenses on medicines brought from the Kingdom of New Spain regularly rise to 10,320 pesos."[36]

Closer insights into the contents of these consignments are offered by two rare surviving full inventories of drugs shipped on board the Manila galleon, one dating from 1642 and the other from 1717. The former list included 82 different items that weighed almost 110 *arrobas*, or ca. 1,200 kg (table 2.1).[37] Around 90 percent of the identified substances were proven Galenic remedies, both simples and compounds. The 1717 shipment comprised nearly 300 items in 32 crates, and its overall weight increased to ca. 1,500 kg. More than 90 percent of the drugs were Galenic.[38] According to the records, some remedies had arrived spoiled, damaged, or missing—perhaps they had been used during the trans-Pacific voyage or embezzled. In terms of both quantity and weight, the two shipments were dominated by quintessential Galenic medicines: for example, drugs prepared using roses represented nearly 20 percent of the combined weight alone.

Another inventory, produced in 1718 as part of an official audit undertaken amid the reforms, listed all the drugs, instruments, and books present in the Hospital Real's stock prior to the delivery of the 1717 consignment.[39] The register enumerated 137 medicines, with a total weight of around 315 kg. More than 90 percent of the identified substances were proven Galenic remedies. There were also four American and several chemical drugs, discussed in the final section. The pharmacist had at his disposal six books associated with the Spanish pharmaceutical tradition, including the works of Mesue, which provided a conduit for Islamic knowledge to the Latin West; *Pharmacopoeia Valentina*, first printed in 1601 to regulate apothecary practices in Valencia; and the popular handbook *Tyrocinio Pharmacopeo* by Jerónimo de la Fuente Piérola, which went through numerous editions following its publication in 1660.[40]

In addition to substances present in the pharmacy stocks, the audit recorded those that had recently been used up or discarded (table 2.2).[41] Among the sixty-five identified items, sixty-one came from the traditional Galenic corpus, with the remainder consisting of three American remedies and one Paracelsian one. All but eight of these substances had been listed a century before, in a catalog of goods that Spanish pharmacists were to have in their stocks, signed by Andrés Zamudio de Alfaro, the Protomédico General of Castile from 1592 to 1599.[42] Therefore, it seems that if one visited Manila's Spanish royal hospital pharmacy around the year 1700, the remedies on offer would have been akin to those found in apothecary shops in peninsular Spain. These records highlight the remarkable stability, resilience, and adaptability of the Galenic pharmaceutical tradition within the Spanish Empire.

The provenance of the drugs present in the hospital stocks is unclear, although their composition suggests that they most probably arrived from New Spain with previous shipments. The audit stated that small amounts of supplies were also regularly procured from the royal warehouse (*Almacenes Reales*) in Manila, especially locally produced goods such as palm liqueur (*vino de coco*, or *lambanóg*), and vinegar from the Ilocos region.[43] In addition, the hospital stocked around ten substances native to Southeast Asia, including cinnamon, cubeb, sandalwood, rhubarb, and zedoary. These items had probably been acquired from regional traders, rather than New Spain, but remained in the minority compared to trans-Pacific shipments. Apart for palm liqueur, the substances had long been part of European pharmacopoeias, indicating a near total absence in pharmacy stock of remedies foreign to Europe, whether native to the archipelago or to the surrounding regions. Such medicines were commonly consumed in Manila, and we cannot rule out that some may have found their way into the Hospital Real. Yet the

TABLE 2.2. INVENTORY OF THE RECENTLY CONSUMED OR DISCARDED MEDICAL SUPPLIES FROM THE PHARMACY OF THE HOSPITAL REAL DE ESPAÑOLES IN MANILA (1718)

Spanish term	English equivalent	Quantity
goma armoniaco	gum ammoniac	0.5 *libra*
cascarilla del Perith [= Perú?]*	cinchona bark*	6 *libras*
oropigmiento	orpiment	1 *libra*
piedra lipis	lapis lazuli	0.5 *libra*
caveza de adormideras	poppy heads	1 *libra*
cañosuelas [= *cañas sueltas?*] *de rosas*	roses, loose-stemmed	2 *onzas*
cortesas de granada	pomegranate peels	6 *libras*
cortesas de zidras y naranjas	lemon and orange peels	0.5 *libra*
semilla de cubovas	cubeb seeds	1.5 *libras*
calamento	calamint	3 *libras*
flor de epitimo	dodder flower	2 *libras*
flor de borrajas	borage flower	1 *libra*
xarave de epitimo	dodder syrup	11 *libras*
conserva de torongil	lemon balm preserve	4 *libras*
sumo de acacia	acacia juice	2.5 *libras*
unguento de a[r]tanita	sowbread ointment	22 *libras*
raiz de pelitre	pellitory (Spanish chamomile) root	2 *libras*
esquinanto	camel grass	2 *libras*
raiz de seduario	zedoary root	2 *libras*
raiz de valeriana	valerian root	1 *libra*
simiente de azederas	sorrel seed	1 *libra*
azeite de dialthea	dialthea	1 *libra*
simiente de peregil	parsley seed	1 *libra*
semilla de apio	celery seed	2 *libras*
salvia	sage	1 *tenate*
tomillo	thyme	1 *tenate*
trebolo	clover	1 *tenate*
parietaria	pellitory-of-the-wall	1 *tenate*
torongil	lemon balm	1 *tenate*
raiz de apio	celery root	1 *tenate*
marullos [= *marrubios*]	white horehound	1 *tenate*
meliloto	melilot	1 *tenate*
raiz de borrajas	borage root	1 *tenate*

raiz de lirios	lily root	1 tenate
raiz de brucio	butcher's-broom[?] root	1 tenate
ruibarbo	rhubarb	6 libras
epildoras	pills[?]	1 libra
trosiscos, de espodia	troches of spodium	1 onza
eligir propietatis	elixir proprietatis	1 onza
semilla de membrillos	quince seed	3 libras
semilla de agras	agraz seed	3 libras
xarave de cantueso	French lavender syrup	10 libras
xarave de zarsa*	sarsaparilla syrup*	10 libras
xarave de echicoria	chicory syrup	10 libras
frazqueras desquadernadas, sin vidrios, ni llaves	disorderly bottle-cases without bottles or keys	4
caxa sin llave	box without the key	1
raiz de peonia	peony root	1 libra
raiz de brionia	bryony root	0.5 libra
raiz de gencian	gentian root	1 libra
eufracia	eyebright	0.5 libra
raiz de azaro	asarabacca root	4 onzas
raiz de erebolo blanco	white hellebore root	1 libra
raiz de yndivia	endive root	3 libras
cortesas de palo de guayacan	guaiacum bark*	4 libras
eleboro negro	black hellebore	0.5 libra
raiz de bruzco	butcher's-broom root	4 libras
semilla de lechugas	lettuce seed	2 libras
palo de balsamo	xylobalsamum (probably American)	2 libras
mirabolanos de belericos	beleric myrobalan	8 libras
mirabolanos quebulos	chebulic myrobalan	3 onzas
mirabolanos emblicos	emblic myrobalan	6 libras
albayarde	ceruse	1 libra
bleo	blite	[not recorded]
palo de taray	tamarisk wood	2 libras
ojas de sabina	savin leaves	1 libra
cortesa de mejo del sol	common gromwell rind	4 libras
calamo aromatico	sweet flag	1 libra
semilla de eneldo	dill seed	4 libras
semilla de agno casto	chaste tree seed	4 libras

evidence suggests that the institution's management sought to procure predominantly established Galenic remedies from New Spain, in addition to sourcing a limited number of substances locally.

Spanish importation of drugs to Manila at great expense and distance was rooted in a combination of political, economic, religious, and medical concerns. In her survey of pharmacists in early colonial Lima, Linda Newson has ascribed their commitment to Galenic remedies to the colonial institutional landscape.[44] The humoral doctrine was perpetuated by professional training, and backed by the state and church organizations of the Protomedicato and the Inquisition. The situation was similar in New Spain, where drugs for Manila's Hospital Real were sourced.[45] While institutional frameworks certainly contributed to the steady supply of Galenic drugs to the Philippines, the case of the St. Ignatius bean highlights that the deeply entrenched humoral system could facilitate as much as obstruct the adoption of foreign substances. Rather, I suggest that the continuous provision of Galenic remedies reflected the underlying and entangled social and natural orders of Spanish colonial regimes.

The contents of the consignments were partly shaped by the power dynamics within the Spanish colonial establishment. The lucrative exportation of medical supplies was supported by both state and medical institutions in New Spain, which leveraged their power to generate income and reinforce control over the Philippine colony. By overseeing the provisions, authorities in New Spain provided local suppliers with access to lucrative contracts. These deals not only guaranteed more secure earnings than bioprospecting projects, but also ensured that some of the silver intended for export stayed in Mexico. For the Philippine colony, by contrast, trans-Pacific consignments incurred significant costs. As Ramírez argued in his complaint, such high expenditure was unnecessary, since many remedy supplies for the Hospital Real could be sourced locally.[46] In response to such pleas, the crown eventually decided that drugs for the hospital should be procured using more proximate networks. The 1717 shipment was the last one formally received by the Philippine government. Authorities in New Spain initially showed disregard for the new royal policy, and continued to dispatch medicines to Manila until 1720. But local officials returned the unsolicited cargo, and complained to the king.[47] In 1739, Philippine authorities reported that the reforms had resulted in savings of nearly 120,000 pesos.[48] The reform of the medical provisions for the Hospital Real in Manila therefore offers insights into interactions and frictions between different branches of the colonial state.

Furthermore, the continued supply of drugs to Manila reflected the anxieties of the colonial regime over the humoral vulnerability of the colonizers' bodies, and the perceived threats posed by non-orthodox practices peddled by those who were alleged to be *hechiceros*. Colonial authorities favored Galenic remedies because they were certified by centuries of tradition, sanctioned by the establishment, and considered better suited to serve the needs of Spanish bodies, due to the connections that humoralism posited between environments, substances, and bodily constitutions. In contrast, ingesting foreign and unproven substances posed medical and spiritual risks for the colonizers' bodies and souls. Offering an orthodox, Spanish alternative to local medical pluralism was central to the wider colonial efforts to manage health, discipline bodies, and assert epistemic authority.

The colonizers were constantly reminded of the dangers of the local environment by encountering unfamiliar plants and animals, as well as Indigenous bodies considered inferior in terms of their humoral makeup. As the Augustinian friar Gaspar de San Agustín wrote in 1720, the constitution of Indigenous Filipino communities "is cold and humid, because of the great influence of the moon. . . . This disposition and influence make them fickle, malicious, untrustworthy, dull, and lazy; . . . they have little courage, on account of their cold nature, and are not disposed to work."[49] The unhealthful Indigenous bodies were seen as the product of the deleterious Philippine climate, which also persistently threatened the lives of Spanish settlers. The official Hernando de los Ríos Coronel complained in 1594, "Due to the difference of the climate from that of our Spain, many people die."[50] Still a century later, when the Jesuit pharmacist Kamel struggled to cure the newly arrived *oidor* (judge) Juan de Ozaeta y Oro, he advised the sufferer to "seek another dwelling, more appropriate to his constitution."[51]

By importing large quantities of proven Galenic medicines, the crown therefore sought above all to supply its colonists with drugs suitable to their constitutions, which would protect their bodies from unsanctioned healing practices, and unfamiliar climates and substances. The trans-Pacific shipments reflected crown policies of ethnic segregation, which spilled over into medical care. Since access to the Hospital Real was officially restricted to Spaniards, the provision of Spanish drugs was meant to protect Spanish male bodies—primarily soldiers and officials, the bedrock of the colonial apparatus. Access to these remedies was a privilege of being Spanish which was closely tied to, as well as reinforcing, Spanish identity. At the turn of the eighteenth century in Manila, the consumption of drugs was therefore a marker of ethnic identity and cultural and bodily supremacy, as well as a means of preserving the superior Spanish humoral complexion. The channel supplying these drugs kept the Philippines Spanish, and tied the archipelago to the distant heart of the empire.

THE PHARMACY OF GEORG JOSEPH KAMEL: INVENTING GALENIC DRUGS

By way of comparison with the Hospital Real, I now turn to the projects of Georg Joseph Kamel, a Jesuit pharmacist in Manila who extensively used Philippine plants in his medical practice. Kamel was born in 1661 in the Habsburg Monarchy, in the city of Brno (present-day Czech Republic), where he attended the local Jesuit school. He entered the Society of Jesus in 1682 as a lay brother, and was trained in pharmacy.[52] In 1687, Kamel was selected by the Bohemian Provincial for the Philippine mission.[53] After arriving in Manila sometime between 1688 and 1689, Kamel opened the first Jesuit pharmacy in the Philippines.[54] His workshop was located inside the Jesuit Colegio de San Ignacio, within the city walls of Manila (fig. 2.1). In addition to working in his dispensary, Kamel also "saved the college the salary paid to the physician, as he himself filled this position."[55] Given the limited number of credited colonial practitioners in Manila, Kamel quickly established himself as a medical authority. As he explained in a letter to his compatriot Šimon Boruhradský in México, "there is no physician here but four friars who know little more than my pair of trousers."[56] Under Kamel's tenure, the humble Jesuit workshop became a major colonial medical establishment.

In his treatments, Kamel extensively relied on materia medica native to the Philippines, which he described in several treatises published in London.[57] These accounts include cases from his medical practice that shed light on his patients and the drugs employed. All patients for whom Kamel provided closer information were men, but he commanded a more diverse clientele than the Hospital Real, largely composed of fellow Jesuits, but also including colonial officials, the poor, enslaved individuals, and Indios. One of the native plants Kamel used in his practices was the rush-like herb called "tanglat" in Indigenous Filipino languages.[58] He wrote that a syrup or decoction prepared from tanglat "stimulates the evacuation of urine, menses, fetuses, gas, watery edema, and renal calculi; the latter I have experienced with Pedro de Sylva," a

FIGURE 2.1. Map of Manila in 1671 (AGI, MP-Filipinas, 10). The Jesuit Colegio de San Ignacio, which housed Kamel's pharmacy, was located in the southernmost tip of the Spanish walled city (letter Y). Reproduced with permission of Archivo General de Indias, Seville.

fellow Jesuit. Kamel also successfully trialed "water distilled from fresh bulbous heads" as a laxative on another Jesuit, Pedro Silvestre Navarro, who "suffered from considerable constipation."[59]

When using Philippine substances in his medical practice, Kamel interpreted them through the prism of Galenic drugs and humoral theory. He adopted tanglat because he had previously identified it with camel grass, a rush-like plant discussed by Dioscorides and Galen, employed for promoting bodily evacuations and digestion.[60] His accounts suggest that Kamel combed Philippine nature for plants cognate with those he already knew, which he could use as substitutes in the treatments he had learned at home. This approach drew on the established pharmaceutical practice and genre of *succedanea* ("replacements") or *quid pro quo* ("this for that") modeled on the treatise *De succedaneis*, attributed to Galen. This text provided practical alternatives for ingredients that were rare, expensive, or simply unavailable at the time of need.[61] In Europe, medical authorities issued lists of accepted substitutes that pharmacists were to follow. But in Manila, Kamel was at liberty to identify and trial his own *succedanea*. His methods followed handbooks of *quid pro quo*, which advised that substitutes should agree in virtues, nature, and sensory characteristics, such as taste and smell. Most of his substitutes, including tanglat, were based on morphology alongside sensory cues. For example, he used the herb called "suganda" in Indigenous Filipino languages in place of oregano, due to their similar odor.[62] Kamel's investigations of Philippine nature, therefore, seem not to be motivated by hunger to discover new cures so much as by the desire to find substitutes for those already known to him.

Kamel explored his vicinity for locally available substitutes, largely because importing drugs was so costly, protracted, and unreliable. Compared to the crown, the funds available to the Jesuit Philippine Province for purchasing drugs were considerably more modest. To finance their medical activities, religious orders relied on combinations of alms and donations, their own coffers, and sponsorship from the crown: the so-called *limosna de medicinas* ("alms for medicines"). The amount annually dispensed to the Philippine Jesuits was a meagre 150 pesos, around 1.5 percent of the sum spent on drugs for the Hospital Real.[63] As the Jesuits regularly complained, this royal subsidy was hardly sufficient to procure adequate medical provisions. In 1686, the Philippine *procurador* Luis de Morales lamented that "the medicines are of such bad quality that they are more likely to harm than help," and, therefore, "the pharmacists fear administering them, so as not to put at risk the health of those who, at such a cost to Your Majesty, have been brought to these missions."[64] It was for these reasons that Kamel "devoted himself to the study of the many medicinal herbs that grow in these islands," as his peer Pedro Murillo Velarde put it.[65] Similar accounts of simple remedies which framed native materia medica in humoral terms, and identified local *succedanea*, were compiled by missionaries across Spanish colonial territories.[66] Practitioners like Kamel therefore adopted indigenous and foreign substances, yet largely out of necessity and as substitutes for the rare, expensive, or outright unavailable Galenic drugs.

These developments offer nuance to accounts of bioprospecting which continue to loom large in histories of colonial medicine. Jesuit participation in the drug trade has featured prominently in these narratives, largely due to their involvement in the extraction and distribution of cinchona, which earned the drug the moniker "Jesuits' bark."[67] The Society of Jesus established a worldwide network of pharmacies and plantations that facilitated the communication of medical knowledge and substances around the globe. Yet evidence from Spanish colonial settings suggests that few "new" remedies became staple features in Jesuit apothecary shops. Where funding and institutional support were available, importation of Galenic medicines remained the preferred option. The Jesuit pharmacy in Córdoba (present-day Argentina), which was subsidized by the crown as a regional distribution center, continued to rely on drugs brought from Europe all through the seventeenth and eighteenth centuries. In the early 1700s, its pharmacist, Heinrich Peschke, wrote to the Society's headquarters in Rome that "almost all medicines come from Europe, at great expense and risks."[68] The pharmacy's inventory, compiled in 1769 after the Jesuit expulsion, was still dominated by Galenic remedies.[69] The practices of both royal officials and Jesuit missionaries therefore reveal the same preoccupation with proven Galenic medicines. The differences between the drugs available in colonial medical establishments can largely be attributed to disparities in funding. It was when access to European imports was unavailable that foreign substances were adopted—and commonly as substitutes for Galenic remedies.

JESUIT BEANS AND VOMIC NUTS: "GALENIZING" PHILIPPINE FLORA

Galenism underpinned European adherence to established humoral remedies, but it also offered a compelling framework for appropriation of foreign substances. The flexibility of Galenism enabled practitioners like Kamel to employ humoral ideas to describe unfamiliar realities in familiar terms and facilitate their translation into frameworks sanctioned by European authorities. Such efforts to Galenize colonial natures fed into the processes of appropriation and commodification of drugs. Here, I focus on the case of the St. Ignatius bean. Kamel's treatise, published in 1699 in *The Philosophical Transactions*, was the first comprehensive account of the drug printed in Europe (fig. 2.2). The plant, called "igasud" in Indigenous Filipino languages, was identified by modern botanists with *Strychnos ignatii*, a tree native to the Visayan Islands, where its strychnine-rich seeds had been used in healing practices prior to the European arrival.[70] The eastern portion of the Visayas fell under the Society's jurisdiction, and it was probably here that the Jesuits learned about the seeds' powers.

The adoption of igasud and its reinvention as a legitimate medicine required its translation from Indigenous contexts into knowledge frameworks recognized as valid by European authorities. Crucially for the Jesuits, there were spiritual concerns over the origins of Indigenous

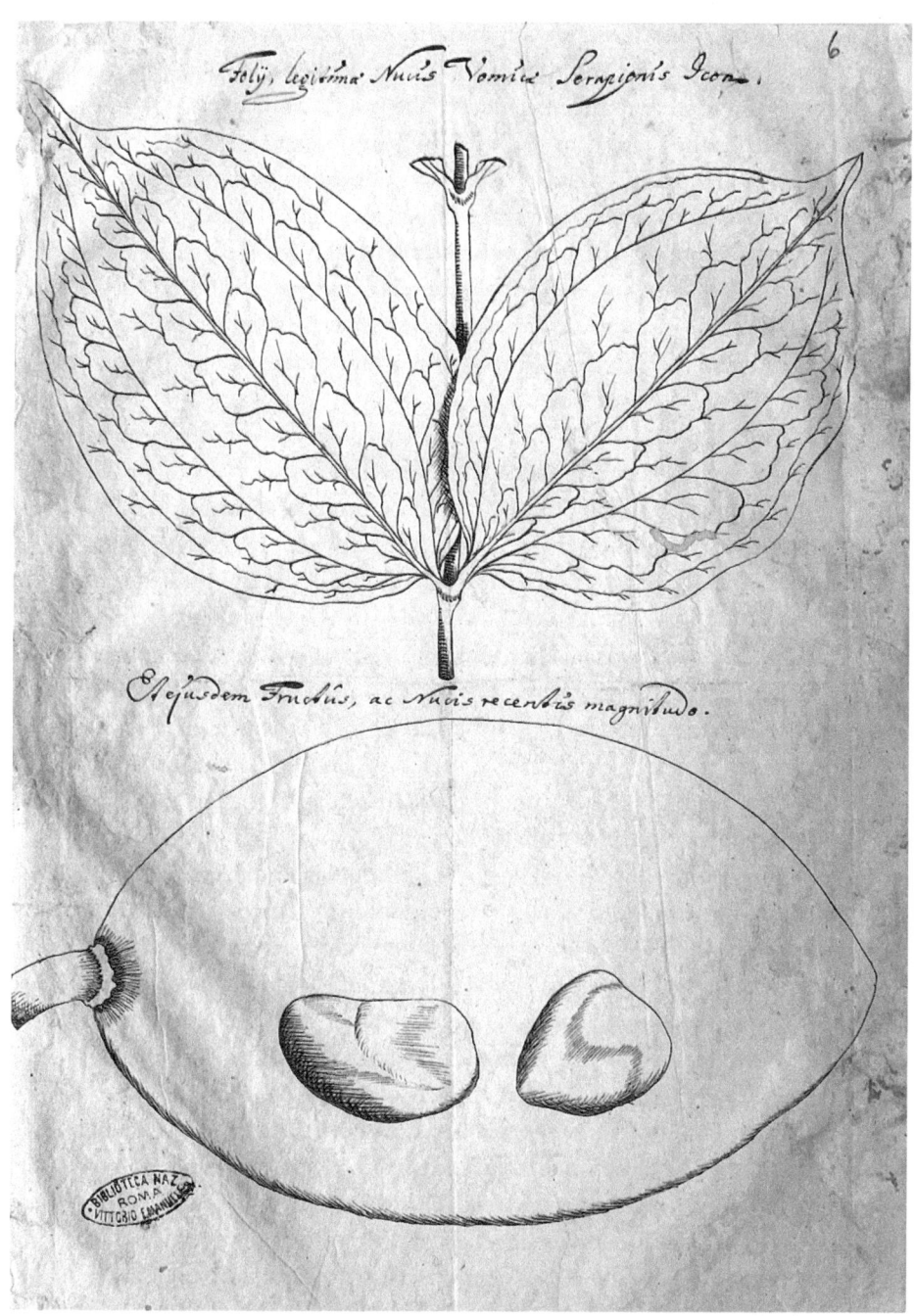

FIGURE 2.2. Kamel's drawing of the St. Ignatius bean (BNCR, 12.11.H.6). Reproduced by permission of the Biblioteca Nazionale Centrale Vittorio Emanuele II di Roma.

knowledge.⁷¹ Since Filipino communities had been pagan prior to the Spanish invasion, their knowledge was regarded with suspicion, for its source and powers could be demonic. The Jesuits could either shun Indigenous healing practices and ascribe them to demonic magic, or—if proven efficacious—incorporate them into frameworks compatible with the missionary effort. To make Indigenous remedies acceptable for European adoption, the Jesuits therefore had to dissociate the source of this knowledge from Indigenous contexts, and instead attribute it to chance, divine inspiration, or natural causes, such as learning through observation and experience. Effacement of Indigenous agency presented the safest way toward adoption.

To disconnect igasud from what they considered unsafe and potentially demonic origins, the Jesuits framed its discovery as the result of divine interference and rebranded the substance as the beans of St. Ignatius, the founder of their Society. Kamel recounted a captivating story of a Jesuit priest saved by the beans when a spiteful Indigene attempted to poison him. It was *"casualiter"*—by chance—that the priest "happened to have a dried bean on him." In fact, this was no lucky accident but a portent of divine will and favor. Kamel presented the incident as "the first occasion on which Spaniards learned about the virtues and powers of igasud."⁷² This "chance" discovery led to further investigations, the results of which Kamel recounted. His story thus inverted the role of Indigenous agents from that of informants to deadly adversaries.

Having detached knowledge of igasud from its Indigenous origins, the Jesuits proceeded to situate the drug within a theoretical framework acknowledged as legitimate by Europeans. Historians have shown how Europeans ignored local theoretical constructs, and pictured Indigenous expertise as mere know-how, raw material that could be used for the production of new and genuine knowledge.⁷³ Galenization enabled the Jesuits to integrate foreign materia medica into the dominant European medical framework embraced by the Society, and thus to present such substances in familiar terms, downplaying their foreignness and novelty. In this way, the Jesuits were able to produce credible and mobile knowledge, as well as to bolster their global project of proselytization. By pointing to commonalities between nature in the Old World and in colonial settings, the Jesuits sought to construct unified histories of the world in defense of the theological concept of monogenism, which posited shared roots of all humankind and thus the universal potential for conversion.⁷⁴ The project of converting Indigenous souls to Christianity and Indigenous nature into European medical systems was therefore part of the same program of universal salvation, in which natural history and medicine readily intersected with religion.

Kamel Galenized igasud by identifying the drug with the *Nux vomica* of Serapion the Younger, based on morphology, virtues, and taste. Serapion was a twelfth-century Christian physician who wrote in Arabic, and whose work later became assimilated into the Galenic corpus. His "vomic nut" was well known to European practitioners and consumers as, above all, a powerful emetic.[75] European experts deemed the plant native to Arabia or the East Indies, but Serapion's brief and fragmentary commentary offered no definitive information. Since the source of true *Nux vomica* was unknown, there was a gap in knowledge and markets that enterprising individuals like Kamel could exploit.

To convince his readers of the correspondence between igasud and *Nux vomica*, Kamel provided a detailed description of the Philippine plant, and performed experiments with its seeds. Accounts of cases from his medical practice attested to igasud's emetic virtues, and integrated the drug into Galenic therapies. Kamel reported that the ailing *oidor*, Juan de Ozaeta y Oro, "regurgitated a great deal of viscous phlegm" after swallowing an entire bean and being relieved with oxymel, a traditional Galenic remedy. Based on additional evidence, Kamel concluded that the bean "frequently tends to induce vomiting."[76] Thus, Kamel presented the knowledge of igasud's virtues as emerging from Jesuit trials, rather than extracted from Indigenous practices.

Igasud's potency matched the reputation of *Nux vomica*, which some European practitioners hesitated to prescribe, due to its powerful purgative effects. Kamel even reported that some of his patients suffered from violent seizures:

> I once diluted one scruple of the powder of igasud and gave it to Vicente Olzina, endowed with a melancholic constitution, to provoke regurgitation. He was troubled with indigestion, diarrhea, frequent nausea, sour belching, and copious flatulence. As soon as he took it, he was seized with a tremor of the whole body, which lasted for three hours, together with an itching and terrible convulsive twitching, so that he could not stand; it was strongest and most troublesome in his jaws, forcing him to a kind of laugh: he was having a seizure. Meanwhile, there was no notable alteration in the pulse, he did not vomit, and there were no other subsequent symptoms. Afterwards he felt somewhat better.[77]

Kamel ascribed these violent effects to the differences between Indigenous and European bodies. His experiments showed that igasud "almost always causes spasmodic convulsions in Spaniards, but not in the Indios."[78] This observation agreed with Galenic theory that posited fundamental disparities between Spanish and Indigenous Filipino humoral constitutions, which had been reared under distinct climates and could thus react to the same substance in divergent ways. Perhaps due to these differences, the bean failed to produce the desired effect in Olzina, who did not vomit. As a disciple of Galenism, Kamel therefore tested the suitability of Philippine substances for European humoral constitutions. Although colonial practitioners could adopt native materia medica as substitutes for Galenic remedies, this step presented risks for unaccustomed colonists' bodies. Like his European colleagues, Kamel thus advised using *Nux vomica*, or igasud, with caution and in small doses.

The reinvention of igasud as a Galenic medicine paved the way for its deployment on both local and global scales and markets.[79] Foreign knowledge and substances were in danger of being regarded with suspicion by European authorities, practitioners, and customers. Yet association with recognized frameworks and authorities, like Galenism and Serapion, facilitated the adoption, appropriation, and commodification of materia medica native to colonial worlds. It was by recourse to the language of tradition that Kamel was able to describe and assimilate novelty. He used canonical authors and their plants essentially as dictionary entries: stripped of their contextual meanings and stabilized through tradition, these terms were fixed enough to translate across different contexts.[80] The language of humoral theory and its authorities afforded some degree of stability and continuity amid the early modern flood of names and objects. Through inscription into the humoral system, local plants could be turned into global drugs, and Indigenous knowledge into legitimate medicine. These practices of translation were thoroughly reliant on non-European knowledge, even as they denied this dependency by erasing traces of foreign agency.

MANILA AT A CROSSROADS OF MATERIA MEDICA

The colonial medical establishment in Manila showed commitment to Galenic remedies, yet both the accounts drawn up by the Hospital Real, and those drawn up by Kamel, mentioned several drugs of chemical and American origin. Such integration of chemical and New World materia medica into Spanish pharmacopoeias reflects broader trends in colonial medical practice. Spanish practitioners had long been familiar with alchemical techniques, due to the enduring tradition of Islamic pharmaceutical alchemy in Iberia, and the seventeenth century saw an increasing influx of chemical remedies into Spanish pharmacies, in both Europe and colonial spaces. In the words of Paula De Vos, practitioners pursued a "chemico-Galenic compromise," or amalgamation of Galenic and alchemical traditions.[81] Indeed, drug inventories for Manila's Hospital Real comprised diverse chemical medicines, including substances discussed by the classical authorities Galen (vitriol) and Dioscorides (alum); preparations touted by Islamic alchemists (spirit of vitriol, flower of sulfur); and medicines associated with the Paracelsian and iatrochemical traditions (refined laudanum, spirit of common salt, cream of tartar). Chemical remedies were also produced locally, with the hospital pharmacy equipped with four alembics and an alquitar. Kamel, too, relied on chemical methods, as shown by his use of a water distilled from tanglat. By the late seventeenth century chemical medicines appeared in the Spanish crown's official shipments, and were prepared and consumed in Manila.[82]

The inventories from the Hospital Real also listed thirteen substances of American origin, native to regions controlled by the Spanish Empire. Included were the purgatives mechoacan, jalap, zacualtipán, and matlalitztic; the febrifuges cinchona and sassafras; the resins tacamahaca and copal; sarsaparilla and guaiacum, used for treating the pox; as well as the antidote cohuanenepili, the mineral salt tequesquite, and two kinds of chili peppers. Although native to the New World, most of these substances had entered Spanish pharmacies and pharmacopoeias by 1600. Nicolás Monardes discussed seven of the thirteen drugs in his *Historia medicinal* of 1574: chili, copal, guaiacum, mechoacan, sarsaparilla, sassafras, and tacamahaca.[83] Among these, guaiacum, mechoacan, and tacamahaca were even listed by Zamudio in his catalogue of goods that all Castilian pharmacists were to stock, written in the 1590s.[84]

It was through recourse to Galenic remedies and ideas that Spanish practitioners were able to incorporate into their pharmacopoeias many of these American substances.[85] Chilies, copal, and tequesquite were considered American substitutes for pepper, biblical balsam, and *Sal nitrum*. Spanish observers identified Indigenous Mesoamerican uses of mechoacan, jalap, and matlalitztic with purgatives, a category of drug from their own Galenic pharmaceutical traditions. Reframing them within humoral theory may thus have facilitated the movement of these drugs between Mesoamerican and Galenic knowledge systems.[86]

Guaiacum and sarsaparilla, the most abundant American items in the consignments, were considered to act specifically against a particular disease—the great pox—which was itself widely considered a new illness originating in the New World.[87] Treatment of American afflictions with American substances reflected the longstanding belief that local diseases required local medicines, rooted among other humoral connections between therapies, illnesses, and climates. Medical practitioners therefore found a use for humoral ideas and practices in translating select New World drugs into Spanish pharmacopoeias—so much so that, by the mid-seventeenth century, colonial authorities in Manila were formally requesting that American substances be included in consignments for the Hospital Real.

The works of Kamel highlight the increasing presence of American plants and drugs in Manila, and their consumption by local communities. In his accounts of Philippine flora, Kamel listed New World crops, including maize, potato, and cassava; fruits and nuts, including pineapple, cashew, and peanut; and medicinals, including mechoacan, cinchona, and cocoa.[88] To make sense of American materia medica and "compare the plants of that [Mexican] kingdom with the local ones [in Manila]," Kamel repeatedly sought—and failed—to acquire the renowned compendium of New World nature compiled in the late sixteenth century by Francisco Hernández.[89] Still, several American imports found their way into Kamel's medical practices, such as the purgative root of mechoacan, and the resin tacahamaca.[90] For the latter, Kamel even strove to locate Philippine plants that would serve him as cheaper and more accessible substitutes, which attests to the close yet unreliable links with America.

Most of the New World drugs discussed here were brought to Manila in their processed form, as medical commodities. However, Kamel's writings suggest that trans-Pacific travelers and migrants had introduced numerous American species into Philippine nature, whether accidentally or deliberately. Many of these plants became vernacularized and incorporated into local healing practices. Sweet potato leaves found their use in treating stingray wounds, while the Mexican acacia, cuahmochitl, came to be locally called "quamochil," and combined with the Philippine plants tangal and ananapla in a compound skin remedy.[91] American plants, and knowledge about them, did not end their journey in the Philippines. Cuahmochitl spread through Asia, gaining folk names, such as Manila tamarind and Madras thorn, and finding new culinary and healing uses in China, India, and Arabia.[92] As the Manila galleon provided the first stable bridge across the Pacific, so the Philippines became a place where Asia met America.

BEYOND BIOPROSPECTING

Patterns of drug consumption in Manila offer nuance to prevailing narratives of colonial bioprospecting and the European hunger for "green gold." Not all European encounters with foreign substances were bioprospective, and colonial sites did not by default yield new drugs for European exploitation, as imperial institutions may have envisaged. While vast quantities of Spanish drugs were flowing into Manila, few medicinal substances traveled back to Europe. The St. Ignatius bean presents an exception rather than the rule. Foreign substances faced the risk of being regarded with suspicion, and their consumption was a matter of authority and trust. Both Kamel's medical practice, and Hospital Real shipments, speak at once to the anxieties of the colonial regime over the consumption of medicine, and to wider imperial efforts to impose authority over colonial worlds by asserting the superiority of European bodies, knowledge, and remedies.

Persisting humoral views on climate continued to inform regimens of the body and cure, as well as feeding into the construction of ethnic hierarchies that shaped access to healing and medicines. Yet Galenism underpinned European commitment to established remedies, just as it also offered a compelling framework for appropriating colonial natures. Kamel's treatment of the St. Ignatius bean reveals how Galenization enabled colonial practitioners to equip local plants for movement beyond their points of origin, and for introduction as legitimate drugs in European markets. To enter the networks of European drug trade, plants native to colonial settings first had to be detached from Indigenous cultural contexts, and then codified within theoretical frameworks sanctioned by European authorities.

This chapter shows that a shift in geographical and cultural viewpoints allows us to decenter European notions of the "exotic." If Spanish remedies were regularly imported in Manila, where do European cultures of consumption end? How was "exotic" defined in these spaces? Despite the extensive presence of Galenic medicines in Spanish colonial territories, their lives and receptions in these contexts are yet to receive closer attention. We now know how substances such as chocolate, tobacco, and tea became global commodities, and how they transformed cultures in Europe.[93] Yet the prevailing focus on the global lives of substances that were new only to Europeans has eclipsed the question of how European medicines were received in colonial spaces. Attention to the emergence of European drugs as globalized goods may yield new insights into the dynamics of transcultural interactions in colonial sites and, perhaps somewhat ironically, contribute to the decentering of early modern histories of medicine and consumption.

Some fifty years ago, Alfred Crosby's pioneering study launched the debate around the ecological consequences of the trans-Atlantic Columbian exchange.[94] Evidence from Manila reveals the Pacific as another stage of dynamic transcontinental exchanges in substances and the associated knowledges and practices. But we know comparatively little about the early modern opening of the trans-Pacific route between Asia and America, and its impact on the environments, communities, and knowledge traditions on both sides of the ocean.[95] How did substances crossing the Pacific enter new cultures of consumption? How were they assimilated into the increasingly global networks of drug trade? How did these processes shape the drugs' availability and use in other areas, including Europe? Like the Spanish outpost in Manila, these developments may at first glance seem distant from European spaces, practitioners, and consumers. Yet they shed light on how colonial consumers conceptualized categories of the "exotic," and interpreted, accessed, and consumed drugs. As such, examining these processes is crucial to our understanding of the changing relationship between Europe and other parts of the world at the turn of the eighteenth century.

3

THERIAC AS A DOMESTICATION TECHNOLOGY

The Indigenous and the Exotic in Galenic Pharmacy

Barbara Di Gennaro Splendore

Most histories of "exotic" drugs focus on non-European drugs. In early modern Europe, "exotic" was often synonymous with rare, foreign, or unknown.[1] Tracing the paths of opobalsam, china root, cocoa, or cinchona bark, historians have asked questions about how botanical commodities and knowledge circulated across cultures, what economic preoccupations influenced consumption, and how political and imperial visions shaped medicine.[2] Shifting the traditional perspective, this essay examines the question of pharmaceutical exoticism through one of the staples of Galenic pharmacy, theriac. Unlike foreign drugs, theriac needed no introduction to early modern European consumers, and it was widely perceived as an indigenous production. Following Galen and other medical authorities, physicians for centuries considered theriac the most powerful antidote and used it to cure all sorts of ailments, from toothaches to the plague.[3] In fact, it was so widespread that the word "theriac" served also as a general term synonymous with "antidote." But theriac's indigeneity was only a part of its character.

Although theriac seemed familiar and "domestic," it contained numerous ingredients sourced from different parts of the world. When a sixteenth-century author, Symphorien Champier, wanted to praise a theriac made in Lyon, he compared the French city to ancient Corinth, "where merchants arrive from all places."[4] For Champier, the provenance of theriac from commercial hubs like ancient Corinth and early modern Lyon vouched for the quality of the final compound. The intricate centerpiece of Galenic pharmacy, theriac stood as a legacy compound passed through the ancient Roman, Hellenistic, and Arab empires, and a shared practice across the Mediterranean.[5] Made from sixty-three ingredients—animal parts, earths, and herbs from far-flung places like licorice juice from Cappadocia, *Cyperus longus* from Africa, opium from Egypt, and *Costus odoratus* from India—theriac could only be produced using well-established commercial networks, and proximity to ingredients' places of origin granted advantages in both identification and supply.[6] Until the end of the sixteenth century, Christian pilgrims traveling from Europe to Egypt and Syria sought to acquire local theriac. Egyptian theriac, though rare in Europe, enjoyed a better reputation than theriac produced on the northern shores of the Mediterranean. In Europe, Italian apothecaries were the major producers of theriac, exporting it to France, Germany, and England since the Middle Ages.[7] In short, theriac was at once exotic and indigenous, foreign and local.

Theriac's incorporation of both indigenous and exotic ingredients offers an unusual perspective on early modern pharmaceutical practices in relation to *exotica*. In the early modern world, I argue, theriac represented a technology of domestication of new materia medica. As I use it here, "domestication" means something more than merely acclimatizing rare plants in botanical gardens or selecting and breeding "wild" species to make them more suitable to humans' purposes. "Domestication" means establishing a shared set of practices to identify foreign drugs, reducing them to known familiar terms, and incorporating them into pharmaceutical practice and even into social life. Because of its history and characteristics, theriac served as a domestication technology facilitating these processes. During the sixteenth century, scholars, physicians, and learned apothecaries in Italy and across Europe researched the local and foreign ingredients of theriac. They believed that knowledge regarding the ingredients of Andromachus's formulation of theriac—*Theriaca magna*, seen as the most effective recipe since its creation in the first century CE—was gradually lost during the Middle Ages.[8] As a result of the attention scholars devoted to theriac and to new studies in materia medica, production of theriac increased significantly over the seventeenth and eighteenth centuries. Apothecaries wanting to compound theriac, and patients eager to buy good quality compound, gained access to foreign materia medica, to knowledge about it, and to the new methods of natural history. Because of theriac, foreign plants entered into both pharmaceutical practice and social life.

Theriac served a similar function around the world. Recent scholarship shows that, despite intensive bioprospecting, in the early modern period Europeans abroad often remained attached to their own pharmaceutical formulations, and were reluctant to abandon the Galenic medical framework.[9] Colonizers, merchants, and missionaries still understood illness and the body in humoral terms. They incorporated new plants into their medical practices following the principle of substitution, one of the central epistemological pillars of Galenic pharmacy. Each materia medica was classified according to the degree of a set of qualities (hot, dry, moist, cold) and ingredients of the same or similar degree could substitute for one another in case of absence or scarcity.[10] On the epistemological level, the possibility of using substitutes granted Galenic pharmacy a wide degree of flexibility. On the practical level, theriac's broad array of ingredients offered a strong motive to find substitutes, as well as a well-established procedure to incorporate them. Through theriac, European pharmaceutical actors both within Europe and abroad expanded their knowledge, redefined their practices, and de-exoticized new materia medica at multiple levels.

From the turn of the sixteenth century, the advent of new chemical remedies and the arrival of drugs from other continents challenged traditional Galenic pharmacy, raising hope for the possibility of novel and more effective therapies. Galenic pharmacy not only resisted these additions but thrived, and its therapeutic promise remained appealing to patients and consumers well into the nineteenth century. Alongside the growth of empiricism in Western science and the rise of bioprospecting and extraction economies in the early modern world, Galenic pharmacy—including but not limited to theriac—showed much resilience. The global success of theriac in the seventeenth and eighteenth centuries epitomizes the internal vitality of Galenic pharmacy in the face of new competition created by the rise of chemical drugs and exotica.

I begin with an analysis of theriac in the Roman pharmacopoeia, in order to understand when and how the new methods of natural knowledge regarding the identification of exotic drugs became relevant to the broader public of apothecaries, not just scholars or natural philosophers. Pharmacopoeias reveal invaluable information about medical practices, the transfer of knowledge, and aspirations of imperial power.[11] In the second section, I demonstrate that by the 1640s, questions about foreign materia medica had won the attention of a wide readership in Rome, taking on political meanings that had little to do with pharmaceutical questions. In the third section, I analyze exports of theriac to the Portuguese, Dutch, and Ottoman empires, and to northern Europe. Theriac's success among European colonizers and missionaries, as well as among non-Europeans, complicates the narrative of a mostly unidirectional flow of materia medica toward Europe.[12]

KNOWLEDGE OF FOREIGN MATERIA MEDICA IN ROME

To trace the growth of apothecaries' interest in theriac and its exotic ingredients over time, I compare different editions of the official Roman pharmacopoeia. Because the pharmacopoeia had several successful editions, I assume that the choices of editors and publishers reflected not only their ideas but also those of their intended readership. First published in Latin in 1583, *Antidotario romano* was the result of a collaboration among several Roman physicians.[13] Its objective, like that of other official pharmacopoeias, was to serve as a guide for apothecaries in compounding medicaments, and as a definitive reference point for physicians in charge of inspecting apothecary shops. The Roman pharmacopoeia was a traditional pharmacopoeia, heavily based on medieval texts, which did not mention exotica already in use, such as sarsaparilla and guaiacum. The book was conventionally organized in the form of *scholia*: editors' comments followed basic information on the recipes.[14] In 1612, the apothecary Ippolito Ceccarelli completed the first translation of *Antidotario romano* from Latin into Italian. Once in the vernacular, the book was republished numerous times in Rome. It became the most popular official pharmacopoeia in Italy, with editions published in Milan, Messina, and Venice.[15]

Comparing several editions of the *Antidotario romano* reveals how Roman apothecaries' knowledge of foreign drugs changed during the seventeenth century. Famous court apothecaries, collectors of *naturalia* and *artificialia*, and learned apothecaries who published on materia medica and corresponded with collectors and physicians all over Europe, often possessed several pharmacopoeias, including the *Antidotario romano*. This particular pharmacopoeia, however, was primarily addressed to non-elite physicians and apothecaries who lived and worked in cities far from commercial hubs. A seventeenth-century physician, for example, wrote that Roman apothecaries "do not possess very many books and only study the Roman *Antidotarium*."[16] Compared to their Venetian colleagues, the *Antidotario romano*'s intended audience had fewer opportunities to come into contact with merchants and travelers who brought specimens from abroad. Venice was the main port where foreign drugs entered the Italian markets. Several Venetian apothecaries were themselves importing drugs through their commercial networks. Their counterparts in Rome and in other Italian cities relied heavily on Venice for necessary foreign drugs.

In 1612, Ceccarelli added his "Treatise on the Apparatus of Theriac and Reasons for Its Ingredients," which described his own production of theriac in 1605, to his translation of the Roman pharmacopoeia.[17] In his introduction, Ceccarelli proudly asserted, "Since we can find many perfect simples for this purpose [making theriac] in our Roman countryside, I made a list, so as to be sure to be able to pick them at the right time and place."[18] In the city of Rome, Ceccarelli gathered green horehound in the Farnese Gardens on the Palatine Hill, while he found Roman turnip rape in the gardens on the Viminal Hill. For other herbs, he needed to leave the city: the best place to find Roman calamint was the hills outside of Porta Angelica, while wall germander grew outside of Porta Appia. Ceccarelli had to supply other ingredients from nearby regions like Tuscia, renowned for its vipers, and Ascoli, famous for its licorice juice. He spent a year acquiring all the local simples outlined in the book. Exotic ingredients, such as opium, cinnamon, and iris, Ceccarelli ordered from ports such as Venice or Portugal, two locations that were European points of arrival for cargo ships coming from other continents.

The arrival of unknown species thanks to colonial and imperial conquests, and to the opening of new intercontinental maritime routes, elicited excitement as well as hesitation among scholars of materia medica and physicians. The rise of bioprospecting in the sixteenth century accompanied the "invention of the indigenous"—the rediscovery and promotion of species indigenous to Europe—as described by Alix Cooper.[19] Used in this sense, "indigenous" is a relative category, designating what is local to a specific place. Ceccarelli valued the indigenous: his botanical knowledge was proudly rooted in his local environment. Among scholars, enthusiasm for the foreign coexisted with appreciation for local plants and knowledge. For example, naturalist Ulisse Aldrovandi declared, "I have always desired to see and know that Moluccan tree called Panacea. . . . These are the simples that Christian princes could have brought here."[20] But he cautioned against the indiscriminate acceptance of foreign materia medica. In 1580, for example, in revising the regulations for apothecaries, Aldrovandi advised that they keep local ingredients in their shops: "Plants that grow in Italy, in different places, whose virtues and faculties are not inferior to the oriental species, since the mountains of the Apennines host the utmost variety of herbs, compared to the rest of Europe."[21]

Ceccarelli's rendering of the provenance of the ingredients he collected for his production obfuscated much geographical information. A mental map of the provenance of Ceccarelli's ingredients would include detailed images of Rome and central Italy, along with only two more distant locations: Venice and Portugal. Ceccarelli did not mention, for example, that the substitute he used for the extremely rare white opobalsam from "Judea" was a black balsam coming from Central America (then New Spain). He only wrote that the balsam arrived from Portugal.[22] In its simplicity, the map that would emerge from Ceccarelli's descriptions is problematic and yet telling. Most likely, his descriptions reflected the apothecaries' mental map, exposing the information they actually had, as well as that which they lacked. If Ceccarelli knew more, he did not deem that information useful for his fellow apothecaries. Indeed, according to a contemporaneous physician, apothecaries were more interested in recipes than in descriptions of "unknown and doubtful simples."[23] Focusing on local herbs and referring to Venice for other ingredients was practical: Roman apothecaries had to rely on their Venetian colleagues for supplies, and, to a certain extent, for the identification of these simples. Furthermore, Ceccarelli's simplification served as a more or less conscious domestication of foreign ingredients. As far as the apothecaries were concerned, what did not come from Rome came from Venice. This state of things was soon to change.

Seven years later, in 1619, the frontispiece of the new edition of the *Antidotario romano* announced that it featured a treatise on Egyptian theriac alongside Ceccarelli's treatise on Roman theriac. The new treatise was a vernacular translation of a chapter on theriac by the Venetian physician Prospero Alpini, in his *Medicine of the Egyptians* (1591). Until the end of the sixteenth century, Egyptian and Syrian theriac were highly regarded by Italian patients. The prestige depended on the ready availability of certain ingredients in Egypt. Alpini explained, "I believe that [the Egyptians] use the most true and legitimate ingredients, because in those countries it is not difficult to supply many rare simples, since Egypt is at the border with Arabia, Ethiopia, India, Greece, where many simples that we desire here grow spontaneously or are easily brought."[24] Alpini praised the production of theriac which Egyptian apothecaries carried out in public with the utmost care, as the drug was made for the Sultan in a "temple" called "Morestan." During his journey in Egypt in the 1580s, Alpini worked hard to obtain the Egyptian recipe for theriac, which Egyptians kept secret. Only with some tricks and the help of a local Jewish acquaintance (who helped him translate the recipe from Arabic) was Alpini able to return home with the cherished recipe, snatching it from a local erudite of *res herbaria*.[25] Alpini's treatise contained long and detailed descriptions of rare ingredients such as "asphalato" and "xylobalsam."[26] He was critical, however, of the theriac recipe used by Egyptians for their Cairo production. This recipe departed significantly from that of Andromachus the Elder, considered the first and best formulation. Though Alpini put the prestige of Egyptian theriac under scrutiny, he still deemed Egyptian theriac superior because of its ingredients.

The addition of Alpini's treatise to the 1619 Roman pharmacopoeia signals apothecaries' growing interest both in theriac and in the sixteenth-century scholarly literature on materia medica and the methods of processing these. Ceccarelli's treatise on Roman theriac, published just seven years earlier, had inadequately addressed physicians and apothecaries' concerns about theriac's foreign ingredients. Even apothecaries lacking scholarly backgrounds would now recognize the importance of comprehending how ancient and modern authors' descriptions of simples agreed or diverged. Observation of real specimens became unavoidable for correct identification of materia medica (this is why collectors, travelers like Alpini, and people living where plants grew had an edge over other herbalists). Finally, apothecaries had to know that they could perform a number of tests on how a plant or a substance reacted to certain conditions (for example when aged, cut, boiled, or squeezed).[27] The incentive to understand these new methods better came from a mix of social and cultural factors: being conversant in the new language of natural knowledge created an advantage in the growing business of theriac. From 1619 onward, Roman apothecaries' knowledge about foreign drugs was expected to increase, and all subsequent editions of the pharmacopoeia kept the "Treatise of Egyptian Theriac" in the frontispiece.

The mental map of Roman apothecaries would expand again with Pietro Castelli's commentary on Ceccarelli's treatise, published in the

1637 edition of *Antidotario romano* (Messina), and promptly republished in Rome in 1639. Castelli (1574–1662) was a prominent Roman physician who published extensively on materia medica.[28] In his commentary, Castelli cited several authors, both ancient and modern, and included lengthy discussions about the provenance and identification of theriac's rare ingredients.[29] A map based on Castelli's comments would encompass Slavonia on the Danube, Macedonia, Cyprus, Crete, Greece, Cappadocia, Spain, the Arabian Peninsula, Ethiopia, India, and the Moluccas. This expanded map resembled that of the Roman Empire between the first and the second century CE, at the height of its power, and included not only the empire's borders, but also commercial routes stretching as far east as southeast Asia and as far south as Ethiopia. During the first century CE, Andromachus the Elder, the alleged inventor of theriac, created a drug which was only possible within a powerful empire with reliable global connections. In this light, theriac was an imperial drug by birth. The product of geographical conquest and colossal wealth, theriac aspired toward universality, both in its claim to be an antidote to many ailments, as well as in its recipe that combined drugs with contrasting qualities, even by Galenic standards. Theriac conquered exotica, incorporating them into itself, and cured everything.

In a matter of twenty years, between 1619 and 1639, the publication of Alpini's treatise and Castelli's description broadened the mental map of Roman apothecaries concerning the geographical provenance of theriac's components. Apothecaries now had ready access to descriptions of the physical characteristics of these plants, their histories, and more. Alpini's treatise, for example, provided apothecaries with information about Egyptian medical practices. Entries about foreign ingredients were abridged versions of the debates of erudite sixteenth and seventeenth centuries physicians. Reading about theriac in their pharmacopoeia, Roman apothecaries familiarized themselves with a new scientific language, while also learning that foreign materia medica required careful assaying. Unlike that of other rare drugs, the therapeutic value of theriac's ingredients was already sanctioned by the Galenic tradition, and thus they were more familiar than other exotic ingredients.

EXOTICA IN THE URBAN SPACE

Castelli's comments about opobalsam in the Roman pharmacopoeia underlay a wide-ranging dispute which broke out in Rome in 1639, demonstrating that an "exotic" drug could thus take on local political meanings. One of the most precious ingredients of theriac, opobalsam carried a halo of mystery and opulence that added fascination and value to theriac. The resin or sap of a bush called balsam, whose supply was extremely scarce in Europe, opobalsam had been known since antiquity, and was allegedly used for anointing kings and for the Christian rite of confirmation.[30]

In 1638, the pope's nephew, Cardinal Francesco Barberini (1597–1679), prompted Roman apothecaries Antonio Manfredi and Vincenzo Panutio to use "true opobalsam" to make theriac following Andromachus the Elder's recipe.[31] Cardinal Barberini wanted to have his name associated with an especially prestigious production of theriac, a form of patronage diffused among other Italian princes and rulers. Manfredi and Panutio procured opobalsam from Venice through a known merchant. Then, following standard practice, the two apothecaries exhibited the ingredients in public in order to gain approval from the college of physicians. The college of physicians, along with four archiaters (physicians to the pope), unanimously approved the ingredients.[32] This certifying protocol for controlling the quality of theriac dated back to the Middle Ages. Similar medical-legal practices were in place in Italy, France, Spain, Germany, England, Egypt, and Syria well before the sixteenth century, when materia medica became a prioritized area of research for scholars.[33] Although there was neither a common botanical classification, nor a national or international system for policing drugs in Europe, the certifying measures in place for theriac provided two things: a widespread reminder of the necessity of scrutinizing materia medica, and an institutional framework to carry on such scrutiny.

In a show of defiance, the Roman college of apothecaries refused to acknowledge the college of physicians' approval, employing legal procedures, circulating letters, and printing pamphlets to make its position public.[34] The college of apothecaries maintained that the opobalsam exhibited by Manfredi and Panutio was artificial, and that therefore their theriac was toxic and should not be sold.[35] Roman apothecaries were not just quarrelling with each other and showing off their feathers to the learned physicians. In Rome, groups including the two colleges of physicians and of apothecaries, the pope's archiaters, and other physicians who were Barberini clients constantly competed for medical authority.[36] When contesting the quality of opobalsam, Roman apothecaries were defending their authority and commercial positions in a crowded marketplace. An observer noted that Roman apothecaries, "envious" of their colleagues Manfredi and Panutio, were most likely retaliating against a recent decision: The previous year, Pope Urban VIII (r. 1623–44), a member of the Barberini family, had set out to "curb the prices of all things belonging to apothecary shops, which for many years had exorbitant prices."[37]

In attacking a production sponsored by a relative of the pope, apothecaries were targeting the pope himself. In 1639, the physician and *protomedico* Paolo Zacchia (1584–1659) wrote, "Showing no respect for what their superiors had prudently approved, some apothecaries maintain that *protomedici* and councilors have approved a fake balsam, almost as if the providence with which His Sanctity rules the world did not provide the city of Rome with men able to evaluate all medicaments made by apothecaries or by human hand. Their temerity is of great prejudice toward the College and toward the public good."[38] By contesting Manfredi's opobalsam, Roman apothecaries were casting doubt on the pope's ability to provide for the city's needs during an especially tense moment.

At the beginning of the dispute, the pope was involved in a conflict with the Farnese family, and the War of Castro (1641) fanned the flames of this conflict. Fought in Lazio, the war drained millions of *scudi* from the church's coffers, and burdened Roman citizens with numerous new taxes. In 1639–40, relations between France—siding with the Farnese—and Rome had degenerated to the point where some feared the interruption of diplomatic ties.[39] In 1643, the pope introduced a tax on milled wheat, tripling its price.[40] Urban VIII's popularity was at its nadir. His health was rapidly declining, and he had already fainted in public several times, revealing his weakness.

Between 1638 and 1644, both Rome's college of physicians and its college of apothecaries sought counsel on the quality of opobalsam from authoritative physicians and apothecaries around Italy and beyond: letters and prints circulated between Rome, Florence, Venice, Padua, Messina, Naples, and Nuremberg. An entire network of apothecaries and physicians became involved in the dispute, which triggered the publication of no fewer than twenty-five texts over six years.[41] In the first years of the dispute, several texts were dedicated to Cardinal Francesco Barberini. While most of the texts were long dissertations written in Latin, several letters and pamphlets written in vernacular survive. Often satirical and anonymous and sometimes quite vulgar, they show that the dispute transcended the medical establishment to reach a wide audience.⁴² Yet it is difficult to believe that in the midst of a war non-physicians and non-apothecaries could find the interest to read and debate about opobalsam.

Filippo de Vivo has shown that printed communication was a form of political action in seventeenth-century Italy. In Rome, strict control of the press prevented the publication of all political information, which included "the pope's health, curial business, foreign alliances, epidemics, harvests, and court intrigues."⁴³ Public political discussions had to find different outlets. Even those with no political power were interested, and played a role, in political events.⁴⁴ Laurie Nussdorfer has argued that pageantries were one way in which political themes found their way to a wider audience.⁴⁵ The dispute over opobalsam, similarly, became a means of destabilizing the higher medical echelons in Rome, and especially the authority of the Barberinis.

In debating opobalsam, Romans were actually discussing the pope and his family. The correspondence of Cassiano Dal Pozzo (1588–1657), Cardinal Francesco Barberini's secretary, shows how Dal Pozzo was carefully orchestrating publications in support of the opobalsam used by the apothecaries Manfredi and Panuzio. Dal Pozzo closely followed the physicians involved in the dispute, and may have either commissioned or written one of the anonymous pamphlets in circulation. In 1643, for example, Johann Vesling (1598–1649), a professor at Padua, reported to Dal Pozzo on his new treatise on balsam. Vesling regretted that, "considering the dispute," he could not dedicate his new book to the cardinal as he had initially planned.⁴⁶ By 1643 Francesco Barberini realized that, while he had once wanted to be associated with a prestigious production of theriac, it was now wiser to step back from this backfiring dispute.

Urban VIII died in 1644, and the opobalsam dispute came to an abrupt end. In a long and stormy conclave, Barberini's candidate to the Holy See lost the vote. The first act of the new pope, Innocent X (r. 1644–55), was to investigate Francesco Barberini's handling of finances during the War of Castro. He then confiscated the properties of members of the Barberini family. Cardinal Francesco fled to France, where he spent several years.⁴⁷ The balsam dispute shows that in Rome, an apothecaries' conflict about the authenticity of a foreign drug of uncertain supply became a scientific debate about medical authority, and a way to challenge the pope's family and its power. By means of theriac, an exotic drug—opobalsam—blended into the Roman medical landscape, and was eventually domesticated into a local political conversation. Through theriac, opobalsam was now part of the Roman urban space.

EXPORTING THERIAC

In addition to the opobalsam dispute and changes in the Roman pharmacopoeia, theriac's popularity in seventeenth-century Europe is evident in the astonishing number of publications containing the word "theriac" in their titles: over two hundred eighty in total, including medical treatises, pamphlets, broadsheets, and even non-medical texts.[48] By the 1630s, apothecaries held public productions of theriac across Europe, typically under the purview of local collegiate physicians and state and ecclesiastical authorities. People of all social levels flocked to enjoy the music and decorations accompanying the exhibition of the sixty-odd ingredients of theriac. In an expanding pharmaceutical market, theriac and its sister remedy, mithridate, represented the height of Galenic pharmacy, which had become the official pharmacy throughout Europe.[49]

The long and complex list of theriac's ingredients could have posed an insurmountable obstacle to its exportation. Instead, Jesuit friars, physicians, apothecaries, and state officials took theriac, along with their pharmaceutical knowledge of Galenic medicine, throughout the world. Moving from Rome to other regions, the examples below demonstrate that *Theriaca magna* helped apothecaries domesticate new materia medica and perpetuate Galenic pharmacy. The Jesuits first established an apothecary shop in the college of Bahia (Brazil) and in other cities in the Portuguese empire in the sixteenth century, and were active there until their expulsion in 1759. Jesuit apothecaries began producing theriac, importing all the original materials for theriac from Europe. But the expense of this practice, along with the challenges of obtaining the original ingredients, eventually prompted the Jesuits—who were gaining a deeper understanding of local flora working closely with Indigenous healers—to adapt the formula using substitution. Over time, they created a *Triaga brasilica*, mainly composed of native plants. Among other plants, they used local variations of *Aristolochia*, *Piperaceae*, and tubers, including locally cultivated European species. Like *Theriaca magna*, *Triaga brasilica* was used as an antidote. While in Europe a theriac with too many substitutes was deemed low-quality or even fake, an eighteenth-century author reported that "if [*Triaga brasilica*] is not better than theriac from Europe, it is not inferior to it in any respect."[50] The need to find substitutes in a local context led to the creation of a new kind of theriac, *Triaga brasilica*, which the Jesuits considered neither fake nor inferior to the original, and which was produced in Brazil for two hundred years. This novel compound freed the Jesuits from importing materia medica from Europe.

A similar example comes from Asia. From 1656 to 1796, Sri Lanka was under Dutch rule. Within their fortified settlements, the Dutch established hospitals equipped with apothecary shops, to serve soldiers, government personnel, civilians, and travelers.[51] Starting in the mid-1660s, the Dutch East India Company (VOC) began employing physicians and chemists to study indigenous medicinal plants and identify alternatives to remedies imported from the Netherlands. This initiative aimed both to reduce the expense of importing materia medica from Europe and Batavia (modern-day Jakarta) and to collect botanical specimens for affluent Dutch patrons, who were the primary beneficiaries of these bioprospecting efforts.[52] In time, VOC medical professionals succeeded in finding effective substitutes. In 1679, for example, Hermanus Nicolaas Grimm published *Insulae Ceyloniae thesaurus medicus laboratorium*, which listed preparations using local ingredients.[53] The earliest known Dutch pharmacopoeia compiled in Sri Lanka, dated 1757, included both a "theriac of Andromachus" and a "theriacal spirit," a distilled version of theriac.[54] However, many of the ingredients in the Sri Lankan theriac of Andromachus differed from those found in the Galenic tradition, for example cubeb, a type of pepper, calumba (*Jateorhiza palmata*) from East Africa, and ekaweriya (*Rouvolfia serpentina*), a native plant commonly used in Ayurvedic medicine.[55]

In the Portuguese and Dutch empires, theriac retained its therapeutic connotations as an antidote against venoms, poisons, and diseases, yet apothecaries transformed the recipe with local ingredients. Theriac, in other words, served as a container for new materia medica and gave colonizers a motive to grow European plants abroad. In Bahia, the transformation of theriac also had social consequences: the recipe of *Triaga brasilica* became a secret.[56] This secrecy served the authority of the Jesuits: it made the treatment mysterious, turning it into a quasi-religious practice. A drug originally born as an ancient imperial creation, theriac served as an early modern colonial domestication technology: a known and respected pharmaceutical drug, familiar to both practitioners and patients, into which the unfamiliar could be placed and used.

In northern Europe, instead, we find another permutation of

FIGURE 3.1. Exhibit of *Theriaca magna*, *Theriaca cœlestis*, and mithridate in the workshop of Frédéric Stroehlin, 1744. *Expositio theriacæ Andromachi et coelestis ut et mithridatii. In officina Stroehliniana. MDCCXLIV.* Coloured engraving, 22.7 cm x 30.9 cm. MS 0 615, Coll. and photogr. BNU de Strasbourg.

theriac. In sixteenth-century Europe, the use and production of chemical remedies increased, alongside the general proliferation of pharmaceutical offerings. Northern European customers sought new remedies but wanted products that made sense to them, rather than unproven foreign remedies. *Theriaca* was both familiar and mutable. The case of "celestial theriac"—a chymical version of theriac—shows that some apothecaries capitalized on Galenic pharmacy, chymical procedures, and exotic drugs the better to position themselves in a crowded market. A 1744 image depicting the display of ingredients for a simultaneous production of *Theriaca cœlestis*, *Theriaca magna* and mithridate blends Galenic, exotic, and chymical. The image shows an astonishing number of ingredients, representations of Europe, Asia, Africa, and America, a black man lying on opium, and several *savants* in ancient and Arab clothing (fig. 3.1).[57]

Frederic Greiff, an apothecary from Tübingen, first made celestial theriac in 1634, using a chymical recipe created thirty years before by Joseph Du Chesne (ca. 1544–1609).[58] Following both Galenic and chymical methods, Du Chesne proposed a reformation of all theriacs, including *Theriaca magna*.[59] Chymical theriacs incorporated "the more substantive virtues" of *Theriaca magna* through "ars spagyrica or true chymistry," procedures that separated good qualities (virtues) from "what is there of filthier and excess material."[60] Du Chesne added new ingredients to *Theriaca magna*—a few apothecary preparations, and a good dose of opium essence—making a new product, celestial theriac. Rather than presenting his novel theriac as a new medicine, however, Du Chesne advertised it as a product of "the ancients" to elevate its prestige.[61] While continuity with tradition granted the product's reliability, innovation responded to evolving consumer demands.

Du Chesne realized that to the eyes of novices, celestial theriac might appear as a very complex preparation, but he reassured readers that this remedy "was nothing but a work for women, which means work that a woman can do." At the same time, he contradicted himself, saying that only "true philosophers and sons of the art" would be able to follow the recipe.[62] Subsequent apothecaries may have found the task a little daunting and the gain uncertain, as it took twenty-five years before Greiff attempted a production of celestial theriac. Physicians and wealthy patients in Strasbourg welcomed the novelty. Greiff produced *Theriaca cœlestis* again in 1641 and 1652, changing the recipe and keeping it secret, with the complicity of the apothecaries of the house of Württemberg.[63] The timespan between each production testifies to the complexity and expense involved in making this drug. By the 1660s, apothecaries in other cities had produced their own chymical theriac: Nuremberg (1664, 1675), Hannover (1668), Hanau (1670), Geneva (1674), Marburg (1674), Kassel (1676), Frankfurt (1680), and Giessen (1680). Between 1641 and 1711, at least thirty-one productions of *Theriaca cœlestis*, synchronous with the theriac of Andromachus and mithridate, were celebrated in apothecaries' publications. Apothecaries produced *Theriaca magna* and *Theriaca cœlestis* in bulk, and stored them in their shops, making supplies for several years. Celestial theriac entered the Strasbourg pharmacopoeia, and by 1760 it cost sixty-two times more than the traditional *Theriaca magna* of Andromachus.[64]

As made evident in the discussions of theriac across northern Europe, and in the Portuguese and Dutch empires, theriac helped Galenic apothecaries innovate by integrating different pharmaceutical traditions and diverse materia medica. The success theriac encountered in the seventeenth and eighteenth centuries, however, also involved the Ottoman Empire. Turks, Egyptians, and Arabs living in the Ottoman Empire needed no introduction to theriac. Galenic pharmacy, and thus theriac, was a Mediterranean tradition. In the sixteenth century, Egyptian theriac enjoyed great prestige in Italy, and Turkish theriac similarly figured as a valuable present for Christian princes, dukes, and ambassadors.

Both in Egypt and Turkey, local production of theriac persisted in the seventeenth century. The Ottoman traveler Derviş Mehmed Zillî (1611–1682), better known as Evliya Çelebi, left a detailed account of the grandiosity of an Egyptian theriac production (*tiryaki faruk*) performed in the Qalawun hospital in Cairo. The elaborate and spectacular manufacture involved capturing and subsequently killing thousands of extremely venomous snakes. Çelebi reported that Egyptians exported their theriac to Anatolia, Persia, Arabia, and Europe, and also sent as it as a gift to the "king of Dunkarkiz" (possibly the king of France?).[65] In the 1630s, Jean-Baptiste Tavernier (1605–1689), a French merchant and traveler who visited Istanbul, reported that at the imperial palace it was "in the Cup-bearer's Apartment that the Treacle is made, which the Turks call Tiriak-Faruk and there is a great quantity of it made, because they use it as a Universal remedy, and charitably bestow it on all sorts of people, as well in City as Country."[66]

When Tavernier traveled through Turkey thirty years later, on the way from Italy to Persia, he carried with him twenty-four vases of theriac. The heat made the drug ferment, causing all the vases to crack.[67] Why would Tavernier bring theriac with him to Turkey, where he could find high-quality local theriac? Perhaps he did so because it was a gift from the Grand Duke of Tuscany, or because the balance of prestige between Italian and Levantine theriac had started to change.

Expensive drugs became more and more fashionable in the Istanbul medical marketplace toward the end of the seventeenth century, a market oriented more toward medicines than services.[68] In 1687, when the English ambassador arrived in Istanbul, among the presents he brought for the Ottoman court was Venetian theriac. "One could say that the world of William Trumbull was held together by Venetian theriac," one historian wrote in accounting for the importance of gifts at the Ottoman court.[69] Several accounts from the Venetian ambassadors in Istanbul (*baili*) reported that Ottoman subjects valued Venetian theriac highly. In the 1720s, Venetian officials brought theriac, along with velvet and mirrors, as presents to lubricate their relationships with Ottoman officials and grant them favors. In 1719, for example, the *bailo* Emo brought six vases of theriac as part of his embassy.[70] Venetian theriac was "greatly valued [in Istanbul], beyond all belief," affirmed the *bailo* Foscari in 1757, also lamenting that reception of Venetian goods in Turkey was declining, apart from theriac, "which was a singularity of Venetian production."[71] The case of theriac presents some analogies to the better-known case of coffee, which went from being a luxury imported from the Ottoman Empire to a colonial commodity exported to the Ottoman Empire.[72] Exports of theriac to Istanbul and Aleppo continued until the end of the eighteenth century.[73]

The value Turks placed on Venetian theriac raises several questions, yet historical records offer only peripheral clues. Until the end of the sixteenth century, theriac moved in one direction, from Turkey to Venice. Why did the direction reverse? Furthermore, Turkish and Egyptian theriac were presented as gifts for high-ranking people in Europe, such as grand dukes, popes, and ambassadors. In the seventeenth century, theriac in Italy went from being a luxury product for the wealthy to a much more common drug, its price ever more accessible. Similarly, in Turkey, theriac became a present for mid-to-high-ranking Ottoman officials, such as the household of the vizier (not the vizier himself), dragomans, kiajas (lieutenants), and the *Reis Effendi*.[74]

How does the reduced value of theriac factor in the change of direction of the drug? It is possible that Egyptian- and Turkish-made theriac remained more valuable, but was in short supply. Ottoman patients may have therefore turned to Venetian theriac because it was cheaper.[75] We still do not know enough about the market for medicines and the eighteenth-century Ottoman Empire medical marketplace to understand how medical commodities and practices influenced one another. The success of the Venetian drug in Turkey, however, provides evidence that the strategies Venetian apothecaries put in place to contest growing competition from foreign drugs and spices were highly successful, not only in Europe but also in the Ottoman Empire.

The examples from Rome, the Dutch and Portuguese empires, and northern European *Theriaca cœlestis* demonstrate that, by relying on Galenic pharmacy, apothecaries were able to integrate previously unknown *materia medica* and new chemical procedures into their pharmaceutical practices. Theriac itself—the most symbolic compound of Galenic pharmacy—worked as a domestication technology and a transformative vessel, capable of incorporating the unknown, and innovating within the tradition. Analysis of different editions of the *Antidotario romano* makes evident the escalation of interest in foreign materia medica among apothecaries. Between the 1620s and the 1640s, their mental map expanded. By the mid-seventeenth century, ordinary apothecaries had to have the skills to identify and know the provenance of foreign simples, in order to carry on their production of theriac. Using new approaches to natural knowledge, they could protect their business, both from the inspections of collegiate physicians and *protomedici*, and from competing apothecaries. The balsam dispute in Rome shows that foreign drugs could became a bone of contention between competing institutions. The debate over exotic opobalsam acquired a local quality when it catalyzed political and popular discontent against the ruling pope. Through theriac, foreign ingredients and new procedures became medically and politically domesticated.

European physicians and apothecaries relied extensively on theriac to respond to the global challenges Galenic pharmacy faced in the seventeenth and eighteenth centuries. Theriac's versatility reflected the same characteristic in Galenic pharmacy. Physicians, apothecaries, and Jesuit friars in different locations used theriac to market their productions and increase their authority. Studies show that Europeans often clung to traditional Galenic pharmacy, and the flow of drugs from Europe to colonial settlements was significant.

"Exotic," like "indigenous," is a relative category. Compared to bioprospecting and imports of materia medica to Europe, we still know too little about European pharmaceutical exports to fully understand early modern global exchanges of drugs. Venetian exports of theriac in the Ottoman Empire are understudied, and many facets of this history need further investigation. Why was the direction of commerce reversed? Was the value Turks attached to Venetian theriac in some ways related to an *exotic* fascination that Venetian treacle exerted on them?

PART II
MATERIA MEDICA
Substances and Texts

4
COMPETING MEDICAL SUBSTANCES DURING AN EPIDEMIC

Causes and Consequences of the Interference of Peruvian Bark and Cascarilla, 1720–1740

Wouter Klein

Tracing the convoluted trajectories of drugs in the early modern period presents the historian with significant challenges. As a subject of inquiry, drug trajectories denote the developmental processes of remedies, such as the growing or declining economic importance of a drug; shifts in scientific interest, due to ongoing research into therapeutic efficacy and side effects; and acceptance or rejection of drugs in the public domain, for financial, religious, ethical, or other reasons. The early modern period is not generally thought of, even among historians, as a period for which a lot of big, serial data exists. Therefore, drug trajectories are often portrayed in an episodic way, by tracing significant places, people, or contexts where drugs made the impact that helped them find their way into bodies of codified knowledge, such as medical or pharmaceutical manuals. But heterogeneous collections of big, digital data can provide more detail about drug trajectories, by revealing the parameters for their success or failure as medical commodities.

Big data for this period (here, the early eighteenth century) has a different character than it does for later periods. The quantitative amount of available data for the global flow of medical commodities in the seventeenth and eighteenth centuries is substantially smaller than for the nineteenth and twentieth centuries. Qualitatively speaking, however, a rich and diverse body of material emerged during this period, including newspapers and trade records, where drugs were presented not only as therapeutic substances but also as commercially valuable products. Such sources can help historians study the trajectories of drugs in new ways, including exotic substances. It can be argued that there are only three non-European medicinal substances for which a substantial amount of serial historical data is available in quantitative form: opium, cinchona bark, and rhubarb. Accordingly, it is these that have received most attention from historians. These were arguably the "big three" among exotic medicines, with the largest impact on early modern society, both in terms of their commercial volumes and therapeutic applications, and as topics of interest in contemporary medical literature.[1]

Early modern apothecary shops, however, were packed with other exotic substances that were much "smaller," whether in terms of commercial volume, value, or use. Many of these eventually became staple products in pharmaceutical practice, although explanations for their sustained availability and use are difficult to provide, beyond their purely medical—often retro-identified—efficacy. In this chapter, the commercial trajectories of a "big" remedy (cinchona bark) and a "small" one (cascarilla bark) are interconnected by means of a case study: the fever epidemic that occurred in Dutch cities during 1727 and 1728. This chapter starts with the early history of cascarilla from the late seventeenth century onward. It then discusses the fever epidemic of 1727–28, as it presented itself in advertisements for secret fever remedies, and in firsthand accounts, especially those of the famous Dutch physician Herman Boerhaave (1668–1738). Finally, the commercial trajectories of cinchona bark and cascarilla are correlated through the lens of the epidemic.

As this chapter will demonstrate, the historical interconnections between cinchona bark and cascarilla are most conspicuously revealed during an epidemic. Epidemics are much less visible in primary sources dating from the eighteenth century than in those dating from the seventeenth century. But they can be discerned clearly by combining three sources. Mortality figures, a conventional source for tracing large demographic shifts, are combined with two types of Dutch newspaper advertisements to unveil the supply of remedies during the fever epidemic of 1727–28: advertisements for irregular fever remedies, i.e., secret remedies beyond the purview of regular physicians and apothecaries; and advertisements for public auctions of drugs in Amsterdam, where cinchona bark and cascarilla can be found as crude medical commodities for the domestic market of the Dutch Republic.[2]

Newspaper advertisements were one of several new media that promoted medical goods and services in the early modern period, alongside handbills and almanacs, among others. The oldest of these formats was the almanac, which began to appear in the late sixteenth century. According to Louise Hill Curth, almanacs had three major benefits over newspaper advertisements and handbills. First, they were a true mass medium, with hundreds of thousands of copies printed each year. Second, they had the relatively long lifespan of a whole year, which meant that they would be read by the same reader multiple times. Thirdly, they could potentially reach readers from local to international levels.[3] These characteristics, however, were certainly not unique to almanacs, which continued to appear alongside handbills and newspaper advertisements throughout the early modern period. In fact, newspaper advertisements also had the ability to reach multiple readers per newspaper copy, and while the lifespan of a newspaper is very short, many medical practitioners in the eighteenth-century Dutch Republic advertised their remedies week after week in one or more newspapers, over years or even decades.[4] This assured a similar longevity of readership to that of almanacs. The impact that medical advertising in newspapers could have has been a fruitful topic of research for several years.[5] But mass digitization has opened up entirely new prospects for research, such as the possibility of studying day-to-day changes in advertising practices, something which is impossible for almanacs. Handbills, although sharing the essential brevity of newspaper advertisements, which immediately captivated the reader, cannot reveal such short-term developments.[6] A systematic analysis of newspaper advertisements requires us to reevaluate Curth's third argument. Many advertisements for remedies promote their unproblematic distribution to the far corners of the globe, demonstrating that a locally produced remedy could reach an international audience.[7] Thousands of advertisements, both for secret fever remedies and for drug auctions, are to be found in Delpher, the portal for digital resources of the Dutch National Library in The Hague. Newspaper advertisements are typically brief, often no more than two or three lines of text. This should not be regarded as a disadvantage: brevity implies that every word served a purpose. In advertisements for remedies, this means that any remark about something other than the drug's essential characteristics (such as its name, inventor, virtues, price, or distribution location) is likely to catch the reader's attention, both the early modern patient and the present-day historian. In trying to trace the trajectories of drugs, therefore, newspaper advertisements offer a suitable lens through which to glimpse shifts in understanding and acculturation.[8]

THE RISE OF CASCARILLA: A NEW TYPE OF PERUVIAN BARK?

Even before the epidemic, the histories of cinchona bark and cascarilla were closely related. Both were American febrifuge barks, and the history of cinchona bark offers many avenues for writing about the lesser-known cascarilla, as a parallel narrative to that of its "big brother." Although cascarilla had a history of its own before 1727–28, the epidemic consolidated cascarilla's position as an accepted exotic remedy for fever, alongside cinchona. The history of cinchona, then, offers a platform for historians to address the trajectories of "historically related" substances like cascarilla.

The reason for correlating the trajectories of cinchona bark and cascarilla bark does not emanate, in the first place, from cascarilla's "intrusion" into the European market for exotic febrifuges between 1720 and 1740. There is a more obvious, linguistic reason: cinchona bark was referred to as "cascarilla" in the Spanish Atlantic empire, well before the discovery of what would come to be known as cascarilla bark. And long after cascarilla bark was successfully acculturated in European medicine, the term continued to be used to refer to cinchona in the Spanish-speaking world.[9] The very word "cascarilla," meaning "little bark" in Spanish, was initially a synonym of the bark derived from "true" cinchona trees (i.e., *Cinchona* species, retrospectively), until it was realized in Loja, the center of bark harvesting in Peru, that "true" cinchona bark should be distinguished from similar species. "False cinchona bark" became the name of such similar types of bark, derived mainly from *Iva frutescens* (marsh elder) and *Croton* species.[10]

But acknowledgment of the existence of a new type of "cascarilla," similar to but distinct from "true" cinchona bark, preceded the linguistic confusion between the two types. The French apothecary Pierre Pomet (1658–1699) was the first author to recognize cascarilla as a distinct substance. Pomet added an appendix that included several newly discovered substances to his *Histoire generale des drogues* (1694), including a section entitled "Du Kinquina Femelle." The substance was described as pale, initially tasteless but later unpleasantly bitter, with a cinnamon-like appearance. Pomet equated it with the "Falsa-Kaskarina" of the Indians, which might suggest that the distinction between "true" (*Cinchona*) and "false" cascarilla was recognized sooner by native Peruvians than by Europeans. According to Pomet, the substance had first become known in France around 1670. His own specimen had been given him by Pierre Bonnet-Bourdelot, physician in ordinary to Louis XIV. But more than two decades after it arrived on the European scene, Pomet was still unable to say much about it[11]

This brevity is typical of early authors who wrote about cascarilla. In the third edition of his *Dictionaire ou Traité universel des Drogues Simples* (1716), Nicolas Lémery (1645–1715) included cascarilla for the first time, by the name of "Eleaterium." Lémery's description is even shorter than Pomet's, and calls the substance similar to, but of lower quality than, cinchona bark.[12] Lémery also mentions that cascarilla is often adulterated with tobacco. This curious fact was later confirmed in a letter and lengthy description by Hendrik van Raat (d. 1731), a merchant from Rotterdam. The Amsterdam physician Willem van Ranouw (ca. 1670–1724) included this letter in its entirety in a series of articles about cinchona bark varieties that he published a few years after Lémery's description. In his letter, Van Raat admitted that the visual differences between samples of cinchona bark and cascarilla were easy to overlook, but this did not discourage him from studying the details of various samples himself.[13]

Van Ranouw maintained that, although the barks of many trees were very similar, the colors of the inner bark in cinchona bark and cascarilla were clearly different, and the moss that grew on these two barks was also dissimilar.[14] He illustrated these and several other findings about his samples on a copperplate that showed various samples of bark to which the text alluded (see fig. 4.1). The images of these samples, which look similar enough to the average observer, only made sense with the accompanying text. Van Ranouw made it seem as if his descriptions of

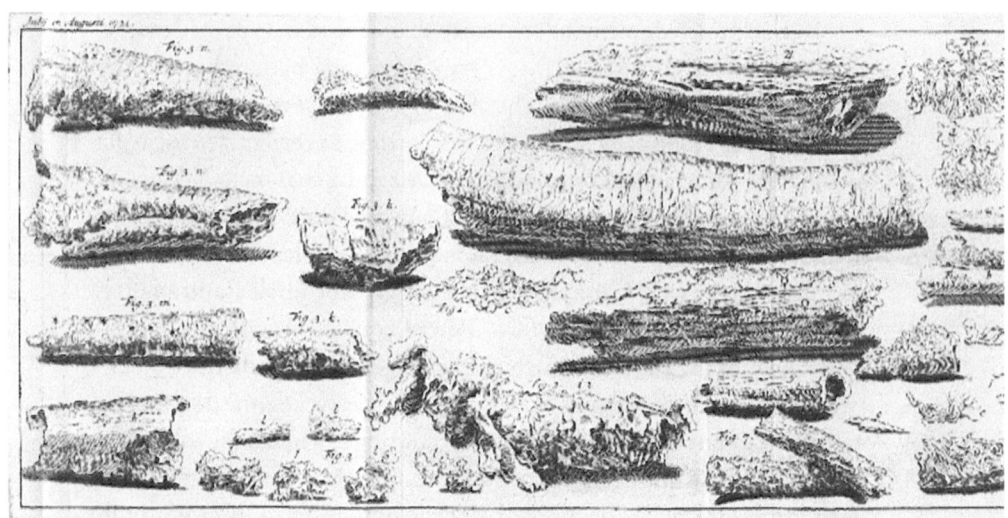

samples were identical to the observations made by Hendrik van Raat, but both men drew their conclusions on the basis of different sets of samples. Although Van Raat had promised Van Ranouw in his letter that he would send bark samples to Amsterdam along with his descriptions, in the end both Van Ranouw's observations and his images were based on samples he received from the apothecary and collector Albertus Seba (1665–1736) in Amsterdam.[15] By connecting Van Raat's descriptions to the images of Seba's samples, Van Ranouw presented the information that was available to him as a unity. In other words, once Van Ranouw had compiled all the information he could about cinchona bark and cascarilla, he fused input from various trajectories: printed information from medical works by authors like Pomet and Lémery; commercial information from Hendrik van Raat; and Albertus Seba's knowledge as a pharmacist and collector of naturalia. All of these men were experts in their respective fields, and in Van Ranouw's mind at least, the fragmentary bits of information they provided could be brought together to form a coherent picture of both cinchona bark and cascarilla.

There were, however, three fundamental and closely related problems that produced a great deal of contradictory information about cascarilla in Van Ranouw's day: the name of the substance; the plant that produced it; and the place from whence it came. Lémery's "Eleaterium" was still an *herba nuda*: a substance devoid of cultural connotations, and without an etymological explanation of its name. The first publication that was exclusively devoted to cascarilla, written by the French chemist

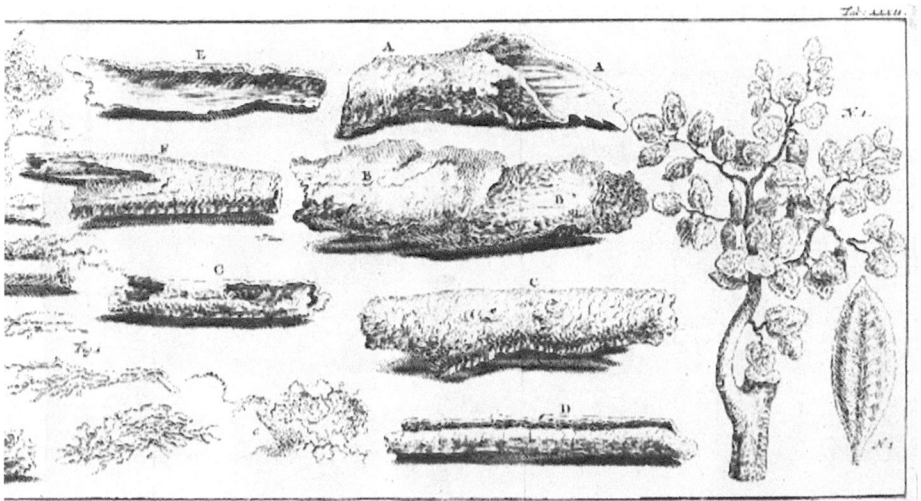

FIGURE 4.1. Illustration of samples of Peruvian bark accompanying the articles of Willem van Ranouw. Van Ranouw, "Vierde Verhandeling," tab. XXXII; National Library of the Netherlands, The Hague, KW 223 M 55[-63].

Gilles-François Boulduc (1675–1741), asserted that the name "Elaterium" for cascarilla was a reference to the squirting cucumber (*Ecballium elaterium*).[16] Another explanation for the name was given by Hendrik van Raat: "Elaterium" could refer to the island Eleuthera in the Bahamas. But Van Raat doubted that the name was chosen because cascarilla actually came from this island. Instead, he argued that the name could also have resulted from a mash-up of an earlier name for cascarilla, "Cortex de lateribus."[17]

Whatever the case, the association between cascarilla and the Bahamas persisted. The publication of the English naturalist Mark Catesby's work on American nature would do much to consolidate the dissociation of cascarilla from cinchona bark. In the second volume of his *Natural History*, published in 1743, there is a description of cascarilla or "The Ilathera Bark," derived from shrubs which "grow plentifully on most of the Bahama Islands."[18] With Catesby, then, linguistic confusion produced a shift in knowledge about cascarilla's geographical origins, from Peru to the Caribbean.

Catesby's recognition of cascarilla's Caribbean origins resolved an ambiguity that is only apparent to modern observers, but was not to Catesby's contemporaries. Cascarilla had always been regarded as somehow similar to cinchona bark, including in its Peruvian provenance.

FIGURE 4.2. Illustration of cascarilla in Catesby's book, behind a copper-belly snake. The image is on the unnumbered page preceding the description of both snake and bark. Catesby, *Natural History*, vol. 2, after 46; Wellcome Collection, London; CC PDM 1.0; https://wellcomecollection.org/works/un67cvxt.

After Catesby, both of these assumptions had to be rejected. Nowadays, cascarilla is regarded as the species *Croton eluteria*, a plant native to the Caribbean, but naturalized in other parts of tropical America. *Croton* species are part of the *Euphorbiaceae* family, characterized by tricapsular fruits. Catesby's description certainly concerns a *Croton*, as he mentions "tricapsular pale green Berries." Cinchona bark, however, derives from *Cinchona* species, from the very different *Rubiaceae* family. Although Linnaeus would be the first to define the species of cinchona bark and cascarilla in the 1750s, a definitive botanical separation of the two was already evident with Catesby several years earlier, particularly as Catesby was the first to add an image of cascarilla to his description. Half-hidden behind the image of a copper-belly snake, a branch of cascarilla is visible (see fig. 4.2). A patchwork image, clearly based on Catesby, is already to be found in a dissertation by Philippus Adolphus Boehmer dating from 1738, which offered the first extensive description of cascarilla (fig. 4.3).[19] Catesby's and Boehmer's images differ

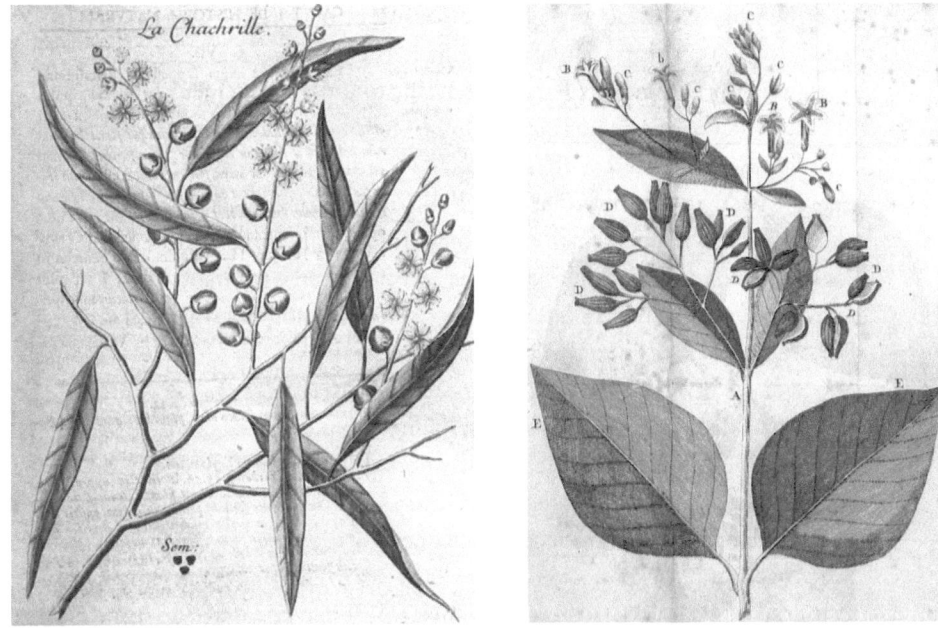

FIGURE 4.3. (LEFT) Illustration of cascarilla, in the 1738 description by Boehmer. Philippus Adolphus Boehmer, *De Cortice Cascarillae*, 17; Bayerische Staatsbibliothek, Munich; CC NC 1.0; http://reader.digitale-sammlungen.de/de/fs1/object/display/bsb10822926_00021.html.

FIGURE 4.4. (RIGHT) Illustration of Peruvian bark ("Quinquina"), accompanying the 1738 description by La Condamine. Charles Marie de La Condamine, "Sur l'Arbre du Quinquina," after 244; Wellcome Collection, London; CC PDM 1.0; https://wellcomecollection.org/works/wvjzue8k.

significantly from the first realistic image of cinchona bark, contained in a description by Charles Marie de La Condamine of 1738 (fig. 4.4).[20]

Meanwhile, the history of testing cascarilla bark for its medicinal properties followed a trajectory that seems to have been largely independent from the quest to disambiguate its name, plant of origin, and geographical provenance. Successful medical applications of cascarilla probably dated back as far as Pomet's day. Boulduc, for instance, relates how Louis XIV's personal physician, Guy-Crescent Fagon (1638–1718), confided to him that he had successfully resorted to cascarilla for treating intermittent fevers at a time when "true" cinchona bark was in short supply.[21] Later research in German territories laid the basis for cascarilla's widespread use as a medicine in Europe. The various steps were recorded in quite some detail by the French physician and chemist Étienne François Geoffroy (1672–1731), in his *Tractatus de Materia Medica* of 1741.[22]

The story so far suggests that a few assumptions may be made about the European acculturation of cascarilla in the late seventeenth and early eighteenth centuries. First, the botanical and geographical origins of cascarilla were very unclear to scholars before the publications of La Condamine and Catesby. These publications disambiguated "true" cinchona bark from the new cascarilla bark, and assigned a clear, distinct place of origin to both. Such disambiguation was highly necessary because the characterization of drugs as "exotic" often went hand in hand with their geographical indeterminacy, as the contributions by Hjalmar Fors and Clare Griffin in this volume make clear.

Second, however, these ambiguities had little effect on the appropriation of cascarilla in European medicine. No protracted conflict like the *querelle de quinquina*, which greatly perturbed medical debate about cinchona bark in mid-seventeenth-century Europe (especially France), seems to have affected the acceptance of cascarilla. The separate trajectories of botanical exploration and medical testing may have been manifestations of different research programs in different centers of learning. Botanical disambiguation was a major concern mainly for English and French scholars, while most of the medical testing took place in German academic centers.

In the Dutch Republic, cascarilla was already gaining importance by the 1720s, as Van Ranouw's publications make clear, and the fever epidemic of 1727 did much to consolidate its significance as a medical commodity. In Amsterdam, the medical and commercial trajectories of cascarilla became intertwined thanks to the efforts of Albertus Seba. Occasionally, we encounter his name in archival records of public auctions of drugs.[23] During one auction, in October 1727, Seba bought batches of cinchona bark along with several other druggists, but he also managed to get his hands on a batch (three seroons) of "Sacorille."[24] The spelling, apparently a Dutchification of the French *chacrille*, suggests unfamiliarity with the substance on the part of either the seller who sold the goods, the broker who organized the auction, or the printer who produced the catalog. This was not the first auction of drugs that included cascarilla: one barrel of "Cascarille" had been sold the previous year.[25] The significance of Seba's purchase of cascarilla in 1727 would not be obvious, had it not been for the fever epidemic in the same year. The fact that Seba had already communicated with Van Ranouw about cascarilla some years earlier demonstrates his continued interest in cascarilla bark.

THE FEVER EPIDEMIC OF 1727-28

Zooming in on Dutch newspapers, we can find cinchona bark and cascarilla in advertisements during the fever epidemic that raged in various cities during 1727 and 1728. In general, Dutch newspapers reveal very little about the existence of epidemics, as they mainly discussed foreign news. Most news about fevers between July 1727 and June 1728 concerned individual cases of feverish kings, queens, princes, and dukes throughout Europe.[26] The same goes for yearbooks like the *Europische Mercurius*, which, however, does associate the high death rate in Amsterdam in 1727 with the fevers raging in the city in the same year.[27] A more obvious source where historians might discern an epidemic are mortality figures, which generally register substantial increases. Since no personal accounts of the epidemic are known for Amsterdam, we only have the mortality figures to inform us of the severity of the disease there.

The yearly death rate in Amsterdam between 1720 and 1740 shows the great impact of fever in 1727, as can be seen in figure 4.5. The year 1727 was the deadliest year for Amsterdam during these two decades: there were 13,755 burials, i.e., an increase of 49 percent from 1726, and more than double the figure for 1725. The *Europische Mercurius* related that 672 people died in the final week of October 1727, or one every fifteen minutes, the highest death rate so far.[28] It has been calculated that during the course of 1727, 6.3 percent of the population of Amsterdam died, with the highest percentage (3.3 percent) dying in the last four months of the year.[29] As the figures for Amsterdam show, this elevated mortality continued into 1728.

The outbreak was not confined to Amsterdam. The exact geographical range of the epidemic is hard to determine, but it seems to have occurred mainly in certain urban centers in the province of Holland.[30] People suffered from fever in Hoorn, where the epidemic lasted until April 1728.[31] There was a mortality peak in Enkhuizen in the same period (September 1727 to June 1728), which may have been related to the same outbreak.[32] Figure 4.5 also shows the death rate for some other cities in Holland for which yearly figures are available. Besides Amsterdam, there were also high death rates in Leiden and the cities of the "Noorderkwartier" to the north.[33] Other cities, however, show no high death rate around 1727–28. Cities like Delft and Rotterdam experienced relatively high mortality in the second half of the 1720s, but if these peaks are relatable to disease, it was probably not epidemic fever.[34] Therefore,

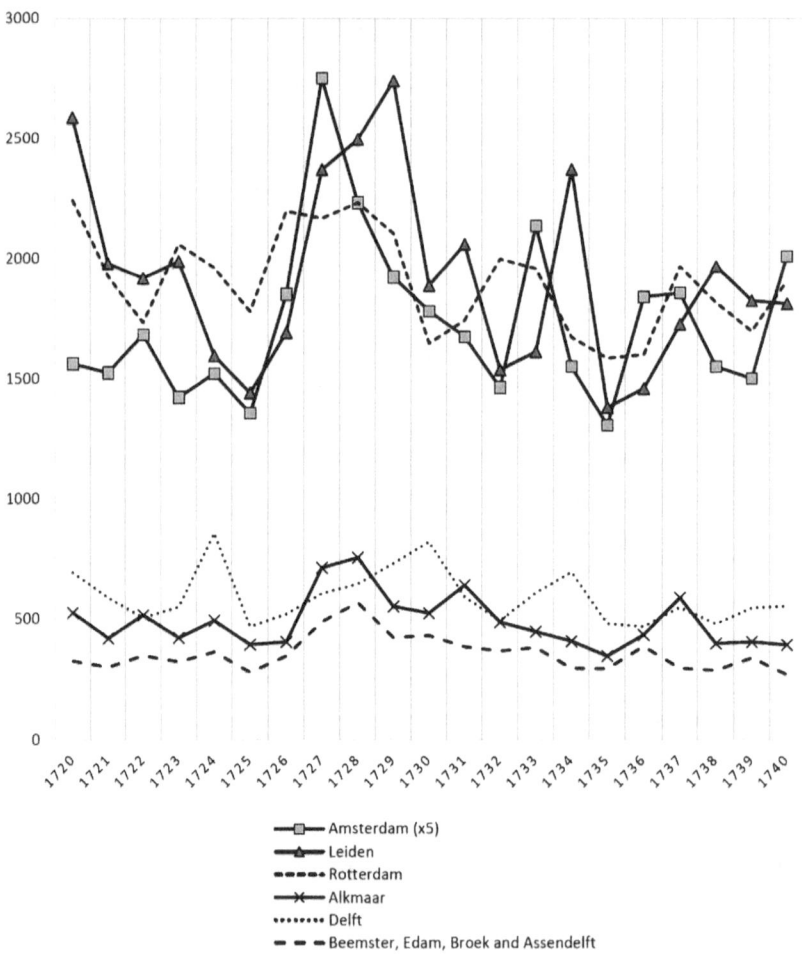

FIGURE 4.5. Number of burials in cities in Holland, 1720–40.

the fever epidemic seems to have struck Amsterdam, Leiden, and the cities of the "Noorderkwartier" in particular.

In addition to mortality figures, the years 1727 and 1728 also saw a sharp increase in the number of advertisements for fever remedies in Dutch newspapers. This points to a growing market for irregular remedies in the wake of the epidemic. In this way, these advertisements provide a good starting point for finding out more about the fever epidemic of 1727–28. Furthermore, the epidemic can be said to have had a lasting impact on advertising practices for secret fever remedies.

Whenever we encounter advertisements for remedies in Dutch

newspapers from the eighteenth century, the remedies to which they refer can generally be classed as medical secrets. Although "secrecy" in early modern science and medicine is nowadays often highly conceptualized, the secrecy of advertisements can be understood from a practical point of view for the purposes of this chapter, both commercially and legally. Those who invented a remedy, or those who distributed a remedy on behalf of an inventor (or as middlemen, i.e., *in commissie*), had a stake in keeping the contents of the remedy a secret. Local entrepreneurs could have good medical knowledge and competence—or none at all—but their remedies were usually produced and distributed outside of regular apothecary shops. Such remedies were often not controlled by the medical profession, and their contents were not known to anyone but the inventor. Disclosure of the ingredients of secret remedies could create a dual problem: forgery by other irregular medical practitioners, and thus loss of income; and/or legal proceedings from regular physicians, who might regard such remedies as fraudulent.[35]

It is therefore particularly striking to find references to cinchona bark, with highly negative connotations, in Dutch newspaper advertisements for fever remedies during the 1720s and 1730s. The bark is not mentioned as an ingredient in these advertisements. On the contrary, the advertisements explicitly say that the remedy does *not* include cinchona bark. Proclaiming the absence of an otherwise obvious choice of febrifuge ingredient was apparently regarded as a unique selling point of these secret remedies. This practice disappeared again after 1730. From late 1727 onward, these advertisements thus exhibit an extraordinary advertising feature that requires explanation.

The number of advertisements that mention cinchona bark with negative connotations is fairly modest. Out of 4,861 advertisements for fever remedies found in Dutch newspapers from the period 1680 to 1799, visualized in figure 4.6, only 88 mention cinchona bark at all; 64 of these date from the period 1720–1740, and all mention the absence of cinchona bark from the remedy.[36] The eight remedies that formed the subject of these advertisements were apparently tapping into a market of patients who disliked cinchona bark for some reason. The first advertisement to take such a line, appearing in the *Leydse Courant* in November 1727, described "an expert remedy against the fevers currently prevailing." It could be bought at the bookshop of J. Hayman in Amsterdam, for 20 *stuivers*. The advertisement ends explicitly: "Everyone can be assured that there is no cinchona bark in it."[37]

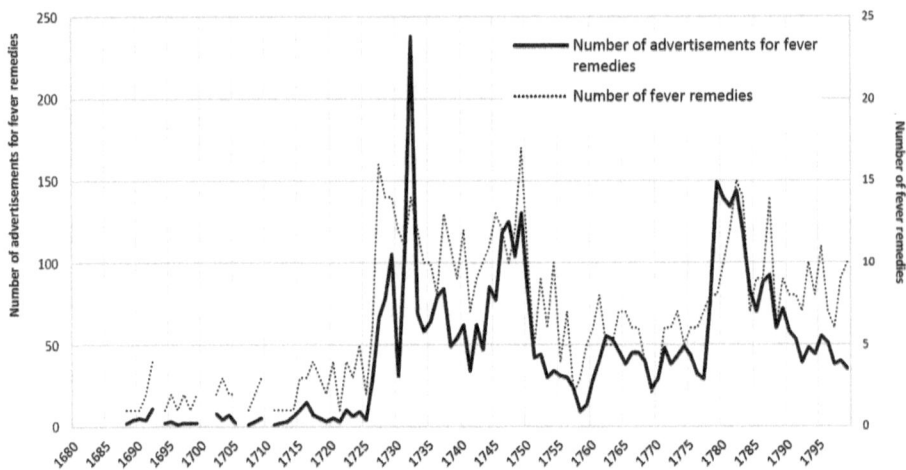

FIGURE 4.6. The distribution of advertisements for fever remedies across the period 1680–1799, and the corresponding number of fever remedies, in Dutch newspapers.

Another remedy, the "Miraculous Antifebrile Salt" (*Sal Mirabile Antifebrile*), invented by an English "physician" (*geneesmeester*), was advertised from October 1727 onward. The intent may have been to compete with the other remedy, for the same audience. If so, then the "Miraculous Salt" probably won the battle. Its usefulness for fevers, again "especially in these times," was characterized in much the same terms as the earlier remedy.[38] For 18 *stuivers* (slightly cheaper than its competitor), it could be purchased from the bookshop of the widow of Jacob van Egmond, in Amsterdam. Forty-one advertisements for this remedy were found, thirty-seven of which appeared between October 1727 and January 1730. This may not indicate greater commercial success, but at least it demonstrates greater public visibility of this remedy. In any case, the producers of the "Miraculous Antifebrile Salt" quickly expanded their business, setting up a distribution network of vendors throughout the province of Holland, as may be seen in later advertisements.

As these two remedies illustrate, advertisements for fever remedies around 1727 might address one or both of two new themes: the absence of cinchona bark from their ingredients, and their applicability in cases of "currently prevailing fevers." Given the novelty of these remedies, it is reasonable to assume a correlation between the two themes. Excluding cinchona bark from fever remedies apparently appealed to an audience whose experiences with that substance had been bad or inconclusive, probably during the fever outbreak of 1727–28. Advertisers for secret remedies were quick to respond to this sentiment. The negative image that arises from advertisements in this period is unique: one rarely comes across sources that exhibit negative publicity for a remedy with such a canonical status as cinchona bark had acquired by the early eighteenth century. Therefore, retracing the epidemicity of fever in this period sheds light on the trajectories not only of cinchona bark, but perhaps also of competing substances like cascarilla.

First, some remarks should be made about the nature of the epidemic: an ambiguous category like "fever" in the eighteenth century does not lend itself to an obviously medical or cultural interpretation, so an analysis of the disease from the actors' perspective will help to clarify the propensity to use certain remedies. The absence of contemporary news about the fever epidemic of 1727–28 is as striking as the epidemic's apparent severity. Very few modern studies mention a fever epidemic in these years at all. J. Z. Kannegieter gives mortality figures, and briefly discusses the social impact of the epidemic on public life in Amsterdam.[39] Jan de Vries and Ad van der Woude suggest that this may have been the earliest recorded malaria epidemic in the Netherlands.[40] It would be valuable to consult first-hand accounts that could offer a glimpse of fever experience and treatment during this episode. Luckily, there is one excellent source that contains both. Herman Boerhaave, the most famous physician of his era, describes the epidemic in two letters to Joannes Baptista Bassand (1680–1742), court physician to the Duke of Savoy and the imperial family in Vienna. The first letter does not mention the word "fever," but it clearly concerns an epidemic disease: "An autumn epidemic [*autumnalis epidemicus*] of a rather dangerous nature and fatal to many has attacked our part of the country so that there are sufferers everywhere. Never have I had more patients to treat, although I have long since given up my practice (doing rounds); I have seen all of them cured by an appropriate emetic, diluting medicines, Polychrest salt, Contrayerva [root], and opium. Finally I fell ill myself, but far worse than the other sufferers I had seen."[41] The second letter to Bassand goes into more detail about both the nature of the disease and its treatment: "It broke out after a long period of hot weather and extreme drought. The fever was atypical [*anomala*], and was accompanied by continual vomiting, most violent headaches, and a cerebral inflammation which was soon fatal, or that, while seeming to abate, quickly rose again, attended by obvious swelling of the abdomen, vomiting and subsequent decease. Others have been livid, pale and dropsical throughout the entire winter."⁴² Although Boerhaave's treatment was successful, other patients evidently died from fever. Boerhaave initially thought that he would not succumb to the disease himself. When he eventually fell ill as well, he applied his treatment to himself:

> I treated myself in the same way as all the others. On the first day I took an emetic, on the second an ounce of cream of tartar, on the third the same, and on the fourth two enemas. Throughout this period I drank a simple decoction of barley mixed with a large quantity of elder-berry juice and Polychrest salt. By means of this treatment I first purged the intestines and cleared the head. But I was troubled by great weakness, almost to the point of swooning; then I took copious draughts of Rhine wine with lemon juice, sugar, and toast; and on the fourteenth day I underwent three purges, and subsequently revived my weakened body by drinking Peruvian bark, made like tea, for several days in the daytime; thereupon I recovered completely, rose from my bed and seem to feel stronger than previously.⁴³

Boerhaave had more trouble curing himself than others. After ridding himself of the febrile matter which caused the disease, he was troubled by weakness for quite some time. Cinchona bark features as the final step in his treatment, which prompted his full recovery.⁴⁴

Boerhaave's attitude toward cinchona bark was ambivalent. On one occasion he confided to his student Julien Offray de La Mettrie (1709–1751) that cinchona bark had done more harm than good, and should never have been discovered.⁴⁵ When he wrote about fever treatments to Bassand, he argued briefly that cinchona bark "is rarely or never necessary."⁴⁶ But other letters show that Boerhaave used the bark himself on several occasions. He had suggested a number of remedies with cinchona bark to Bassand in 1714, to treat a violent disease suffered by the Princess of Savoy.⁴⁷ Many years later, in 1733, Boerhaave suggested a similar remedy to Willem Roëll (1700–1775), professor of anatomy in Amsterdam.⁴⁸ Once, in 1717, Boerhaave even sent twelve pounds of cinchona bark to Bassand himself.⁴⁹

While historians tend to think of cinchona bark as an essential early modern remedy against fevers, in Boerhaave's use, it appears far less important. Cinchona bark was always part of a longer treatment process, and can hardly be regarded as the focal point of Boerhaave's fever therapy.⁵⁰ This was the case during the epidemic of 1727, but the epidemic did not seem to change his opinion of cinchona bark in any way. If patients had begun to distrust cinchona bark after the epidemic of 1727 (as referenced in advertisements for fever remedies), no evidence for increasing distrust on the part of physicians like Boerhaave can be found.

Boerhaave was an illustrious figure, but can his account be regarded as representative? Some other sources, published long after the epidemic took place, add context to Boerhaave's experience.[51] These include the *Constitutiones Epidemicae* by his student Gerard van Swieten (1700–1772). Van Swieten's observations include his own illness in November 1727. He got better quite soon after having consulted Boerhaave, who suggested similar remedies to those he had taken himself.[52] Another account can be found in Salomon de Monchy's (1716–1794) study of medicine at sea. De Monchy recalled a lecture from his student days by Herman Oosterdijk Schacht (1672–1744), a colleague of Boerhaave at the time of the epidemic, who taught theoretical and bedside medicine. Oosterdijk Schacht maintained that purgative remedies had been used with success during the epidemic of 1727.[53] There is a striking contrast between these positive accounts and the one by Theodore Tronchin (1709–1781), another of Boerhaave's pupils. In 1757, Tronchin recalled the colic that fever patients suffered in 1727 as a result of treatment with cinchona bark:

> Being soothed by the bark, while the putrid bile continued to work, she [i.e., the colic] affected the mesentery and the membranes of the intestines to such an extent that the accident [*toeval*] of colic pain was an indubitable sign of preceding fever. The remnants of that disease, which would still be active years later, accustomed the physicians to that disease: oh, if only that had been the case with the remedy as well. One could often see the poor sufferers, who resulted from it, passing on the street like ghosts; bloodless, deadly pale; with crooked hands, flabby arms, their voice hoarse and faint; yes, some were even speechless.[54]

This vivid description of the effects of cinchona bark reveals the impact such a treatment might have on patients. Tronchin's description is reminiscent of the ways in which producers of fever remedies described the negative effects of cinchona bark in their advertisements. Patients who recognized Tronchin's description would undoubtedly be receptive to fever remedies without cinchona bark.

In sum, the medical trajectory of cinchona bark in Leiden presents us with ambiguous information: the bark seems to have been used to treat sufferers of epidemic fever in 1727–28, but not always to good

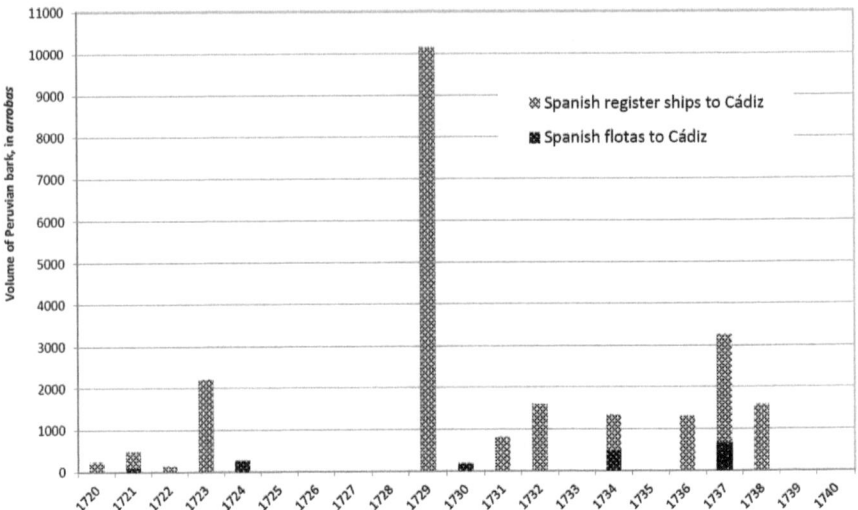

FIGURE 4.7. Global influx of Peruvian bark into Spain in *arrobas* (1 *arroba* = ca. 11.5 kg), 1720–40.

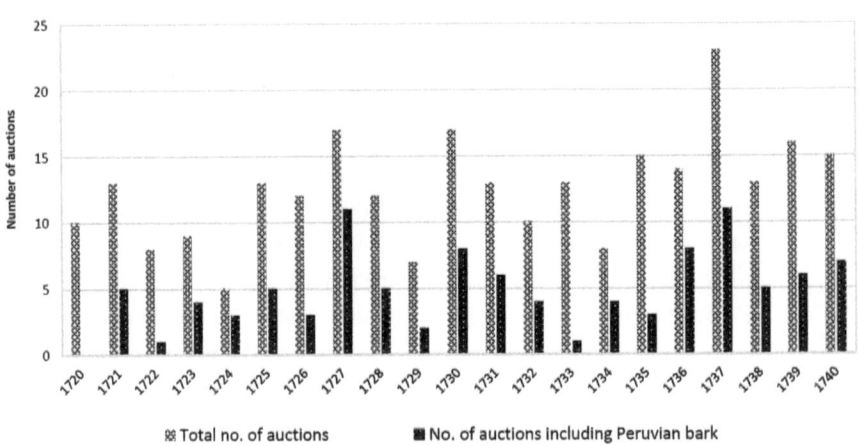

FIGURE 4.8. Auctions of drugs in Amsterdam, and the number of auctions including Peruvian bark, 1720–40.

effect. At the same time, cinchona bark was not the focal point of treatment. Therefore, it would be useful to know whether the epidemic had an effect on the availability of cinchona bark. In the next section, therefore, we shift our attention from the domain of medicine to commerce. This is where we first encounter cascarilla bark as a possible alternative medical commodity.

AVAILABILITY OF, AND COMPETITION BETWEEN, CINCHONA AND CASCARILLA DURING THE EPIDEMIC

We are particularly well-informed about trade in cinchona bark in Amsterdam in the early decades of the eighteenth century, especially when compared to the fragmentary Spanish records of transatlantic trade in this period. Cinchona bark arrived in Cádiz each year with the Spanish galleon fleet (*flota*), which brought the bulk of goods from Spanish South America to Europe: mainly gold and consumer products, like tobacco, cacao, and indigo. As fleets were no longer arriving every year by the early eighteenth century, the system deteriorated. Therefore, Spain allowed so-called "register ships" from other countries to trade directly with the New World, adding to the volume of goods imported by the fleets. No systematic records were kept during this period, so the cumulative imports from *flotas* and register ships is impossible to determine in detail, as can be seen in figure 4.7.

Much more coherent commercial data can be found for public auctions of drugs in Amsterdam. Several members of the Brokers' Guild there were starting to develop the drug trade into a niche market in the early decades of the eighteenth century. The public auctions of goods organized by these brokers were sanctioned by the city authorities, to distinguish official brokers from unofficial ones.[55] As a result, auction activities have left a substantial amount of data for scholars to investigate. For the period 1720–40, the archival record is incomplete, but it can be supplemented by data from newspaper advertisements, in which these same auctions were announced.

Altogether, cinchona bark appeared frequently at public auctions of drugs, as can be seen in figure 4.8. Each year, a fair number of auctions of raw drug materials took place, mainly bought by local druggists in Amsterdam. A substantial number of auctions included cinchona bark, which indicates that the bark was generally available to medical practitioners and patients in the city during the period 1720–40. Figure 4.8 demonstrates the fluctuating availability of bark from year to year, with an expected peak in 1727, when there would have been more demand than usual for the bark. But the data is more pertinent if we zoom in on the auctions for which we know the actual number of units sold, shown in table 4.1. In archival records and advertisements, these units are given as trade units, i.e., "manageable" batches like chests, barrels, and seroons.[56] While little more than 80 or 90 units were auctioned per annum

TABLE 4.1. PUBLIC AUCTIONS IN AMSTERDAM INCLUDING PERUVIAN BARK, 1720-40

Year	A. Number of auctions that include Peruvian bark[1]	B. Auctions in column A that specify amounts[2]	C. Units of Peruvian bark listed in auctions included in column B	D. Specified units of Peruvian bark in auctions included in column C
1721	5	2	76	76 seroons
1722	1	1	8	8 seroons
1723	4	3	95	95 seroons
1724	3	1	10	10 seroons
1725	5	4	96	82 seroons, 14 barrels
1726	3	3	80	68 seroons, 6 barrels, 6 small chests
1727	11	11	244	241 seroons, 1 barrel, 1 chest, 1 small bag
1728	5	3	93	65 seroons, 28 barrels and seroons
1729	2	2	106	106 seroons
1730	8	7	270	259 seroons, 11 bales
1731	6	5	128	105 seroons, 22 barrels, 1 small bag
1732	4	2	44	40 seroons, 4 barrels
1733	1	1	14	14 seroons
1734	4	2	113	113 seroons
1735	3	2	80	80 seroons
1736	8	8	129	51 seroons, 32 barrels, 4 small barrels, 9 chests, 30 small bales, 3 bags
1737	11	11	114	108 seroons, 5 barrels, 1 small barrel
1738	5	5	119	73 seroons, 38 barrels, 8 bales
1739	6	6	122	116 seroons, 5 barrels, 1 bag
1740	7	7	165	133 seroons, 9 barrels, 20 small barrels, 1 box, 2 small chests

Notes: 1. Original diminutive forms in Dutch have been retained, i.e. both barrels (*vaten*) and small barrels (*vaatjes*) have been included in the table.

2. For the auctions where the amounts have been specified in archival records and/or newspaper announcements (column B), specifications are given in columns C and D.

Source: Stadsarchief, Amsterdam, 5069, inv. nos. 1–13.

in earlier years, the number of units tripled in 1727, to 244 units. Only 86 (35.3 percent) of these were sold in the first half of the year; the rest was auctioned in late summer and fall. This points to a quick commercial response to the fever epidemic.

In all auctions dating from the second half of 1727, the prices mentioned before and after bidding did not diverge very much. Apparently, little bidding took place to drive up the price. This seems to indicate that, regardless of the large amount of cinchona bark that was on offer in Amsterdam in the midst of the epidemic, no gold rush of any kind seems to have taken place, and prices were not driven up. In other words, more cinchona bark was available (and probably desired by consumers) during the epidemic than in other years, but auction records do not indicate either scarcity or saturation in the market.

All in all, the trade records for cinchona bark do not offer a strong explanation for why cinchona bark was disliked in 1727. The amount of cinchona bark that was available seems to have been adequate. The bark was also of acceptable quality, because damaged batches were usually flagged at auctions, but no such indications were made in this year. The arrival of shipments of cinchona bark in Spain was usually mentioned in Dutch newspapers, but only rarely is there any hint in the newspapers about quality issues.[57] According to one author, the bulk of the cinchona bark that was shipped from Loja, Peru, to Europe came from an inferior variety.[58] The best varieties were destined for the Spanish Empire, especially the motherland itself.[59] Rather, the consistency with which batches of cinchona bark sold at public auctions in Amsterdam suggests that there were no great concerns about the quality of cinchona bark among brokers and buyers.[60]

Although no pronounced aversion for cinchona bark can be observed in the domain of trade, the epidemic of 1727 did nonetheless consolidate an upward commercial trajectory for its competitor, cascarilla. It might at first seem odd to argue that cascarilla might have gained some of the ground that cinchona bark had lost as a result of the epidemic of 1727. Advertisements for the "Miraculous Antifebrile Salt" frequently mentioned that the remedy did not contain cinchona bark "or anything like it" (*of iets diergelyks*).[61] Surely cascarilla—another bitter, astringent febrifuge from the New World, which partly shared a history with cinchona bark, in terms of naming and provenance—would have counted as a similar substance. But apparently, alternatives for remedies could be acceptable even if their characteristics were similar to those of the remedies they were supposed to replace.[62]

In auction records, the new febrifuge substance recurred frequently after 1727. The occurrence of cascarilla in auctions is shown in table 4.2, which shows data similar to that shown in table 4.1 for the case of cinchona bark. Auction records demonstrate that cascarilla was quite regularly available, though in much smaller quantities than cinchona bark. Cascarilla was apparently a popular product: there are names of buyers for all the units in table 4.2 for which an archival record has survived. A peak in trade volumes was reached at the auction of August 7, 1737, when 104 units of cascarilla were sold, at the price of 12–14 guilders per 100 pounds.[63] The codification process of cascarilla in pharmaceutical handbooks (pharmacopoeias) occurred roughly around the same time. Alkmaar was the first city to include cascarilla in its pharmacopoeia, in 1723.[64]

Evidence from the domain of trade, therefore, indicates that cinchona bark was not clearly disliked during the epidemic of 1727–28, but there was still enough room for the new febrifuge cascarilla to position itself as a possible alternative. Where does this leave us with regard to the status of cinchona bark after the epidemic? Some advertisers continued to claim that they *excluded* cinchona bark from their remedies, but in doing so they still *included* the bark in their advertisements. As late as 1737, the producers of the "Miraculous Antifebrile Salt" advertised that their remedy cured all sorts of fever and always had, "as can be confirmed by a great many people who have used it, [and as] its healing powers have been found to be abundant in the year 1727, etcetera."[65]

Thus, the negative connotations that began to surround cinchona bark in advertisements especially appeared *following*, not *during*, the epidemic. Perhaps a negative recollection of cinchona bark treatment lingered on in people's minds, encouraged by the sight of those who still suffered from its effects, as mentioned in Tronchin's account. Still, as the foregoing discussion has aimed to show, various trajectories of cinchona bark can be correlated to demonstrate that it was not generally disliked in commerce, medicine, or society, either during or after the epidemic. The insistence that cinchona bark was not an ingredient of a given drug can be regarded as a theme that was used in advertisements to address a specific market segment of patients. Some patients would have been receptive to the language of advertisements because of their own bad experiences in fever treatment, but the evidence we have suggests

TABLE 4.2. PUBLIC AUCTIONS IN AMSTERDAM INCLUDING CASCARILLA, 1720-40

Year	A. Number of auctions that included cascarilla	B. Auctions in column A that specify amounts	C. Units of cascarilla listed in auctions in column B	D. Specified units of cascarilla among entries in column C
1721–25	0	0	0	0 units
1726	1	1	1	1 barrel
1727	1	1	3	3 seroons
1728–29	0	0	0	0 units
1730	2	2	25	5 barrels, 20 bales
1731	2	2	37	23 bales, 14 barrels
1732–35	0	0	0	0 units
1736	4	4	5	4 barrels, 1 small barrel
1737	3	3	108	97 *scavassen*,[1] 11 barrels
1738	2	2	5	5 barrels
1739	1	1	1	1 barrel
1740	0	0	0	0 units

Note: 1. It is unclear if and how *scavassen* (also spelled as *schavassen* or *cabassen*) should be translated. The word refers to a type of basket, sometimes with a handle. See http://gtb.inl.nl/iWDB/search?actie=article&wdb=WNT&id=M029363&lemma=kabas.

Source: Stadsarchief, Amsterdam, 5069, inv. nos. 1–13.

that this was probably not a collective sentiment. During and after the epidemic, there was enough cinchona bark available, which was used moderately by physicians, and there were possible alternatives, like cascarilla. It is only in the recurrence of cinchona bark in advertisements in the aftermath of 1727 that we can observe the imprint of the encounter with epidemic fever on collective memory.

It is possible that, when advertisers assured their audience that they did not use cinchona bark, they were distancing themselves from other advertisers who still included bark in their remedies but kept silent about it. In that case, we could be dealing with secret remedies opposing *other* secret remedies, rather than "regular" remedies prescribed by physicians and composed by apothecaries. In other words, the cinchona bark theme might reflect *internal* competition within the market of irregular practitioners and remedies. It is reasonable to assume that the inclusion of the bark in secret remedies might sometimes have produced negative experiences among patients. Including or excluding cinchona in advertisements could therefore help to differentiate fever remedies that were advertised on the same newspaper page.

Although it is impossible to prove, many producers of secret fever remedies must have included cinchona bark in their products: the bark was available in large quantities, and in reasonable quality. Toxicological side effects might therefore have been a real danger, even if these were not generally known in the eighteenth century. If so, then the advertisers of secret remedies *without* cinchona bark attempted to attract patients whose bad experiences had been caused by similar secret remedies *with* cinchona bark. This is an example of secrecy's boomerang effect: the intended safeguarding of the producers' interests worked against them when patients became the victim of their surreptitious practices.

CONCLUSION: EPIDEMICITY DEMONSTRATES THE REGULAR SUPPLY OF DRUGS

The fever epidemic of 1727–28 must have caused much more suffering in Holland than the few sources from the medical and public domains suggest. Alongside mortality figures, however, commercial data from newspapers is one of the best sources—though still largely unexplored—for studying the occurrence and impact of epidemics. This chapter has argued that epidemics can offer a fruitful way to study the acculturation of drugs. Many remedies that are generally hidden in primary sources might reveal themselves when supply and demand increased as a result of an epidemic. This is the case for "big" remedies like cinchona bark, "small" exotic remedies like cascarilla, and secret remedies like those that were advertised for the treatment of fever.

The few personal experiences that survive in writing demonstrate that the disease must have left a considerable imprint on society as a collective, in the form of many and diverse experiences with epidemic fever. These experiences are also reflected in the trajectories of cinchona bark and cascarilla, which had been connected before but became even more strongly intertwined as the epidemic progressed. In the process, various agents handled these substances or formed opinions about them. Patients may have had an aversion toward cinchona bark, perhaps based on negative experiences during the epidemic. Producers of secret fever remedies saw this as a commercial opportunity to address a specific segment of patients by including the issue of cinchona bark in their advertisements. A physician like Boerhaave dealt with epidemic fever as best as he could and pragmatically applied or renounced cinchona bark as the situation required, both for his patients and himself. Merchants and brokers traded in cinchona bark and cascarilla during the epidemic, as they increasingly did in the early decades of the century anyway, thus consolidating their practice and position as intermediaries in the global chain of commerce. Both knowledge and goods were shared, as Albertus Seba did when he provided samples of bark to Willem van Ranouw. Therefore, the trajectories of cinchona bark and cascarilla were greatly affected by these various attitudes, which intertwined during and after the epidemic. As more and more people encountered remedies during an epidemic, out of pure necessity, the acculturation of these substances accelerated. They acquired a more solid character as commodities, compared to a time when they were still little-used exotic novelties.

The epidemic of 1727–28 can thus be regarded as the glue that tied the different trajectories of cinchona bark and cascarilla together, albeit implicitly most of the time. The advertisements that mention cinchona bark hint at raging fevers during 1727 and 1728, but give no details as to their causes, course, or ending. Herman Boerhaave in Leiden discussed the epidemic in some detail, but only twice, with one of his most trusted correspondents, Joannes Baptista Bassand. For Amsterdam, no personal accounts of the epidemic are known, so mortality figures are the most important source we have to trace the epidemic there. Meanwhile, cascarilla had already been acculturated to a large extent in Dutch medicine, judging from its inclusion in pharmacopoeias before the epidemic. The significance of cascarilla, however, increased once the substance itself began to be traded regularly, which happened around the time of the epidemic. Investigations and publications by men like Van Ranouw reveal the interconnection between commerce and research in the process of disambiguating knowledge about cinchona bark and cascarilla. There was a clear and widespread interest in these substances and in knowledge about them, and it would only be a couple of years after the epidemic that key publications about both substances would be published. As such, the epidemic of 1727–28 helped to solidify both cascarilla and cinchona on the medical market.

5

THE MEDICAL RECEPTION OF SASSAFRAS IN EARLY MODERN ENGLISH PRINT

Katrina Maydom

How was knowledge about New World drugs presented, contested, and consumed in early modern print? The Spanish and Portuguese were the first Europeans to encounter American drugs directly, and recent research has examined how these empires investigated and understood the medicinal properties of new drugs, particularly cinchona.[1] In this chapter, I explore how sassafras was understood and recommended for use in medical practice, consulting 114 medical texts published in English over a hundred-year period between 1577 and 1680.[2] By providing a century-long survey, I can examine change and continuity in the medical reception of a New World drug. I find that sassafras was not regularly referred to immediately after its introduction in English print, but rather after a concentrated effort to commercialize the drug and a substantial increase in imports, resulting in a delay of nearly seventy years for sassafras to become frequently discussed. From 1650, the number of references to sassafras, the range of diseases that it was endorsed to treat, and the diversity of preparations described increased significantly in scale. By the 1670s, sassafras was advocated for medical practice in a greater number of medical texts than many drugs with a long tradition of cultivation in England and those primarily obtained through trade with Southeast Asia, including angelica, mastic, cloves and nutmeg.[3] Once sassafras became more prevalent, it was co-opted into contemporary debates, such as those between Galenic and chymical physicians, which had little to do with its status as originating from the New World.

Sassafras is a deciduous tree that can grow up to one hundred feet tall with aromatic leaves, bark, and roots. It has green apetalous flowers and dimorphous leaves that have a citrus-like aroma.[4] The origin of the European name for this plant—"sassafras"—is uncertain but became shared among European languages to spread knowledge about a plant that they became increasingly curious about.[5] As Clare Griffin has recently argued, the Europeans who turned what indigenous peoples knew as *wissoe, winauk,* and *pauame* into "sassafras" were, in the process, constructing their own New World.[6]

Sassafras was first described in English in a translation of work by the Spanish physician Nicolás Monardes (1493–1588). The English gained substantial first-hand experience of the plant in the seventeenth century because it grew abundantly in England's American colonies, including Virginia and New England, and became an important colonial export.[7] The importance of sassafras was influenced by the commercial and political imperatives of England's early empire. It was a natural commodity that was immediately available and required little investment, and it was already known as a medicinal plant through Monardes and earlier Spanish and French accounts. These European colonizers had learned of the medicinal qualities of sassafras from Native American communities, particularly the Timucua in Florida.[8] Groves of large sassafras trees could be easily harvested in the English American colonies and exported as substitutes for sassafras from Spanish and French American colonies.[9] As their roots and bark were stripped, however, sassafras trees died back, creating a boom-bust cycle in exports of sassafras from the English American colonies. When early colonial settlements grew in size and population, they could cover a larger radius of territory in the search for new sources of sassafras, but these new forests were eventually exhausted by unsustainable harvesting.

The earliest book printed in England about sassafras was John Frampton's 1577 translation of Monardes's publications on New World drugs, entitled *Joyfull Nevves Out of the Newfound World*.[10] Monardes's work has been discussed in the historiography as one of the earliest sources to introduce sassafras to Europeans.[11] But locating an entry point for a drug or its associated knowledge does not mean that its further diffusion or permanence within the new culture are inevitable. We need to explore the trajectory by which New World drugs were made available, and how they were experienced and became acculturated. For example, Wouter Klein and Toine Pieters have shown that while cinchona entered the European market in the 1640s, it was only decades later that it became an accepted medical remedy for fevers.[12] This chapter contributes to the emerging literature on drug trajectories within the wider context of the global circulation of things.[13]

Sassafras was unknown in English experience before the sixteenth century, and it therefore allows us an entry point for studying the reception of knowledge about a specific drug in early modern print. While these medicines from the New World had not yet been appropriated into European medical culture, there was a long-standing precedent for incorporating drugs outside the Mediterranean-centric texts of Galen and Hippocrates into medical theory and practice.[14] Sassafras makes for a particularly useful case study for investigating the development and dissemination of knowledge about an exotic drug. Firstly, it was not mentioned in English print prior to the late sixteenth century, and so had not accumulated layers of past understandings from direct experience and from biblical, classical, and medieval writers.[15] Secondly, the number of references to sassafras in early modern English printed works is sufficient to examine wider trends without compromising an attentive, qualitative study.

SASSAFRAS IN MEDICAL TEXTS

I constructed the collection of texts under study from a search of all extant transcribed English printed materials published up to 1680. The medical texts consulted included pharmacopoeias, practical guides, and books of receipts written for both practitioners and lay people, theoretical treatises, discourses about specific diseases, and herbals.[16] These texts were found using a search of Early English Books Online (EEBO).[17] Using "sassafras" and its alternative spellings and abbreviations as search terms, I paralleled the manner in which information was

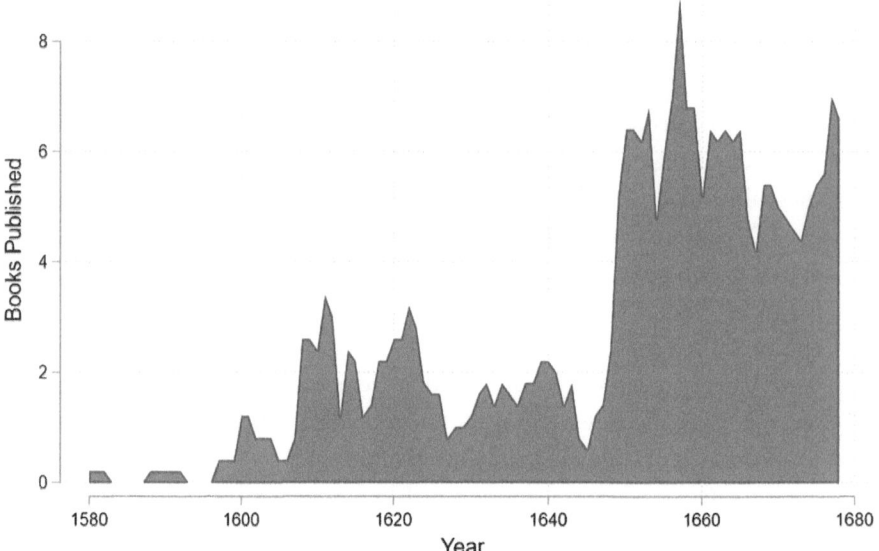

FIGURE 5.1. Medical books referencing sassafras, 1580–1680 (5-year moving average).

organized in the early modern period. Many of the texts contained "alphabetical tables," in which herbs were ordered by their "common English name" so that they could be easily found by readers.[18] Each search result was manually checked and verified as a reference to sassafras through reading the text.

Figure 5.1 presents four periods of sassafras discussions in English print between 1580 and 1680. The first period, between 1580 and 1610, saw a limited number of sassafras references. This can be seen as the "exploratory phase," coinciding with English and Spanish explorations of the Americas, when sassafras was identified as an early commodity for export. Between 1610 and 1630, there was an increase in the number of texts discussing sassafras, reflecting a wider knowledge of and engagement with the medicine: the "expansion phase." During this phase, England established permanent settlements in Virginia and New England, where sassafras trees were abundant. Between 1630 and 1645, there was a decline in sassafras references: the "dormancy phase." This phase reflects the disruption in colonial interest during the period of the English Civil Wars. After 1650, there was a shift in scale of the total number of texts published that mentioned sassafras, with the largest increase being

THE MEDICAL RECEPTION OF SASSAFRAS IN EARLY MODERN ENGLISH PRINT

FIGURE 5.2. Prevalence of sassafras, guaiacum, and mechoacan in English medical texts, 1550–1680.

in medical texts: the "abundance phase." This phase corresponds with an increased interest in England's overseas empire during the Interregnum and after the Restoration. Cromwell's Western Design aimed to revitalize England's imperial ambitions by capturing Spanish possessions in the Americas, and Charles II's imperial policies aimed to consolidate English control in North America between Virginia and Spanish Florida.

The number of medical texts published in English increased over the time period under study, so it is also important to consider the proportion of medical texts that discussed sassafras. In the 1590s, 33 percent of medical texts mentioned sassafras, but by the 1670s 73 percent of medical texts did so.[19] The opposite trend can be seen in geographic and travel texts: 67 percent of these texts published in the 1590s mentioned sassafras, compared with 11 percent in the 1670s. The increase in medical references to sassafras was therefore not merely a function of the increase in the number of medical publications.

How important was sassafras to early modern medical discussions,

FIGURE 5.3. Prevalence of sassafras, cloves, and nutmeg in English medical texts, 1550–1680.

compared with other New World drugs? In figure 5.2, I present a comparison of the prevalence of three New World drugs—sassafras, guaiacum, and mechoacan—in English medical texts from 1550 to 1680.[20] I find that, by the 1660s, sassafras was the most commonly referenced New World drug in the English medical corpus. Guaiacum, a sudorific wood, was the most important drug in medical texts prior to the 1660s, and it was referred to three or four times as often as mechoacan, a purgative root, and sassafras in the sixteenth century. It became less dominant in the early seventeenth century as the importance of other American drugs grew. Sassafras became increasingly important across the seventeenth century, eventually displacing guaiacum as the leading New World drug in the 1660s. This change was likely due to the increasing supply of sassafras in England from colonial trade; by 1663, the supply of sassafras to England exceeded domestic demand such that 2,240 lb. were reexported to European countries.[21] While guaiacum was ubiquitous in the Spanish possessions in the Americas, sassafras was more prevalent in the English American territories. Guaiacum had

FIGURE 5.4. Prevalence of sassafras, angelica, and mastic in English medical texts, 1550–1680.

been imported into Europe since the early sixteenth century from the Spanish American colonies, and the uptake of sassafras may have been eased by the preexisting market for a sudorific American wood. Sassafras did not displace guaiacum; indeed, they were often prescribed together to give a more potent effect. Guaiacum and sassafras were both associated with the treatment of the pox through their hot and sudorific qualities in the Galenic system.

How does the prevalence of sassafras in early modern medical writings compare with drugs from other regions? In figure 5.3, I display the trajectories of three exotic drugs in English medical literature, from 1550 to 1680. I contrast sassafras with two drugs primarily sourced from Southeast Asia: cloves and nutmeg. Cloves were the dominant drug across the period of study until the 1670s, when sassafras became the leading drug. Until the 1620s, both Southeast Asian drugs were more prevalent than the New World one. By the 1630s, however, sassafras was discussed twice as often as nutmeg. Over the seventeenth century, sassafras became increasingly important compared with these Southeast Asian drugs.

How does the use of sassafras compare with drugs cultivated extensively in Europe? In figure 5.4, I evaluate the trends in medical discussions of angelica and mastic in relation to sassafras. Mastic was the dominant drug in the sixteenth century, and angelica became referenced on the same scale as mastic in the early seventeenth century. This shift between European drugs indicates that the rise and fall in importance of drugs was not a case of "Old" versus "New" Worlds. The European drugs were much more prominent than sassafras until the 1620s, and in its turn, sassafras overtook angelica and mastic in the 1660s.

The period when references to sassafras overtake other drugs in these examples—whether sourced from the Americas, Southeast Asia, or Europe—occurred between the 1660s and the 1670s. In the mid-seventeenth century, there was a boom in drug exports from the English American colonies, and sassafras became an important medical commodity.[22] As greater quantities of sassafras became available on the English marketplace, medical writers began to debate the medicinal qualities and uses of sassafras more frequently in their publications. Sassafras also came to be recommended for a wider variety of diseases in this period, as will be discussed in more detail later in this chapter.

PROPERTIES

The physician Gideon Harvey (1636/7–1702) described the medicinal properties of sassafras in the context of curing gout. He described sassafras as hot, dry, aromatic, a sudorific, discutient, and aperitive, and its effect in treating gout as miraculous.[23] Many of the writers agreed with Harvey in recommending sassafras to cure the gout, which was the second most common disease to be treated with the plant. For example, the German physician Daniel Sennert (1572–1637) praised sassafras's "opening, discussive, and attenuating faculty," which was commended for curing gout, among several other diseases.[24] In 1621, the physician Tobias Venner (1577–1660) combined two American drugs, tobacco and sassafras, in his medical treatise on the medicinal benefits and dangers of consuming tobacco smoke. He viewed tobacco as especially potent in drying out cold ailments and claimed that, when sassafras roots were mixed in with the tobacco, even better results could be achieved, especially in opening obstructions in the body.[25] Nearly all the authors presented a similar assessment to Harvey of how sassafras affected the body.

Sassafras was incorporated into prevailing models of European medicine with little resistance. For example, in the 1657 translation of

Jean de Renou's *Dispensatorium Medicum*, de Renou promoted the use of sassafras, dedicating a chapter to its medical uses, and classified it as an "exotical calefactive."[26] While most of the plant was determined to be in the second degree of Galenic heat, the rinds of its roots were often believed to be more powerful and in the third degree of heat and dryness.[27]

In the early modern period, sudorific medicines were understood as more powerful than diuretics in their purging abilities, thanks to their hot and tenuous nature. Sudorifics could "penetrate into the farthest parts of the body and cut humours, they attenuate, rarify, and turne into exhalation" and "drive malignant humours to the superficies of the body," according to a translation of Daniel Sennert.[28] Across the seventeenth century, sassafras was widely agreed to be one of the principal sudorifics, or medicines that provoked sweats, alongside guaiacum, sarsaparilla, and china root. These four medicines were often used in conjunction, as we will see later in this chapter. Sassafras was generally used in all diseases that came of cold, raw, thick, and corrupt humors, such as the pox or diseases of a "foul nature." In these "chronical and contumacious diseases," the French physican Lazare Rivière (1589–1655) justified the use of sudorifics to heat, melt, attenuate, and evacuate the humors of deeply fixed diseases, including "Epilepsies, Palsies, obstinate Catarrhs, Dropsies, Gouts, and any cold affections, and especially the Pox, [which] require more powerful Medicaments to eradicate them."[29]

Amongst Galenists, there was little debate over how sassafras worked in the body, or the properties with which it was imbued. As William Coles explained, the Galenic humoral properties of sassafras could be "manifestly perceived in the decoction," and could be known by anyone from their personal experience of the drug.[30] In his chapter on "sudorifick medicaments" in his *Universal Body of Physick*, translated into English in 1657, Rivière favored the use of sassafras, guaiacum, sarsaparilla, and China root over Old World remedies. Rivière explained that his recommendations differed from "ancient Physicians" because the more effective sudorific New World medicines had been unknown to early medical writers.[31]

SASSAFRAS AND THE GALENIST-CHYMIST DEBATE

The first concentrated resistance to the use of sassafras in English medical print arose in the 1650s with the Galenist-chymist debates. In the Galenic tradition, individuals were treated by addressing their constitutions and restoring the balance of the humors within the body. The chymical approach was to treat the disease with knowledge gathered from experience rather than ancient texts, and many of its supporters rejected Galenic practices such as purging and bloodletting.[32] European encounters with American plants had not prompted change in the fundamental theories of medicine. Medical understandings of the properties of sassafras had been formed within the humoral system, and this angered chymists who were in competition with Galenists for philosophical recognition and status.

Walter Charleton (1619–1707), a president of the Royal College of Physicians and one of the leading iatrochymists in England, was a strong advocate of Helmontian chymical theories. He complained bitterly of attacks on his views by traditional physicians in the College. In 1650, he translated the section entitled "Deliramenta Catarrhi" from Jan Baptista van Helmont's *Ortus Medicinae* (1648).[33] Here, Van Helmont argued that drugs from the New World were used according to classical theories, but since these theories were faulty, physicians were merely treating the symptoms and not the disease: "[Let D]rinks of China, Zarza [sarsaparilla], Sassafras . . . and other deceitfull remedies of the same order, be wholly layed aside, which are brought into use by Physicians, that they might not appear to have received their fees for nothing. . . . They never release the sick out of their hands: but perpetually oblige them, like purchased Bondslaves . . . while Physicians remain ignorant of the fundamentals and Causes of the disease."[34]

It is important to emphasize that it was not New World remedies as such that were rejected in Helmontian medicine, but rather the manner of their application. The chymists bemoaned the interpretation of American drugs by recourse to Galenic medical theories rather than chymical experimentation. For example, the physician George Thomson (1619–1677) dismissed the common Galenic practice of prescribing "Diet-drinks of Guaiacum, Sarzaparilla, Sassafras, out of an intent of drying up superfluous moisture, and imaginary Catarrhs in the Body."[35] He proclaimed that "these things Helmont hath plainly shewed to be ridiculous."[36] The chymists sought to develop new medical theories and validate them through trials and experimentation. They saw the incorporation of New World drugs into the old medical system as a missed opportunity. In the chymists' view, sassafras might alleviate certain symptoms, but it could not truly cure disease without physicians' understanding the fundamental causes of disease. Writers in the chymical tradition insinuated that apothecaries and physicians had a vested interest in having their patients return again and again to manage the symptoms of uncured diseases; a patient who was cured would have no need to pay for subsequent treatment.

The chymists' criticism of the prescription of costly exotic remedies was an ongoing line of attack against Galenists. In Paracelsus's *Herbarius*, he called for greater attention to be paid to medicines that could be gathered in Europe, rather than expensive foreign remedies.[37] He attacked the use of guaiacum as a treatment for the pox, advocating the use of mercury instead, which was cheaper and, Paracelsus argued, more effective. As Andrew Wear has stressed, the conflict between the Galenic and chymical physicians was as much a social issue as an intellectual one. The chymists criticized contemporary medicine for not taking care of the poor, demonstrating outrage in their writings against the Galenists for privileging exotic drugs over the cheaper, local remedies chymists used in their own preparations.[38]

Writers in the chymical tradition were not opposed to American drugs such as sassafras and guaiacum because of their foreign origins. The medical practitioner and writer Everard Maynwaring (1628-ca. 1699) observed that "they themselves, which despise Chymical medicines for their novelty do use Rhubarb, Mechoacan, Cassia, Guajacum, Sassafras, Sarsaperilla, Bezoar stone, and many more which were unknown to Hypocrates and Galen."[39] Maynwaring argued that Galenic medical practitioners could not attack physicians for using chymical remedies because they were new medicaments, since Galenists also used remedies that were unknown to classical writers. Maynwaring himself had a commercial interest in promoting new remedies, because he sold his own chymical preparations.[40]

Despite these criticisms by chymical physicians, the prevailing conceptualization of sassafras throughout the period of study remained grounded in Galenic theory. The medicinal properties of this drug continued to be understood within the humoral framework. Sassafras was hot, dry, aromatic, sudorific, discutient, and aperitive, and it cured diseases through these properties. While the chymists expanded the diversity of compound preparations of sassafras, their arguments concerning the method by which sassafras cured disease ultimately failed to displace the Galenic interpretation of how the drug worked.[41]

METHODS OF PREPARATION

Between 1577 and 1680, sassafras was prepared in increasingly diverse ways, although the most common method of preparation—decoction—remained constant. Other preparations included electuaries, powders, pills, salts, and oils. Changes in preparation reflected the increasing importance of chymical approaches to medicine from the mid-seventeenth century onward. This provides further evidence that, while chymical physicians often attacked exotic drug commodities like sassafras on the basis of their high price, they also made significant use of such drugs.

The earliest, longest-standing, and most common preparation of sassafras was in the form of a decoction, with thin pieces of sassafras root being steeped in boiled water, a technique learned from the Amerindians. Monardes recommended two primary ways of preparing sassafras: a decoction of the roots and a mash made of its leaves.[42] He recommended the decoction as a treatment for many diseases, including the pox. He advised that the mashed leaves be applied directly to external injuries, such as bruises and wounds. Around twenty years later, Richard Hakluyt also discussed the internal and external applications of sassafras. He recommended that the bark and leaves be made into a decoction, rather than the roots.[43] He also suggested that the dregs of the decoction should be applied to external wounds rather than mashed leaves. Over time, directions for preparing the different types of sassafras decoctions became more detailed. For example, in 1633 John Gerard included a color indication that the sassafras wood should be steeped until the decoction was the color of claret wine.[44]

The decoction of sassafras became so famous and widely consumed that it was developed as a base for the consumption of other medicines. John Hartman's translation and enlargement of the German alchemist Oswald Croll's (1563–1609) *Basilica Chymica* recommended that sal ammoniac should be taken in a decoction of sassafras.[45] The translation included a much-enlarged section on the preparation of sal ammoniac; Croll's original text, published in 1609, did not include any references to sassafras.[46]

Sassafras was often a key ingredient in "diet drinks." By the late sixteenth century, the use of New World drugs such as sassafras, guaiacum, and sarsaparilla in diet drinks was already referred to as ubiquitous. Some writers assumed that readers would know both the preparation method of these diet drinks, and how to consume them. An example of this can be seen in Richard Surphlet's 1599 translation of the French physician André Du Laurens's (1558–1609) *A Discourse of the Preservation of the Sight*. He wrote, "We shall make them [diet drinks] with Guaiacum, Zarza-perilla [sarsaparilla], the roote China, and Sassafras, the maner of the setting downe of such, as also of the using of them is sufficiently knowne unto every one."[47] This diet drink was being used for "universal evacuation," which provoked sweats and dried up "the superfluous moisture which is within the bowels" while avoiding "all the waterish parts which are conteined in the veines."[48] After a universal purge of sweating, specific purges could be given to afflicted body parts.

The claimed medicinal qualities of sassafras often served an author's agenda by promoting a desirable solution to a current problem. For example, some writers asserted that sassafras had nourishing and strengthening qualities, which appears at odds with the purgative properties of sassafras. Reports of sassafras causing men to "grow fat" in a colonial environment, where food was periodically scarce, offered writers promoting the colonization of North America a tool for assuaging the fears of potential settlers. The acclaimed virtues of sassafras were at times more a reflection of writers' interests than the drug's inherent properties.

Sassafras was consumed as a pottage in times of desperation by early colonists, and it became a staple in diet drinks to encourage people to gain weight. Monardes was the first to describe sassafras as restoring health in addition to curing specific diseases. He wrote, "The use of this [sassafras] water doeth make fatte, and this is certainly knowen, for we have seene many leane and sicke, that have taken it, and have healed of their evils."[49] These claims were repeated throughout the period of study, for example by Hakluyt in 1599, the natural philosopher Francis Bacon (1561–1626) in 1638, and German physician Daniel Sennert (1572–1637) in 1660.[50] The last text noted that previous authors had found a decoction of sassafras, guaiacum, sarsaparilla, and China root to be "no less nourishing than chicken broth," and that patients consuming decoctions of these drugs had gained weight.[51] In order to reconcile this with sassafras' purgative qualities, however, Sennert argued that the mechanism by which bodies grew fat upon sassafras was not the "nutritive power" that other writers had attributed to sassafras, guaiacum, sarsaparilla, and China root. Rather, Sennert viewed the drug as acting in an indirect manner by taking away the "cause of leanness;" it had "be[en] found by experience, that in the ptisick [wasting disease], Veneral disease, scab [scabies] and other diseases, bodies extenuated, have been restored again."[52]

Among its invigorating properties as a diet drink, sassafras was also claimed to help the memory, quicken the senses, and restore natural heat and strength in the body.[53] It was the first ingredient in the receipt for "The Ale of Health and Strength, by Viscount St. Albans" in the anonymously authored *The Queens Closet Opened* (1659).[54] Francis Bacon suggested that diet drinks made of sassafras, sarsaparilla, China root, and guaiacum could even help to prolong life. He wrote, "In the declining of Age, such Dyets are good to bee kept once in two yeeres, there by to grow young againe, as the Snake doth by casting his skinne."[55] The search for enhanced longevity was an intellectual puzzle for Bacon, and he viewed sassafras as one of the substances that could aid in its achievement.[56] He believed that following a regime with diet drinks had a restorative and refreshing effect on the body by removing its blockages and impurities.

The New World was often represented as a place of healthfulness, one aspect of which was the alleged greater longevity of its inhabitants. Baconian science was influential in shaping the early Royal Society's investigation of nature in the New World, which included the assessment of older accounts and the devising of questionnaires to gather and interrogate knowledge.[57] For example, John Evelyn considered Walter Raleigh's report of a Virginian king who lived for over 300 years to be plausible.[58] Further corroborating accounts were required, however, for such information to be confirmed.[59] In the 1660s, the Royal Society sent out its own questionnaires to collect more information on the longevity of peoples in Brazil and the Bahamas.[60] By consuming sassafras and other New World drugs in diet drinks, there was a perceived possibility that the virtues that would maintain health and prolong life could be imbibed.

In a work translated into English in 1657, the Montpellier physician Pierre Morel discussed the celebrity and use of various diet drinks in France.[61] He noted that diet drinks were often composed of three New World drugs and one exotic drug from Southeast Asia: guaiacum, sassafras, sarsaparilla, and china root. When using sassafras in diet drinks, Morel recommended fresh and newly imported wood mixed with thin slices of bark. He warned that "the decoction of the Wood Sassafras if kept wil [l]ose his grateful smel."[62] Alternatively, the prescription for sarsaparilla was to take its root, "cut in slices, together with his hairy strings"; the prescription for guaiacum was to use its resin; and the prescription for china root to use "the weightiest . . . not worm eaten or rotten."[63] A common concern in the consumption of exotic drugs in the early modern period was obtaining fresh material to ensure the drug's potency.[64]

In the mid-seventeenth century, sassafras oil began to be discussed in medical texts. Culpeper's 1649 translation of the *London Dispensatory* states that sassafras oil should be made like the oil of cinnamon because it was a well-known method for which sassafras should be substituted.[65] The distillation of sassafras oil was conducted in the same manner as for other aromatics, including Old World plants such as lignum rhodium.[66] References to sassafras oil in the medical literature increased throughout the 1660s and '70s. Sassafras oil required significant amounts of plant material in order to extract the oil, suggesting that there was increased availability in the London market at this time. The production of sassafras oil also coincided with the period in which chymical medicines became more popular. The oil was applied externally to the body; Robert Bayfield recounted a case of a gentlewoman afflicted with "tortura oris" (writhing of the mouth), who had "her neck . . . often anointed" with the oil of sassafras as part of her treatment.[67]

Powdered sassafras was used in diet drinks, electuaries, and cordials, and became more prevalent in the mid-seventeenth century. In his 1657 *Adam in Eden*, William Coles promoted the use of sassafras powder in diet drinks to treat "all diseases that come of cold raw thin and corrupt humors," like the pox.[68] He also recommended the smell of the root and the wood to "expelleth the corrupt and evill Vapours of the Pestilence."[69] Powdered sassafras was also mixed with honey or syrup to make electuaries. The medical practitioner John Tanner (1636–1715) recommended the electuary of sassafras as a treatment for a variety of diseases in his 1659 *The Hidden Treasures of the Art of Physick*: "It opens obstructions of the Liver, Spleen, and Kidneys, and is good against cold Rheums and Defluxions, from the Head to the Lungs, Teeth, Eyes; and helps Diseases in those parts, occasioned by such Defluxions: it provoketh the Terms, dryes up the superfluous moisture of the Womb, and all raw thin Humours, and breaks Wind. The dose is half a drachm in the morning."[70] The natural philosophers Robert Boyle (1627–1691) and Nicaise Le Fèvre both included sassafras in their receipts for Sir

Walter Raleigh's Cordial.[71] Raleigh himself produced receipts for several different cordials based on his investigations on New World plants while imprisoned in the Tower of London.[72] Boyle discussed in 1663 how the cordial should be prepared with "the Rinds of Sassafras of Virginia."[73] Sassafras was the only ingredient in the cordial for which Boyle specified its place of production. The cordial could be used for "Feavers, Want of Spirits, violent Fluxes, and several other distempers, where Diaphoreticks and Antidotes are proper."[74] Le Fèvre included sassafras as a required ingredient in his receipt for Sir Walter Raleigh's Cordial, whereas other ingredients could be substituted or were optional. He recommended that both the bark and wood of sassafras should be used to ensure sassafras's potency after its long sea travel.[75]

Near the end of the seventeenth century, sassafras became used in an even more extensive range of preparations. For example, Gideon Harvey recommended sassafras as an ingredient in a decoction, liquor by infusion, and a distilled spirit, all of which were involved in the treatment of scurvy in his 1675 treatise on the disease.[76] Sassafras was prepared in several manners, reflecting its appropriation as a cure under both Galenic and chymical theories and as an increasingly ubiquitous substance through the seventeenth century. The diversity of diseases that sassafras was recommended to treat also expanded in the latter half of the seventeenth century, suggesting that it was becoming more familiar and less exotic as a medical drug.

DISEASES

In sixteenth- and seventeenth-century print, sassafras was primarily discussed within the Galenic conceptualization of disease by both medical practitioners and lay people, such as playwrights and colonial settlers.[77] The body was viewed as porous and interconnected with its environment, and disease could enter the body as a contagion or develop within the body through malignancy and putrefaction. Galenic disease narratives centered around an imbalance in the humoral system, and the body required an evacuation of this excess to restore health.

In table 5.1, I assess how important sassafras was as a remedy for various diseases, organized by the number of times medical works recommended it as a part of the treatment. This table includes all diseases which were referred to in at least five texts. A further forty-seven diseases were discussed as being treated with the use of sassafras in fewer than five texts.[78]

TABLE 5.1. THE USE OF SASSAFRAS IN DISEASE TREATMENT, 1580–1680

Disease Name	Number of Medical Texts Referencing this Disease	Number of Medical Texts Recommending Sassafras as a Treatment	Percentage of Medical Texts Recommending Sassafras as a Treatment
pox	99	39	39%
stone in the reines	35	10	29%
loathing of meat	38	10	26%
king's evil	34	8	24%
barrenness	57	11	19%
agues and fevers	141	26	18%
dropsy	199	34	17%
windiness	76	12	16%
obstinate catarrhs	50	7	14%
obstructions	186	25	13%
rheumes	97	13	13%
epilepsy	78	8	10%
coughs	99	10	10%
gout	193	19	10%
pestilence	118	11	9%
palsy	132	12	9%
rickets	74	6	8%
scurvy	174	12	7%

Source: The "Sassafras Text Collection" in Katrina Maydom, "New World Drugs in England's Early Empire." (PhD diss., University of Cambridge, 2019), 94–96, 233–50.

The diseases that sassafras was used to treat cannot be systematically classified into "disease categories" across the early modern period, because definitions were fluid and varied across time and place. Furthermore, disease was perceived to change in response to the environment and other factors, intensifying, compounding, and transforming organically over time. Even looking at an individual medical treatise reveals many inconsistencies and disagreements about the organization of disease. While at first glance these complicated explanations of disease causation may appear contradictory, they were not necessarily problematic for early modern readers. For example, some writers who

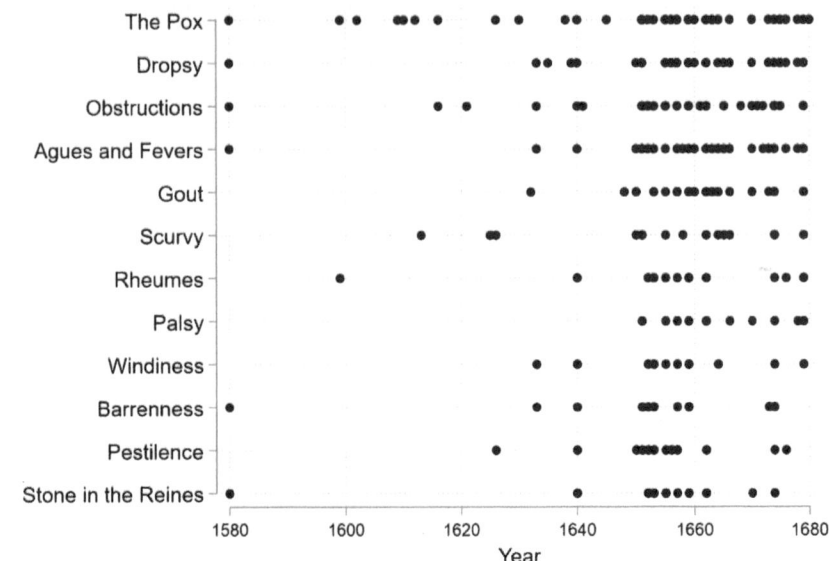

FIGURE 5.5. Top twelve diseases that sassafras was used to treat by year, 1580–1680.

believed that miasma could cause pestilence found the aromatic, sweet smell of sassafras a successful deterrent, while others found it ineffectual in treating the pestilence, if it was deemed to be caused by a divine retribution for human sin. No disease was a distinct entity in early modern England, but instead shifted its nature and form, disappeared and reemerged through the centuries.[79] Table 5.1 thus refers to diseases as conditions or symptoms described by the writers of medical texts.

Sassafras was heavily endorsed as a remedy for the pox. The close association between sassafras and venereal diseases began with Monardes and continued throughout the seventeenth century.[80] More than one-third of the medical treatises on this disease recommended sassafras as a treatment. Several diseases specific to women were emphasized in the medical literature on sassafras, including barrenness, green-sickness, the whites, and the mother. Sassafras was understood to be particularly effective at heating and drying moist and phlegmatic constitutions and generally helpful for women with excessive moisture, which was perceived to be the underlying cause of many of these diseases. Only one of the diseases was specific to men, when sassafras was recommended as part of the treatment for "venereal impotency" arising "from some defect of the Yard in men" in a translation of Giovanni Benedetto

THE MEDICAL RECEPTION OF SASSAFRAS IN EARLY MODERN ENGLISH PRINT

Sinibaldi's *Rare Verities, The Cabinet of Venus Unlocked* (1658).[81] In this text, sassafras was claimed to treat impotence by increasing lust through its heat.

Figure 5.5 displays the distribution over time of references to the top twelve diseases sassafras was recommended to treat, ordered by frequency. Monardes promoted sassafras as part of the treatment for six of these diseases: the pox, dropsy, obstructions, agues and fevers, barrenness and stone in the reines.[82] The expansion of sassafras as a treatment for many of the diseases cannot be seen solely as stemming from its initial introduction into the English medical literature via Monardes. It was in the 1650s that sassafras came to be regularly discussed in English print sources, and by that point the virtues associated with sassafras had developed significantly beyond Monardes's original recommendations. Most of his suggestions for the use of sassafras took around fifty years to become established, with few mentions between 1580 and 1630. Other early seventeenth-century authors promoted the use of sassafras as a treatment for other diseases such as scurvy, rheumes, and pestilence that were not included in Monardes's discussion. In the 1650s, when sassafras became more frequently discussed in the medical literature, writers also suggested using it to treat gout, windiness, and palsy. There was no steady accretion of new diseases with which sassafras was associated. Each time sassafras was evaluated by an author, some previous treatments were included, some new diseases were added, and other suggestions which had been excluded sometimes reappeared.[83] During the 1650s, however, many of Monardes's original recommendations resurfaced, and there was a rapid increase in the number of new diseases for which sassafras was recommended as a treatment.

Sassafras was recommended to treat the pox consistently in the texts throughout the period of study, while most other diseases were discussed much more after the 1650s. Indeed, the familiarity with sassafras as a treatment for the pox in English medical texts was so well-known by the 1630s that the idea that medical students would need instruction in that use was laughable. The physician Alexander Read (ca. 1586–1641) included one such humorous reference in his teaching of surgery. In his *The Chirurgicall Lectures of Tumors and Ulcers* (1635), he wrote, "As for diaphoreticke medicines, the decoction of Guajacke [Guaiacum], Sarsaparilla, Sassafras, and the China roote with Agrimonie, Betony & Coriander, sweet Fennil-seeds, & Annise-seeds carrie away the bell. How effectuall these medicaments are, being judiciously used, not in this griefe only; but in moist ulcers also, and other diseases contagious, I need not to labour to perswade, seeing there are few of this company, who have not often made triall of them."[84]

Read suggested here that his audience of listeners and readers would have known from first-hand experience the effectiveness of sassafras as a treatment for "other diseases contagious," meaning venereal diseases.

From the 1650s onward, sassafras was regularly recommended to treat a wide variety of diseases due to its increasing prominence in the English marketplace. This was the same period in which sassafras became more widely referenced in medical texts, and when chymical physicians began to include sassafras in new forms of preparation. At the same time, colonial projectors of sassafras were advocating for an increase in its trade, and the more frequent discussion of sassafras in the medical literature may have supported this endeavor. Thus the period in which sassafras became more readily available was when it changed from being primarily associated with the pox to more a general medicament, including the uptake of many of Monardes' recommendations for many kinds of diseases.

This chapter speaks to two forms of consumption of a New World drug. The first is the consumption of knowledge about sassafras, and the second is the consumption of the physical drug. Sassafras rose to fame in early modern England for both economic and cultural reasons, as part of a diversified colonial drug economy that expanded in the latter half of the seventeenth century. Medical texts presented sassafras as an efficacious remedy for a growing number of diseases, and informed readers how to prepare and consume the drug.

The changing medical landscape across the seventeenth century—in terms of debates between medical philosophies, the rising volume of texts authored by unlicensed medical practitioners, and the increasing number of colonies from which American drugs were supplied—all influenced how sassafras was represented in early modern English print. Over the course of the century, sassafras became more familiar, less exotic, and was incorporated into medical practice in an increasingly diverse range of preparations. The significance of sassafras in the medical corpus accelerated with a shift in scale from the 1650s. References to sassafras became increasingly common across the seventeenth century, and by the 1660s it had become the American drug most often discussed in English print. By this time, sassafras was also more commonly referred to in medical texts than other drugs such as guaiacum, mechoacan, cloves, nutmeg, angelica, and mastic.

Medical interest in sassafras was spurred by economic development in the English American colonies, which were a ready source of sassafras for the English market. Sassafras was prevalent within England's North American territories that began to export greater quantities of drug commodities in the middle of the seventeenth century after the downturn in the colonial tobacco economy and a change in government, which allowed a rethinking of colonial policy.[85] As a sudorific wood, sassafras bore similarities to guaiacum, which had been imported into Europe from the Spanish American territories since the early sixteenth century. Both sassafras and guaiacum were imported into England from its American colonies in the seventeenth century, and they were often prescribed together, particularly in diet drinks. Sassafras became prevalent in England and overtook guaiacum as the most discussed New World drug in the latter half of the seventeenth century.

The use of sudorific woods such as sassafras and guaiacum as simples and in diet drinks reflected their incorporation within the traditional Galenic framework; they were perceived as hot and dry medicaments that could cure diseases that required sweating to restore balance to the body. Within medical texts, it was often noted where sassafras could be found in North America and the West Indies. Its curative effects, however, were perceived to arise from its sudorific, hot, and dry properties rather than its place of origin. The increase in discussions of sassafras in the mid-seventeenth century reflected the greater availability of the drug through colonial trade networks.

There was little contestation of sassafras's medicinal properties prior to the rise of chymical medicine in the mid-seventeenth century, when its method of curing was debated along with the question of whether it merely treated symptoms, or the underlying causes of disease. The range of preparations of sassafras also expanded at this time to include more chymical preparations, in the form of oils and salts. The chymists' challenge to the Galenists was that exotic drugs were being prescribed for the purposes of profit, duping customers into paying higher prices for novelty, and not because they were necessarily the best curatives in a given situation. Chymical physicians included sassafras in their preparations. They were not opposed to the use of exotic drugs, and they also took advantage of the lucrative sale of New World ingredients in their patent medicines.

The one-hundred-year survey of sassafras in early modern English medical texts presented in this chapter has shown how the plant sassafras was prepared for consumption, the diseases it was recommended to treat, and the medical debates into which it was co-opted. Future research could use the method of long-run print surveys to investigate the trajectories of other drugs in early modern medical literature.

A key advantage of this method is that it allows the identification of long periods of continuity and turning points across the late sixteenth and early seventeenth centuries, for example how the variety of diseases for which sassafras was recommended changed over time, and when chymical preparations of sassafras were introduced. There is a danger of placing undue emphasis on the "first" mention of a drug in print or in a "significant" text. It is important to consider these texts in relation to the wider corpus to research their reception. By examining a printed corpus in this way, it is evident that the assimilation of exotic drugs into English medical culture was not a smooth, linear process but was influenced by commercial forces and medical debates. At the same time, it is important to recognize that print texts were published for profit and for burnishing the credentials of the author, and so discussions of sassafras do not necessarily reflect the availability of sassafras, or how it was prescribed by medical practitioners and consumed by the sick. The evidence from such print surveys should therefore be used alongside that from analyses of trade records, medical practitioners' manuscripts, and household recipe books, to construct a rounded account of how these medical commodities were understood and consumed.

6

ACCUMULATION OF THE EXOTIC IN THE *PALESTRA PHARMACEUTICA* (MADRID, 1706) OF FÉLIX PALACIOS

Paula De Vos

By the beginning of the eighteenth century, there were many new and "exotic" substances in use in the European pharmaceutical tradition through the global drugs trade. Various authors have documented the commercial and medicinal significance of materia medica from the Americas, Asia, and Africa that were brought to Europe as a result of imperial expansion in the early modern period. Building on Columbian Exchange literature, and with a focus on long-distance trade, these scholars have traced the transit of ideas and commodities through networks over oceans and continents.[1] Scholars of the Atlantic world, for example, have focused on the introduction of drugs as part of the new commodities introduced in transatlantic trade: in addition to tomatoes, potatoes, maize, tobacco, and chocolate, purgatives like jalap and mechoacan root, anti-venereals like guaiacum and sarsaparilla, and febrifuges like cinchona began to enter the world market.[2] These, and other medicinal plants from the Americas and Asia, were described and recorded in works by Francisco Hernández, Nicolás Monardes, García d'Orta, and Carolus Clusius that were, in turn, translated into many different languages and published in multiple editions throughout Europe. Patrick Wallis has shown the resulting increase in trade volumes of these long-distance products in apothecary shops of London, demonstrating the global nature of this burgeoning trade in pharmaceuticals.[3] Its effects were also evident in the early modern herbals of Dodoens, Fuchs, Gesner, Gerard, and others who incorporated these substances into the traditional materia medica, or collection of medicinal substances, that had been in place for centuries. New medicinal plants were also cultivated in botanical gardens and incorporated into cabinets of curiosities and natural history museums across Europe—part of what Brian Ogilvie has labeled the "information overload" engendered by imperial expansion.[4]

There is no doubt that these trends were intensifying in early modern Europe; but the European pharmaceutical tradition already had a long history of incorporating medicinal substances from far-off lands. This tradition, referred to as Galenic pharmacy, after the Greek physician Claudius Galen (whose writings formed much of its core theory and practice), developed in areas arguably foreign to Western Europe.[5] Galen himself was from Pergamon in Anatolia, and the Greek basis of Galenic medicine and pharmacy was initially formed in the ancient Near East and Asia Minor. It developed further in schools, hospitals, and pharmacies of the medieval Islamic cities of Baghdad, Gondeshipur, and Cairo, and its key texts, ideas, and practices came to al-Andalus through Islamic expansion, later spread to the Latin West through translation centers in Spain and Italy. From its earliest beginnings, Galenic pharmacy, reflecting the wide reach of the Roman Empire, had incorporated drugs from all over Afro-Eurasia—from India, China, and Southeast Asia; from the Arabian Peninsula and the Levant; from Europe and central Asia; and from Africa, both north and south of the Sahara. In addition to the drugs themselves, this tradition also utilized techniques and ideas that evolved over time and place, constituting another facet of European pharmacy's foreign origins and development.

FIGURE 6.1. Title page to Palacios's *Palestra pharmaceutica chimico-Galenica* (Madrid: Viuda de Joaquin de Ibarra, 1792).

Using the eighteenth-century Spanish meaning of "exotic" as signifying things "outside, foreign, or wandering," this was an "exotic" tradition imported, domesticated (figuratively and literally), and appropriated from the Eastern Mediterranean, the Middle East, Africa, and later, the Americas.[6]

Galenic pharmacy in Europe at the turn of the eighteenth century, then, constituted a tradition of ideas, substances, books, and practices that had developed over centuries and moved far from its places of origin. It consisted of layers of development that began as exotic imports, but by the early eighteenth century had accumulated and been wholly adopted, in a process that one scholar has labeled a kind of cultural naturalization.[7] This chapter aims to trace the various layers of that accumulation as evident in a highly significant Spanish pharmaceutical text, the *Palestra pharmaceutica*, written by the Madrid apothecary Félix Palacios y Bayá (1677–1737). This landmark text, first published in 1706 with multiple editions throughout the century (1716, 1724, 1737, 1763, 1778, and 1792, see fig. 6.1), served as the basis for the official formulary of the Spanish Empire, the *Pharmacopea Matritense* (later called the *Pharmacopea Hispana*), throughout the eighteenth and early nineteenth centuries. The first edition of the *Palestra* alone sold two thousand copies, and by the time of the 1724 edition, it had, together with Palacios's Spanish translation of Nicholas Lémery's *Cours de Chimie*, sold more than seven thousand.[8] There is little secondary treatment of the *Palestra* among Anglophone scholars; Spanish historians of pharmacy have outlined Palacios's biography, and noted the *Palestra*'s important role in the compilation of a standard formulary for the empire.[9] The *Palestra*'s most defining characteristic for these scholars, however, is that it contained medicines of the Galenic tradition as well as a full complement of alchemically prepared "chemical medicines," a separate pharmaceutical tradition that had developed parallel to the Galenic one since the late Middle Ages. The two traditions merged in the seventeenth century, resulting in a "chemico-Galenic compromise," evident in a number of contemporary European formularies and pharmacopoeias, of which the *Palestra* was a prominent example.[10]

However, the significance of the *Palestra* in the history of Spanish pharmacy goes beyond its inclusion of alchemical pharmacy. This text in fact represents centuries of layers of pharmaceutical development, in which the larger history of Galenic pharmacy and its union with chemical pharmacy are visible. In this chapter, I argue that the *Palestra* represents the culmination of centuries of pharmaceutical knowledge and practice that had developed in and around the Eastern Mediterranean, spread throughout the Near and Middle East and entered Western Europe through Spain and Italy during the High Middle Ages. Thus the pharmaceutical tradition of early modern Spain stemmed from Spanish incorporation of "outside" influences—materials and practices not indigenous to Europe but adopted from "foreign" ideas, practices, texts, and materials of the eastern Mediterranean and, later, the Americas. This essay traces the different layers of "exotic" pharmacy as evident in the kinds of medicines included in the *Palestra pharmaceutica*.

In the first place, Palacios listed a series of over two hundred "simples," a term used in Galenic pharmacy to identify natural substances that were derived from plants, animals, and minerals with known healing powers. These simples made up Galenic "materia medica," the collection of medicinal materials employed within Galenic medical practice. First codified in the Eastern Mediterranean by Galen and an important predecessor, Dioscorides, Galenic simples came mainly from the Indo-Mediterranean, but a significant minority were also imported from Africa and East and Southeast Asia. Additional long-distance substances were added to Galenic materia medica under the Islamic empires, and a third grouping of American materials joined the collection of Galenic simples following the Columbian voyages. The various layers of development in Galenic pharmacy are also evident in the kinds of compounded medicines that Palacios included in the *Palestra* and the evolving techniques utilized to prepare them. In Galenic pharmacy, simples were processed and incorporated into various kinds of compounds, remedies such as ointments, plasters, lozenges, and syrups that were composed of multiple processed simples combined with a substrate, usually oil, sugar, honey, gum, or wax, to bind and preserve them. The types and classification of compounds formulated and the techniques used to do so changed over time as well. Classical and early medieval compounding typically employed relatively mild processing techniques and types of compounds that were classified in various ways—according to effect, method of application, and method of preparation. By the early modern period, Galenic compounds in the Spanish tradition were almost universally classified according to method of preparation only, and by the late seventeenth to early eighteenth centuries had begun to incorporate alchemically prepared compounds using alchemical methods. All of these categories of compounds, and the techniques for their preparation, are represented in the *Palestra*, making it an encyclopedic work that incorporated many foreign and exotic elements.

THE *PALESTRA PHARMACEUTICA* AS AN EXOTIC ENCYCLOPEDIA

The *Palestra* is a unique and highly significant work in the Spanish pharmaceutical tradition. The basis for subsequent official formularies of the

Spanish Empire, this work and its contents represent the accumulation over centuries of layers of pharmaceutical materials, learning, and practice that developed largely outside of Europe. Its author, Félix Palacios y Bayá, was a prominent apothecary in Madrid and member of the Royal Medico-Chemical Society of Seville, later named pharmacy inspector to several bishoprics in Spain.[11] As noted in its title, the *Palestra* represented a union of two different traditions in pharmacy, the Galenic and the chemical. Galenic pharmacy, first of all, was the tradition that guided early modern pharmaceutical theory and practice in the West from the first centuries of the common era well into the eighteenth century.[12] It consisted of a set of ideas and practices that began with classical Greek teachings, and developed in stages through the ancient and medieval periods. They stemmed from a series of core ancient texts, including the Hippocratic Corpus (fifth and fourth centuries BCE), Dioscorides's *De materia medica* (first century CE), and various works by Galen (second century CE). Arabic authors such as Ibn Sahl (d. 869 CE), Al-Rāzī/Rhazes (865–925 CE), Ibn Sīnā/Avicenna (ca. 980–1037 CE), and Ibn al-Tilmidh (1073–1165 CE), in the Eastern caliphate; and al-Zahrāwī/Abulcasis (936–1013 CE), Ibn Juljul (ca. 944–994 CE), Ibn al-Wafid/Abenguefit (1008–1075 CE), Ibn Rushd/Averroes (1126–1198 CE), and Ibn al-Baytar (1197–1248 CE) in the Western caliphate translated and built upon these works and those of their Byzantine redactors, Oribasius (320–400 CE), Aetius of Amida (502–575 CE), and Paul of Aegina (625–690 CE), producing important herbals and the first codified formularies. This tradition was brought to Europe in the later Middle Ages through translation centers in Salerno and Toledo, in which major works of the Greco-Arabic medical tradition were translated into Latin. These works were utilized in newly established medical schools of medieval universities throughout Europe, and further disseminated with the advent of print in the fifteenth century. Key to this process with regard to the transmission of knowledge and practice of Galenic pharmacy was the work of a pseudonymic author using the name of a ninth-century Arabic physician and medical author Yūḥannā Ibn Māsawaih (anglicized to John Mesue).[13] The writings of Pseudo-Mesue, whose earliest extant sources appeared in northern Italy around 1285, acted as a conduit for Arabic advances in pharmacy to the medieval Latin world, responding to increasing demands for drug therapeutics and compounding. In the Galenic tradition, Pseudo-Mesue codified the main operations to be used in processing simples and standardized the kinds of compounds included in its formularies. These operations and compounds generally excluded formulation through alchemical means, mainly distillation, sublimation, and calcination, which often involved the use of high heat. For Pseudo-Mesue, these kinds of operations would damage a substance's healing powers, and thus he advocated processing techniques that were arguably more mild, including trituration (grinding), levigation (washing), coction (or decoction, cooking), and infusing.

The *Palestra pharmaceutica* was one of the first major Spanish formularies to include recipes for compounds of both the Galenic and the alchemical traditions. Indeed, Palacios gained his reputation largely based on his embrace of chemical pharmacy and associated chemical pharmacy with modern advances, having great faith in both its explanatory potential and its healing powers. But by no means did he deny the validity or efficacy of the Galenic tradition. The *Palestra*, he noted, combined the traditions of the "ancients" with those of the "moderns" and served to "unite and arrange with great acuity the Chemical remedies with the Galenic, in order to overcome [even] the most difficult illnesses."[14] In order to accomplish this task, he argued that "it is indispensable to know perfectly all the operations of the one and the other Pharmacopoeia [Galenic and chemical]."[15] Indeed, the structure of the *Palestra* faithfully included full treatment of both traditions. Book 1 laid out the basis of pharmaceutical materials and techniques (including diagrams and drawings of alchemical utensils and apparatus); books 2 and 3 dealt with Galenic compounds; and books 4 and 5 dealt with chemical preparations. In fact, the *Palestra* went even further than most works, not only incorporating the latest developments in Galenic and chemical pharmacy, but also providing encyclopedic treatment of simples, compounds, and processing techniques going back to the ancient period. An examination of the simples and compounds included in the *Palestra* reveals the various layers of pharmaceutical development since the first centuries of the common era, describing techniques, processes, and recipes, some of which had become obsolete, and also including the very latest in chemical operations, chemical equipment, and methods of explaining how and why medicines worked. Thus the *Palestra* was encyclopedic in character and demonstrated the accumulation and naturalization of a long history of foreign or exotic ideas, materials, and practices.

PAULA DE VOS

EXOTIC ACCUMULATION: THE LAYERING OF MATERIA MEDICA IN THE *PALESTRA PHARMACEUTICA*

The materia medica in use in Galenic pharmacy (and later utilized in alchemical pharmacy as well) consisted of hundreds of simples derived from plant, animal, and mineral sources, of which plant substances made up the great majority, about 80–90 percent.[16] Plant simples in the ancient period were native to areas throughout Afro-Eurasia, the products of long-term trade in materia medica that arguably developed alongside the trans-Eurasian exchange in the first era of food globalization, around 7000 BCE. That exchange eventually developed into the Silk Roads and other trade routes that linked major world regions of Afro-Eurasia for centuries.[17] Using archaeological and historical evidence, researchers have documented the origins of plant domestication in early human societies of Afro-Eurasia, and traced their dispersal throughout the Old and, later, the New Worlds. Dispersals have been so widespread that many plants are intensely cultivated, and some are most productive outside their original native habitat. The materia medica of ancient Galenic pharmacy, thus, was the product of thousands of years of development, reaching back to the efforts of our human and hominid ancestors.

Over time, the materia medica of medical traditions in Mesopotamia and the ancient Near East were gradually codified and recorded in the written records of ancient Egypt, Sumer, and Babylonia, and in the Hippocratic Corpus and the writings of Theophrastus in classical Greece. The height of codification of materia medica was reached with Dioscorides's *De materia medica*. Dioscorides hailed from Anazarbus in Anatolia under the Roman Empire and traveled throughout Asia Minor and the Mediterranean, identifying, describing, and testing medicinal plants himself and recording the results in a remarkable five-part treatise that named over one thousand plant, animal, and mineral substances. The majority of these materials derived from plants native to Europe and the Mediterranean, but with significant contributions from the flora of eastern and sub-Saharan Africa, the Arabian Peninsula, and South, Southeast, and East Asia.

De materia medica would have longstanding influence and importance within Galenic pharmacy, supplying the core of Galenic materia medica. Galen, who lived in the century following Dioscorides, relied on the work, and praised it highly in his own landmark work on simples, *On the Mixtures and Powers of Simple Medicines*. More than any other ancient text on the subject, *De materia medica* was copied, redacted, translated, read, circulated, and cited in manuscript form, adopted into the Arabic, Byzantine, and Latin medical traditions, and later published, redacted, edited, and reprinted more than a hundred times.[18] The substances named in *De materia medica* continued to make up an average of 50–60 percent of Galenic simples, even into the nineteenth century, and early modern editions of his work appear to accept the indications Dioscorides provided for these substances.[19] Despite this long tradition of stability and continuity, however, Galenic materia medica—like Galenic pharmacy overall—did not remain stagnant. Subsequent prolonged and intensified exchanges, usually under the guise of expanding imperial connections for trade and exchange, served to bring new materials into the collection. For Galenic pharmacy, this happened in two main stages, the first focused on the Mediterranean world and the Indo-Mediterranean, an area that connected the Iberian Peninsula with trade as far east as India under the Islamic empires, and the second focused on the Atlantic world, a product of the Columbian Exchange and identification and collection of medicinal plants from the Americas under the Spanish Empire. These two phases added two new "layers" to Galenic materia medica. The first occurred with the expansion of the Islamic empires during the Middle Ages, in which a series of aromatics that were added to the pharmacopoeia increased contact with and knowledge of eastern lands, and especially Indian Ayurvedic medicine.[20] Greek and Roman medical writings had discussed a number of aromatics, but others were unknown (or known but not part of materia medica as described by Dioscorides) until Arabic authors began to describe them.[21] The second wave of introductions to the traditional Galenic materia medica began with the Columbian Exchange and the expansion of the Spanish Empire in the late fifteenth and early sixteenth centuries, which led to the addition of several American plant simples, plants that had been wholly unknown to Galenic pharmacy. While these did not displace Galenic materia medica, they did add new substances of importance.

These various layers of Galenic materia medica are evident in Palacios's *Palestra pharmaceutica*. In the *Palestra* of 1706 and later editions, Palacios listed the mineral, animal, and vegetable simples commonly used in pharmacy. A survey of these materials compared with those of Dioscorides shows a 45 percent overlap: of the 220 identifiable plant simples in the *Palestra*, 100 had also been named and described in *De*

TABLE 6.1. ORIGINS OF IDENTIFIABLE PLANT SIMPLES IN BOTH *DE MATERIA MEDICA* AND *PALESTRA PHARMACEUTICA*

Place of origin	Plant simples Total: 100	Percent of total
Europe (west, east, central)	barberry, Celtic nard, fenugreek, hermodactyl, imperatoria (masterwort), oregano, rue, sage, scammony	9%
Indo-Mediterranean (including southern Europe, northern Africa, Arabian Peninsula, Middle East)	absinthe, almond (sweet and bitter), angelica, asparagus, azafœtida, betony, birthwort, bistort, bladder cherry, blessed thistle, borage, camel hay/grass, calamint, caper, caraway, carrot, celery, centaury, chamomile, chaste berry, chestnut, chicory, coltsfoot, cuckoo-pint, cypress, date, dittany, dock (sorrel), elder (berry), endive, fennel, figs, fleawort, frankincense, french lavender, fumitory, gum ammoniac, gum arabic, helenium, hellebore, hemp-agrimony, horehound, hyssop, ivy, juniper, labdanum, laurel, lemon balm, licorice, madder, malabar, mallow, marshmallow, mastic, myrrh, myrtle, opopanax, pellitory, peony, pomegranate, poppy, rose, rosemary, saffron, saxifrage, squill, spurge-leafy, St. John's wort, tormentil (lemon balm), tragacath, turnip, turpentine, valerian, violet, water cress, waterlily, yarrow	78%
Asia (east, south, southeast, central)	cardamom, cinnamon, cassia, costus, ginger, nutmeg, mace, marjoram, pepper, plantain, sarcocolla, sweet flag	12%
Africa (tropical)	tamarind	1%

materia medica more than 1,500 years earlier, showing the longevity and continuity of that work.[22] The materials in common also demonstrate the "exotic" nature of the materia medica, in that most of the plant sources originated outside of Europe. Of the one hundred plant simples in common (see table 6.1), only nine were native to western and central Europe, the rest native to Africa, eastern and southeastern Asia, and the Indo-Mediterranean. Many of these plants were transplanted and transferred throughout the entire region of Afro-Eurasia. Some, as with the trans-Eurasian exchange, became naturalized and flourished in regions far outside their native habitat. Others, particularly the Eastern aromatics and Middle Eastern gum-resins, proved impossible to transplant to other locales due to their specific needs. Nevertheless, as discussed further below, they constituted some of the main staples of Galenic materia medica, known and utilized from the classical period, and codified as highly important medicines from the time of Dioscorides to the time of Palacios. Thus, both plant transplantation and long-distance trade and exchange allowed for and sustained the continuous inclusion of these foreign, exotic materials as part of Galenic materia medica in the Western tradition for millennia—to the point where they became naturalized, both culturally and for many, literally, and appropriated to European medicine in the medieval and early modern periods.[23]

The nine simples in common between the *Palestra pharmaceutica* and *De materia medica* that were native to Europe were nevertheless important staples in the Galenic tradition. Rue and sage, for example, were both highly important antifertility agents that also served to staunch blood, cleanse sores, and stop poisons. Fenugreek served as an emollient and anti-inflammatory, as well as an ingredient in shampoo, and hermodactyl was used to treat gout, rheumatism, and arthritis.[24] Celtic nard was used as a carminative, and scammony served as one of the most important and powerful purges in humoral medicine, a subject of much debate among early modern apothecaries. Many of these simples, further, were prescribed regularly and were ingredients in multiple compound medicines.[25]

Sub-Saharan Africa and South, Southeast, and East Asia were the native regions of another thirteen simples described in both the *Palestra* and *De materia medica*. The pulp of the tamarind, native to tropical regions of Africa, was a common ingredient in Galenic pharmacy, and was used as a laxative that also calmed the stomach and brought an end to vomiting.[26] The twelve simples originating in Asia were made up largely of aromatic spices, long associated in the European imaginary with exotic locales and long-distance trade. They were some of the most expensive, highly valued simples, due not only to the high cost of transporting them, but also to their longstanding importance as medicines, incense, perfumes, and food flavoring.[27] They were among the most ancient of commodities, known and traded throughout Afro-Eurasia for millennia.

Among the aromatics common to both the *Palestra* and *De materia medica* was the all-important cinnamon, an emollient and diuretic native to Indonesia that was thought to be good for digestion. Closely related and with similar properties was cassia, or "Chinese cinnamon," which came from tree of same genus as cinnamon and was native to China. The two were—and are—often confused and mixed together.[28] Cardamom came from the Malabar Coast of India; its small, angular seed pods served as an antispasmodic and were widely used for flavoring. One of the cheaper and most widespread of the spices was pepper, also native to India and valued in medicine for its diuretic and digestive properties, and its ability to "draw down the fetus."[29] From India came costus root (also called putchuk), considered to have diuretic and emmenagogic properties; and sweet flag, an astringent, carminative, rubefacient, and stomachic.[30] Southeast Asia was home to ginger, a digestive aid that was cultivated on board Chinese and Indonesian ships to prevent scurvy.[31] Perhaps most important to the historic spice trade were nutmeg and mace, both native to the Spice Islands (the Moluccas of Indonesia), whose rich, light volcanic soils, with much moisture and a high degree of cloud cover, provided an optimal climate and environment for the growth of these spices.[32] Nutmeg and mace were, respectively, the nut and rind or aril of the nutmeg tree, valued in medicine for their astringency and ability to fortify the stomach.[33]

Despite the longstanding significance of these simples, however, the great majority of the one hundred simples in common between the *Palestra* and *De materia medica* came from the Indo-Mediterranean, a vast area that included the Mediterranean basin, the Arabian Peninsula, the Levant, Mesopotamia, and extending east as far as the Indus Valley. Seventy-eight of the hundred simples were native to this area, constituting almost 80 percent of the simples common to both texts. Such an emphasis makes sense: this area constituted a major zone of both overland and sea transport, with important ports along Indian Ocean coastlines, and major cosmopolitan market towns along the land routes. This area had long been connected through the Silk Roads, especially following the conquests of Alexander the Great and the establishment of the Hellenistic empires that connected the Mediterranean with the Indus Valley. Asia Minor, the birthplace of both Dioscorides and Galen, was a key site for these connections, where the influences of East and West would have met. This area was also part of the Roman Empire, whose connections to the Silk Roads and trade in the Mediterranean, the Red Sea, Persian Gulf, and Indian Ocean further emphasized and facilitated travel and exchange in the area, and would have allowed for codification of a widespread exchange and trade in knowledge and materials of medicine from different cultures and traditions.

This region, then, constituted the heart of the origins of materia medica in the Galenic tradition—and though it included regions of southern Europe, it is better characterized as part of this world of trade and exchange that had such a long history. The fact that 78 percent of the simples in common were still evident in a pharmaceutical text from 1706 shows the region's outsize influence on the Galenic medical and pharmaceutical traditions. Yet they were not European—they were Indo-Mediterranean, from North Africa, the Arabian Peninsula, and the Middle East, extending to the Indus Valley—so in that way arguably "exotic" and "foreign" to the area that adopted these traditions in the later Middle Ages from their Arabic neighbors. Some of the most significant simples native to the Mediterranean area were ivy, sorrel, and peony, serving as emmenagogues, anti-inflammatories, and stomachics. Poppy, the source of opium and as such an analgesic and soporific, and tormentil, an astringent, constituted two other highly important Mediterranean native simples.[34] In addition to these, there were a host of other antifertility agents and emmenagogues, including chaste berry seed, caper, juniper, helenium, white hellebore, madder, chamomile, and absinthe. Other simples aided in childbirth and lactation, such as birthwort, coltsfoot, lettuce, and madder.[35] Still other Mediterranean herbs and flowers treated various ailments of the liver, spleen, and the urinary and digestive tracts, including rose, water lily, asparagus, endive, myrtle, laurel, rosemary, chamomile, pellitory, cuckoopint, and hyssop. Many of these also helped to heal wounds, bites, stings, and various infections. In addition, angelica root served as a stomachic, sudorific, and cordial; and betony aided with diseases of the head and nerves.[36]

A number of gums and resins were native to the Mediterranean as well, including turpentine and mastic, used widely as substrates for compounds, though each was considered to have cleansing, astringent, softening, and diuretic properties on its own.[37] Gum-resins were particularly important as medicines due to their versatility: not only did they possess powerful medicinal virtues, but they also served as binding agents and preservatives for many different topical compounds, such as liniments, ointments, and plasters, discussed below. Other significant gums included spurge for treating cataracts, bone spurs, snakebite, and diseases of the hip; and labdanum, the diuretic exudate of the rock rose, gathered from the beards of goats who grazed on its leaves.[38] Many of these simples also served as perfumes, incense, and ingredients for sacred oils and ointments.[39] Turpentine and mastic, for example, had considerable importance for painting and sculpture as ingredients in pigments and varnishes; hyssop was employed in purification rites in the Temple of Solomon and in Catholic rituals, ground date pits were used as pigment in eyeshadows and rose petals for blush, and mastic provided a convenient glue for attaching false eyelashes.[40]

A range of Indo-Mediterranean simples also came from regions east of the Mediterranean, native to the area connecting the Mediterranean basin with the northern plains of India. The most prominent Indo-Mediterranean simples were the emmenagogue fennel; bistort root, an astringent considered good for diarrhea and menstrual regulation; dittany of Crete to cure of wounds and bites; white dittany for epilepsy, regulating menstruation, and destroying intestinal worms; violets, an emollient and gentle purge, good for headaches and fevers; and watercress, a diuretic also said to remove birthmarks and freckles.[41] Also of note, and used widely in prescriptions, were licorice for treating sore throats, chest and urinary tract ailments, heartburn, and wounds; mallow to ease the stomach and bowel, and as an antidote for poisons; marshmallow for treating all sorts of injuries, inflammations, hip and digestive ailments, and to draw down afterbirth; chicory for treating stomach problems; and St. John's wort for treating burns and hip ailments and to "drive out bilious matter and excrement."[42]

A series of additional gums and resins were native to the Middle East, especially the very dry, hot areas of Persia, Mesopotamia, the Levant, the Arabian Peninsula, and eastern Africa. Environmental conditions in these regions made them home to a variety of dry, thorny, "unpampered plants," that survive due to their ability to conserve water and secrete resinous or waxy substances that are sometimes highly flammable.[43] In general, like the Eastern spices, these plants were not easily naturalized or cultivated elsewhere. Also similar to the spices, the secretions, or exudates, of these plants generally had strong aromatic properties. Perhaps the most famous of the Middle Eastern gum-resins were frankincense (also called olibanum or thus), and myrrh, native to both the Arabian Peninsula and the tip of eastern Africa.[44] Frankincense, a pale green to amber-colored gum resin from the frankincense tree, was good for cleansing and healing wounds and stopping hemorrhages, and is still used as incense in the Catholic and Orthodox Christian churches.[45] Myrrh, the red gum-resin of the thorny, bush-like myrrh tree, was considered healing, agglutinative, soporific, and astringent. Both myrrh and opopanax (or sweet myrrh), also from the same area,

TABLE 6.2. PLANT SIMPLES ADDED TO MEDIEVAL PHARMACOPOEIA IN THE PALESTRA PHARMACEUTICA

Simple	Place of origin
bdellium	Asia (South)
camphor	Asia (East)
cañafistula	Uncertain in this period
clove	Asia (Southeast)
cubeb	Asia (Southeast)
galangal	Asia (Southeast)
grains of paradise	Africa
lemon	Uncertain in this period
myrobalan	Asia (South)
sandalwood	Asia (South; Southeast)
senna	Africa, Mediterranean
tamarind	Africa
zedoary	Asia (South; Southeast)

were employed as contraceptives or abortifacients, due to their emmenagogic properties.[46] Moving further east, several species of trees producing medicinal gum-resins were native to Persia; cultivation attempts elsewhere had little success. These included asafoetida, or giant fennel, a contraceptive and abortifacient that also treated nosebleeds, bowel ailments, and snakebites; gum tragacanth, used to treat sore throats and coughs; and gum ammoniac, native to modern-day Iran, an important antifertility agent that also treated ailments of the chest and bowel. Gum arabic, native to a wide area from Africa to the Arabian Peninsula to India, treated coughs and ailments of the stomach and lungs.[47]

Among the aromatics introduced to Galenic pharmacy by Arabic medical authors and listed in the *Palestra pharmaceutica* (see table 6.2), perhaps the most sought-after were cloves, the dried buds of the clove tree that were legendary both in value and in their inability to be cultivated anywhere but the Spice Islands. They were considered an effective muscle stimulant, good for treating gout and paralysis. Another important addition was camphor, the fragrant crystalline substance found in cavities in the trunk of the camphor tree that was known for its distinctive smell and its anti-inflammatory properties.[48] Galangal was another such addition, a plant of the ginger family whose roots were dug up, boiled, and dried for shipping, but could also be consumed raw, and were thought to be fortifying to the kidneys and good for digestion. Sandalwood was another fragrant simple, a wood mainly native to Indonesia, chiefly the islands of Timor and Java.[49] Sandalwood came in several varieties—yellow, white, and red—valued for their ability to fortify the stomach and heart.[50] Also native to Java and to Sumatra was the cubeb, a diuretic whose unripe berries were also dried and used for sore throat and toothache, and could be rolled into cigarettes for smoking.[51] Cubebs were also used to spice meats in medieval Europe.[52] Another eastern spice added to the medieval pharmacopoeia under the Islamic empires was zedoary, a rhizome native to India and Indonesia that was astringent, diuretic, and good for stomach ailments and sometimes substituted for cinnamon or nutmeg.[53] Bdellium, the gum of a spiny shrub native to India, was an important incense mentioned in the Old Testament and the Vedas, and served in medicine as a diuretic and softening agent used for "drawing out the fetus."[54]

A second wave of introductions to the traditional Galenic materia medica, and essentially a third layer of the simples contained in the *Palestra*, began with the Columbian Exchange and the expansion of the Spanish Empire, which led to the addition of a significant number of American plant simples, plants that had been wholly unknown to Galenic pharmacy (see table 6.3). Early interest on the part of the Spanish crown to investigate and exploit New World materia medica led to inclusion of a number of "exotic" simples foreign to the Galenic tradition.[55] Among those included in the *Palestra* were the South American stomachic ipecac (or ipecacuanha), as well as tobacco, which had a wide reputation as a medicinal panacea. For Monardes, tobacco was "a very ancient and well-known herb to the Indians," with "marvelous medicinal virtues."[56] In Oaxaca, it was "the herb [indigenous people] use most generally," good for healing all manner of afflictions: wounds and sores, headaches, chest ailments, asthma, side pain, gas, hiccups, worms, joint problems, swellings, toothache, snakebite, and carbuncles.[57] Also of widespread use was the antidote contrayerva, known in Nahuatl as *cohuanenepili* or *cohuapatli* ("snake tongue" or "snake medicine"), used for treating fevers, and as an antidote to snakebites, scorpion stings, and poisons generally.[58] Guaiacum bark and sarsaparilla root served as important anti-venereals that were sometimes mixed together to treat syphilis, and as sudorifics and purges.[59] As such, they were in great demand in Europe in the sixteenth century, exported in large quantities,

TABLE 6.3. AMERICAN PLANT SIMPLES ADDED TO THE EARLY MODERN EUROPEAN PHARMACOPOEIA IN THE *PALESTRA PHARMACEUTICA*

Simple	Place of origin
contrayerva	America (Central, South)
guaiacum	America (North; Central; South)
gum (caraña)	America (South)
gum (copal)	America (Central; South)
gum (tacamahaca)	America (South; North)
ipecacuanha	America (Central; South)
jalap	America (Mexico)
mechoacan	America (Central)
Peruvian balm	America (Central)
sarsaparilla	America (Central, South)
sassafras	America (North)
tobacco	America (Central; South)

but by the eighteenth century had largely fallen out of favor. Jalap and mechoacan root, purges of widespread use throughout Mexico, were also exported to Europe, and by the eighteenth century were commonplace items in European pharmacopoeias like the *Palestra*. According to Monardes, "many purgative medicines are brought from different parts of the Indies ... which have a great effect."[60]

A number of American gum-resins also entered the Galenic collection of medicinal materials, many of which, as with the Mediterranean and Middle Eastern gums, served a variety of artisanal, medicinal, and religious purposes. Copal, a tree resin widely used from Yucatán to central and northwestern Mexico, for example, served as incense in religious ceremonies, as an adhesive, and as a medicine.[61] Also from Mexico came tacamahaca, identified as a resin of the poplar, used for toothache and to fortify the stomach.[62] Its smoke was used to revive those who had fainted, and it also helped relieve back and joint pain, as well as irritation in the eyes and ears.[63] In this way, we can see the *Palestra* simples as a Galenic tradition that included about 90 percent of non-native plant simples, to which additional layers were added over time, with Arabic additions of Eastern and African aromatics, followed by American herbs, roots, gums, and resins. It was a foreign tradition that was accumulated, appropriated and domesticated, both literally and figuratively.

PAULA DE VOS

EXOTIC ACCUMULATION IN THE LAYERING OF FORMULATIONS AND TECHNIQUES IN THE *PALESTRA PHARMACEUTICA*

The layers of the development of Galenic pharmacy, its exotic (in the sense of non-native) origins and accumulations, are also evident in examining Galenic compound medicines and the addition of alchemical formulations in the seventeenth and early eighteenth centuries, along with the evolution of techniques used to process them. In the Galenic and alchemical traditions, simples or medicinal substances were processed in particular ways in order to heighten or moderate their medicinal powers, and prepare them for inclusion in a compounded remedy. The *Palestra* included comprehensive discussion of the different techniques utilized by apothecaries and medical alchemists for centuries, and represents the accumulation of two main developments, or layers, in the evolution of pharmaceutical technology. In the Galenic tradition, pharmaceutical processing was referred to as "correction," and involved four main techniques: trituration, levigation, coction, and infusion (see figure 6.2). Trituration, or grinding, involved breaking substances into smaller particles, while levigation, the washing of simples, was either performed by rinsing their exterior, or by shaking ground particles in solution, and decanting the liquid to remove impurities. Simples could be cooked in solution over gentle heat to produce a decoction or *cocimiento*, or they could be fried in oil or dry-roasted. Infusions were produced by submerging a simple in heated or boiling liquid, and leaving it to steep. Practitioners employed these techniques in the preparation of medicines over millennia, and they are evident in the classical tradition, as well as in *De materia medica*, but largely constituted "tacit knowledge"—with a few exceptions, written sources in the ancient and early medieval Galenic pharmaceutical tradition rarely discussed operations and practical techniques. The writings of Pseudo-Mesue once again provided a major watershed in the textual tradition, particularly in the *Canones universales* attributed to him, which outlined best practices and the rationale for each of these processing procedures, collectively referred to as "correction."

Although two of the operations associated with correction—cooking and infusing—utilized the application of heat (and infusions began with liquid brought to a boil), for the most part apothecaries avoided the use of prolonged and violent heat in the processing of simples and preparation of Galenic compounds. That was not the case for the second layer

FIGURE 6.2. An apothecary processing medicines, ca. 1500. M0000459: Interior of a pharmaceutical laboratory, 1500, from Peters, *Pictorial History of Ancient Pharmacy*. Credit: CC BY WT/D/1/20/1/5/41. Online at https://wellcomecollection.org/works/uwgcwxmu.

of pharmaceutical processing that developed among alchemists in the Latin West in the later Middle Ages, and later joined the Galenic tradition in the later seventeenth century in the "chemico-Galenic compromise." Alchemical techniques included a series of "thermo-chemical" operations that involved the prolonged application of heat to process simples. These involved distillation, which allowed for the separation of different components of a liquid and extraction of a concentrated

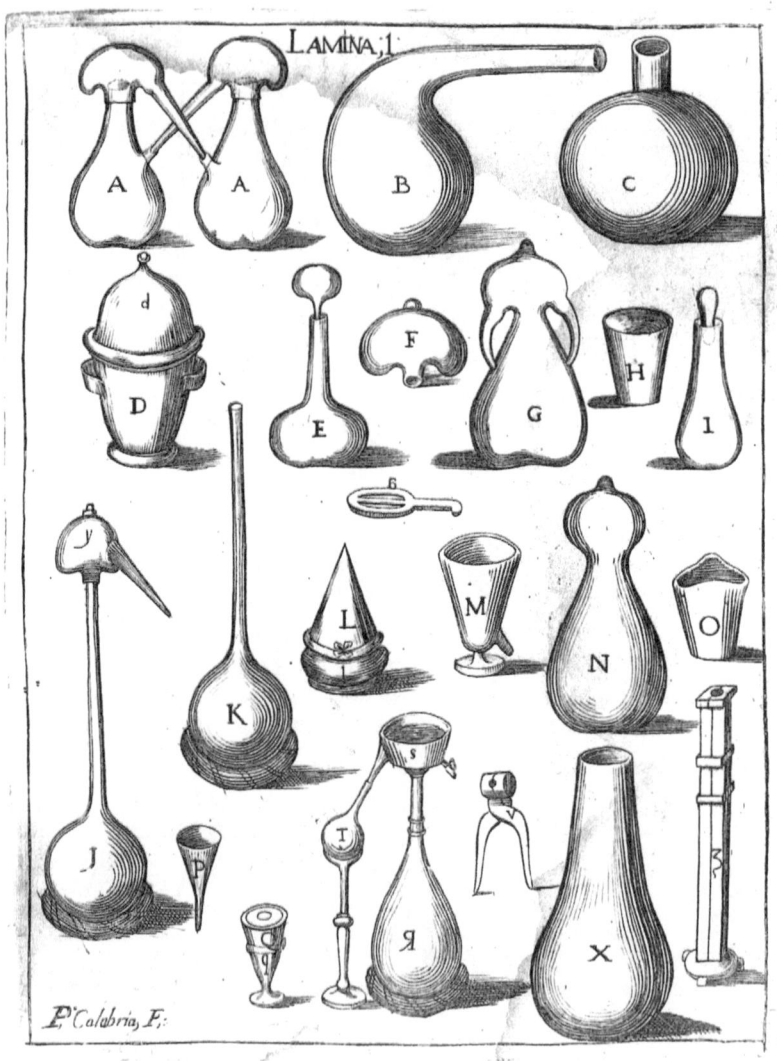

FIGURE 6.3. Alchemical apparatuses in *Palestra pharmaceutica chimico-Galenica* (Madrid: Viuda de Joaquin de Ibarra, 1792).

medicinal power; sublimation, essentially a distillation of a solid material in which the solid turned directly to vapor, and reformed on an upper surface of the apparatus; and calcination, the burning of a material to produce its alkaline ash, or "calx," which was then usually washed in solution and dried.[64] It also utilized methods of solvent extraction, or the repeated mixture of simples in solution, in order to concentrate a power or remove impurities (similar to levigation), sometimes accompanied by evaporation and recovery of a precipitate.[65]

The use of these techniques was rooted in the craft traditions of late antiquity, which provided the basis for early alchemy, whose focus in the Greek and Arabic world was on transmutation of metals, the idea that base metals could be ennobled to become gold or silver. Greek alchemy dealt largely with mineral and metallic substances, and the idea that they could be brought to perfection by correcting imbalances in their material makeup, mainly through the use of thermochemical processes. Medieval Arabic authors expanded this program to include a whole series of organic as well as inorganic substances that were manipulated chemically to separate out impurities and bring about perfection and balance. Like Galenic pharmacy, Arabic alchemy entered the Latin West through the translation movements, upon which medieval Latin authors—among them Arnald de Villanova, Roger Bacon, John of Rupecissa, and Pseudo-Llull—applied alchemical ideas and practices to medicine and the human body. Using alchemical operations and techniques, they began crafting medicines, quintessences, elixirs, and panaceas purported to have extraordinary—even divine—healing abilities. Over the following centuries, these medicines broadened to include an entire formulary, parallel to that of Galenic pharmacy.[66] These two traditions then combined during the seventeenth century to create a "chemico-Galenic compromise." This compromise—perhaps better referred to as an amalgamation—is, again, evident in the *Palestra pharmaceutica*, which included chapters in book 1 devoted to Galenic techniques of trituration, levigation (washing), infusion, and coction (cooking), as well as distillation, sublimation, calcination, fermentation, evaporation and precipitation (i.e., forming crystals) (see fig. 6.2). The front matter of the *Palestra* (see fig. 6.3), moreover, included several diagrams of alchemical equipment, accompanied by descriptions, and a list of alchemical symbols and what they signified.

These techniques, both Galenic and alchemical, resulted in the formulation of several types of compound remedies. In Galenic pharmacy, simples were rarely administered by themselves, but rather underwent substantial processing, as shown above, in order to optimize their medicinal powers and to be included in compounds. Compound remedies had been in use since antiquity, and grew increasingly popular from the Middle Ages onward, as drug therapeutics began to take precedence over traditional emphasis on dietetics and regimen in the maintenance and restoration of health, with widespread recognition that the complexity of disease required a similar complexity of medicinal powers in one remedy, which could only be accomplished by combining multiple simples in that remedy. The formulation and categorization of compounds developed in four main phases, and became increasingly codified over time, so that by 1700 or so, most major European formularies contained a relatively consistent and standardized listing of categories of remedies.

These phases, and the categories of compound remedies developed within them, represent another example of the layering and accumulation of knowledge in the development of Western pharmacy. They included, first, an ancient Greco-Roman phase, in which practitioners employed a more limited subset of the compounded medicines; a medieval Arabic phase, which witnessed a widening of the types of compounds employed; a medieval Latin phase, in which the types of compounds utilized shifted to categories based upon the materials and methods used in their formulation; and finally, an early modern European phase, in which alchemical remedies joined those of the traditional Galenic formulary, resulting in the "chemico-Galenic compromise," of which the *Palestra pharmaceutica chymico-galenico* was an important example. It included a full treatment of compound remedies prepared using Galenic and alchemical techniques.

Compounds were categorized over the centuries in different ways. In the first phase, ancient Greek medicine employed mainly application-based compounds, classified according to place or method of application, including gargles, collyria or eye washes, incense, perfumes, sternutatories or sneeze-inducers, toothpastes, enemas, pessaries, and suppositories.[67] But they also incorporated a number of what I call "method-based" compounds, categorized by the main methods and materials they employed, indicating increasing specialization and emphasis on the formulation of medicines, which would eventually form the basis of a separate pharmaceutical profession. These included honey-based confections, as well as ointments, oils, liniments, poultices, pastilles, and plasters. Ancient texts from Hippocrates to Galen, and including Celsus, Scribonius Largus, and Paul of Aegina's redactions, generally named compounds according to these different classifications. Paul of Aegina, for example, included application-based collyria, perfumes, and pessaries; and method-based liniments, lozenges, plasters, ointments, and oils. The production of medieval Arabic formularies constituted the next step, the second phase, in the development of categories of compound medicines and their formulation. Arabic formularies, from that of Sābūr ibn Sahl (ninth century) to that of Ibn Sīna (eleventh century), expanded the types of method-based compounds, and provided more systematic and consistent categories of the method- and application-based medicines.[68] Method-based categories came to include a series of new compounds formulated using cane sugar, adapted from Indian techniques, and another indication of Indian influence on Galenic pharmacy.[69] A highly versatile sweetener, cane sugar was incorporated into the newly developed syrups, electuaries, confections, and lambatives (very thick, syrup-like medicines, in which sticks or licorice were dipped, and the medicine licked off) that had not been part of the earlier Greek tradition.

This growing emphasis on, and expansion of, the method-based compounds in Galenic pharmacy in the medieval Arabic world continued when Arabic medical texts were translated into Latin, beginning in the eleventh century. As Latin authors sought to assimilate this material, a landmark Galenic formulary by Pseudo-Mesue served as a conduit to the Latin West for Arabic contributions to compounding—another example of his unparalleled influence on the development of medieval and early modern Galenic pharmacy. The formulary, entitled the *Grabadin* (see fig. 6.4) after the Arabic word for formulary—based on the Greek *graphidion*, meaning "list" or "registry"—included recipes for all the added cane-sugar-based compounds, and also signified the consolidation of a shift toward method-based preparations that would dominate the classification of compounds and the organization of European pharmacopoeias from then on.[70] For Pseudo-Mesue's *Grabadin* included only method-based categories, organized into twelve chapters, each covering one of twelve main types of method-based preparation. These included decoctions, infusions, and powders of multiple simples; pressed and infused oils, ointments, and plasters formulated with varying proportions of oil and wax; pills and electuaries bound with honey, gum, or resin; and the cane-sugar-based syrups, preserves, lozenges, confections, and lambatives. These types of compound came to dominate the formulary in late medieval and early modern Spain: my own survey of twenty medieval Latin and early modern Spanish formularies, for example, shows a clear trend toward the method-based compounds in the Hispanic world, and these categories can be seen in virtually every major early modern European pharmacopoeia.

De electuariis.
**ridis & fraudulentis.(Dosis.)Eius est a charatis.vi.vsq; ad tertiam.ʒ.i. vel vsq; ad.ʒ.i.

¶ Ioannis Mesue grabadin. Quod est aggregatio
& antidotarium electuariorum et confectionuʒ:
& aliarum medicinarum compositarum.

Cripsimus in libris explanationuʒ nostraruʒ ex his que experti sumus quedam adhuc re/ memoratione digna. Ex quarum aggregatio ne summam conteximus quod grabadin'no/ strum vocamus:quod est compendium secre torum in quo expertas & preclaras medici/ cinas ex his que acquisiuimus & inuenimus ab eis que medicine artis archana scire faciut tradere dignū duximus.Et ponimus totam latitudinem operis huius sum mas duas. In prima famosas & solemnes medicinas com/ positas tradimus.In secuuda uero medicinas singuloruʒ mē brorum egritudinibus appropriatas distinguimus.
In prima uero distinctione ordinamus.xii.
¶ Prima est de electuariis.
Secunda de medicinis opiatis.
Tertia de medicinis solutiuis.
Quarta de conditis.
Quinta de speciebus lohoc.
Sexta de syrupis & robub.
Septima de decoctionibus.
Octaua de ttociscis.
Nona de suffuf. & pulueribus.
Decima de pillulis.
Vndecima de vnguentis & emplastris.
Duodecima de oleis.
¶ Prima distinctio que est electuariorum subdiuisionem ha bet.Quedam enim ex eis delectabilia sunt quedā uero ama ra:& vtraq; sunt solemnia.Prius tamē ex eis electuaria quo rum est delectationem facere cum titulis utilitatum suarum tradimus:& post de aliis loquemur sicut deo placuerit.
¶ Incipit

FIGURE 6.4. First page of *Grabadin*, 1513 edition of *Opera medicinalia*.

The final phase or layer in the development of compound remedies in European pharmacy was the addition of alchemically prepared formulations. As described above, these medicines were formulated through the thermochemical operations of distillation, sublimation, and calcination, which involved the use of high heat, as well as solvent extraction and precipitation. Following on from the work of medieval Latin authors in the thirteenth century, a series of alchemical compounds formulated using these operations emerged over the course of the following centuries, and were expanded upon with increasing variety and precision. These compounds included distilled waters, distilled oils, essences, extracts, salts, spirits, and tinctures, which employed variations of the alchemical processing techniques described above. Distilled waters, distilled oils, and spirits, for example, utilized distillation to extract medicinal powers from simples (or fermented simples in the case of spirits), while formulation of essences, extracts, and tinctures called for simples to be mashed and placed in a solution, an operation sometimes repeated upon multiple batches. Formulation of salts typically involved calcination of simples, and distillation of the ash in solution, followed by evaporation to recover the crystallized precipitate.

Once again, the *Palestra* bears witness to these different phases of categorizing, formulating, and processing compounds. As with the simples listed in the work, the compound medicines Palacios included also provide a survey of these various layers of development (see table 6.4). Unlike most of his contemporaries, who tended to include only method-based classifications of Galenic compounds, Palacios also included the largely obsolete earlier application-based categories as well—another indication of the encyclopedic nature of the *Palestra*. Chapter 14 of book 2 of the 1753 edition, for example, presented recipes for gargles; chapter 15, chewing gums; chapter 16, suppositories and pessaries; chapters 17 and 18, enemas; and chapter 19, collyria. Each chapter began with a general discussion of the type of compound, its uses and formulation, followed by a series of recipes. Most of the recipes in the *Palestra*, however, in line with most other early modern formularies, were organized under the common method-based categories codified by Pseudo-Mesue, including those in use in classical antiquity, as well as the sugar-based compounds added in the Arabic tradition. As such, chapters 28 to 38 of book 2 treated preserves, syrups (including the thinner versions of syrups and robs), lambatives, electuaries, and powders, while chapters 1 to 4 of book 3 covered lozenges, pills, and pressed

TABLE 6.4. TYPES OF COMPOUNDS IN RECIPES OF THE *PALESTRA PHARMACEUTICA*

Origin	Compound category	Compound type
ancient	application-based	chewing medicines, enemas, collyria, gargles, incense/perfumes, suppositories
ancient	method-based	decoctions, electuaries, oils (cooked or pressed), ointments, pills, plasters, powders, preserves (with honey), infusions/tisanes
medieval Arabic	sugar method–based	confections, juleps, linctures, lozenges, preserves (with honey or sugar), robs, syrups
early modern	alchemically based	distilled waters, distilled oils, extracts, essences, salts, spirits, tinctures

oils. Chapters 9 and 11 covered ointments, liniments and various kinds of plasters, confections, pressed and cooked oils, ointments, pills, plasters, powers, and preserves, and syrups. The *Palestra* also included the alchemical compounds of the chemico-Galenic compromise, distilled waters, extracts, essences, salts, spirits, and tinctures. The work not only brought together these layers but also shows the exotic origins of many of these compound medicines.

In this way, the *Palestra* represents a microcosm of pharmaceutical materials and techniques that developed over the *longue durée*. It demonstrates that the Western pharmaceutical tradition was in fact one that had adopted, appropriated, and arguably domesticated "exotic" elements over the centuries. This study also demonstrates that ideas about what constitutes the "foreign" or the "exotic" in history are highly complex and manipulable—long-distance objects and ideas that might be considered foreign or exotic can be naturalized in both a literal and a symbolic sense, and historical narratives can also be constructed to include or exclude, depending on their purpose. In this case, the role of the Iberian Peninsula as a mediator between the European "West" and the Islamic, Indo-Mediterranean "East" becomes especially apparent, as Islamic culture and society brought Galenic ideas, practices, and plant materials to al-Andalus. For the study of the adoption and use of simples and compounds in Spanish pharmacy, as evident in the *Palestra pharmaceutica,* shows the great importance of the eastern Mediterranean and Indo-Mediterranean to medieval and early modern European pharmacy, both Galenic and alchemical. The connection is so strong that these traditions may be labeled "Western" in terms of their legacies—without the selective exclusion of the Arabic contributions. Yet they originated far from later European centers of power, involving and necessitating long-distance trade and travel to arrive in Europe, and in this way they may be considered "exotic," though without the Orientalist ramifications of what such a term may imply.

PART III
NETWORKS OF COMMODIFICATION

7

A SCIENCE OF THINGS

The Jesuits and the Introduction of American Materia Medica in Europe

Samir Boumediene

1689: Francesco Viva (1656–1702), a Jesuit father from Naples, father superior at the mission of Maynas, Peru, drafts a project for the spiritual and military conquest of the region.[1] In the later sixteenth century, the Spaniards had found significant gold reserves in this area, but they also faced resistance from indigenous communities they called the "Jívaros." In 1599, an assault on the Spanish town of Logroño terminated the expansion of gold exploitation in the region. Almost a century later, the memory of this attack was still strong among religious and civil authorities in Maynas. According to Viva, the province lacked well-trained missionaries and soldiers, and his 1689 project was designed to recruit them. It involved a trip to Spain and Italy of some five years.[2] Usually, such voyages were undertaken by the provincial procurators, those Jesuits sent to represent their province at the Society's meetings in Rome. In Spanish America, most were creoles, i.e., born and raised in the New World. Born in Naples, Viva knew he had little hope of election as a provincial procurator, so he decided to finance his trip independently.

The material basis of Viva's project was to sell American plants in Europe. In 1685, he had deputed thirty Indians to sow around 50,000 vanilla plants, planning to use the profits from these to finance his round trip to Europe.[3] By 1689, the plants were not yet cropping in sufficient quantity. Nonetheless, Viva urged his superiors to let him go to Europe that very year, because "a commodity [had] ended up in his hands by divine disposition": cascarilla, or Jesuits' bark.[4] As this name suggests, the history of cascarilla is closely tied to the history of the Society of Jesus. Introduced around 1640 in Europe, it profoundly transformed medical thought and practice as well as the management of health. By 1689, it was already considered a specific for intermittent fevers, that is, a highly effective remedy against this common fever type. In writing to his superiors, Francesco Viva assured them that he was able to extract fifty mule-loads of cascarilla from the mountains around the town of Loja. According to his calculations, these fifty mule-loads would be worth 50,000 *patacones* (pesos) when sold in Europe. In other words, Viva hoped that trading in the bark would finance a journey he had been planning for several years.

In the middle of the Andean forests, a Jesuit could thus handle the complete process of extraction, exportation, and retailing of a product that would be considered an important good in the European medical marketplace. The example perfectly illustrates the role the Jesuits played in the exotic drug trade during the early modern period. As several scholars have argued, their contribution to early modern pharmacy should be linked to several priorities of the Society of Jesus: the importance it placed on missions outside Europe and on education; the emphasis on assistance to the poor and sick; and lastly, the value it accorded to commercial activities.[5]

DEALING WITH MATERIA MEDICA: A TEXTUAL INTRODUCTION

In light of recent scholarship, the current chapter endeavors to understand the Society's involvement in the European medical marketplace by looking at the American remedies they tried, or managed, to introduce into Europe. At the chapter's heart lies an interrogation into the process of commodification. How were the Jesuits able to convert natural things into commercialized goods? The question has two dimensions: first, the transfer of an object from one continent to another, and second, the introduction of a novelty into a new market, where it had to find its niche. As we shall see, this question of commodification cannot be understood without paying attention to drugs' other roles as samples, curiosities, and gifts. Focusing on the Jesuits allows us to connect all the main aspects of drug importation to Europe: the extraction of materials and knowledge overseas, their movement across oceans, product testing, recipe design, and the sale of medical commodities in European towns.

Jesuits' involvement in introducing American plants into Europe was grounded on their ability to appropriate knowledge. Their contribution can be studied, in the first instance, through the wide variety of writings they dedicated to medicine, natural history, and pharmacy, as well as in other sources like missionary reports and letters.

As with other religious orders, the Jesuit presence in America took several forms: churches and colleges in urban centers, farms, and villages (*reducciones*) on the mission frontier. From Lima to northern Mexico, their interactions with Indians depended heavily on these configurations. Given this diversity, it is impossible to offer a complete account of medical interaction in mission settings. But Jesuit knowledge of remedies used by Indians was never limited to the mere transmission of information. It also involved active inquiries deeply intertwined with evangelization purposes.

Curing the soul implied curing bodies. This conversion-through-medicine strategy implied, first, the creation of a positive attitude toward Christianity. If the missionaries were able to treat diseases among local populations, the latter might be more receptive to the set of beliefs promoted by the former. Healing sick people, affirmed the Moravian Jesuit Johan Steinhoeffer (1664–1716), was a way to "open the doors to their souls."[6] Around 1622–23, in a missionary report (*carta annua*), Father Florián de Ayerbe (1568–1647) recounted how the missionaries fought epidemic diseases suffered by Indians in the Jesuit province of New Granada, especially measles and the pox. The Jesuit mission house, Ayerbe wrote, was not large enough to care for all the sick, nor to help them "die well," so the priests decided to visit the houses of Indians as well Spaniards.[7] Similar news is to be found in most mission reports sent to Rome between the sixteenth and eighteenth centuries. Such documents probably exaggerated the health difficulties of the Indians, and the effectiveness of the medical care provided by the Jesuits, but they illustrate well the prominence given to healthcare in the mission: improvement in material and bodily life was evidence of the goodness of Christianity.

The second dimension of this medical conversion was a negative one. In imposing themselves as the best healers, the missionaries also sought to challenge the authority of indigenous healers. The reason for this was that local healers assumed a wide range of functions, from curing bodies to interceding with "pagan" divinities. Since they were simultaneously "healers" and "priests," challenging their power in one domain, the cure of the body, amounted to challenging their power in the other domain, intercession with the invisible. "The fathers," said an eighteenth-century report from the Mojos missions, "are the doctors and surgeons who cure [the Indians], their barbers who bleed them, their nurses who care for them, visiting the sick of the village twice a day and personally assisting in the application of medicines."[8] This twofold strategy—displaying the goodness of Christianity and supervening over local healers—implied that Jesuits would use remedies coming from the Old World, and so they did, as inventories of their pharmacies show.[9] But this strategy also entailed the study of local remedies.

Issues such as the difficulties of maintaining a supply of European drugs, the need to adapt to local custom, but also the recognition of the medicinal value of local remedies, led Jesuits to use American plants on both themselves and the Indians. In the most remote regions, appropriating local remedies was effectively mandatory for the missionaries' survival, since they could not rely on a stable supply of medicines from Europe.[10] Alongside soldiers, missionaries were thus often the first to learn the properties of drugs still unknown in Europe.[11] This was the case, for instance, for ipecacuanha, a root used to treat diarrhea that became very famous in the Old World during the last quarter of the seventeenth century but that had been described a century earlier by the Jesuits in Brazil.[12]

The use of local medicines was also essential to the "extirpation of idolatries." As a concept, idolatry not only referred to a defined set of beliefs, but rather was embodied in every practice of Indian life: for example, using medicines was a way to interact with invisible entities, and might involve invocations, offerings, divination, and trances. Health was thus an important battleground in the Jesuit war against "idolatry," which further explains the attention missionaries paid to medicinal plants. Based on questionnaires used by religious authorities in the Old World, the numerous handbooks on confession published in Spanish and indigenous languages contained frequent queries about ritual healing, poisons, or abortifacients.[13] However, it was also during this fight against indigenous rituals that missionaries could learn of new remedies they considered useful. In 1621, for instance, an Indian healer named Francisco García Julcapuma appeared before the bishop of Trujillo, Luis de Paz, to make his confession. During his interrogation, he was asked about herbs and potions he might use for love-magic or killing people. Francisco García Julcapuma answered that he knew no such remedies, and only healed people as a Christian. In proof of this, he supplied information concerning several herbs he used to treat stomach pain, bloody flux, and dropsy.[14] As Andrés Prieto has shown, the ritual of confession was an important moment for appropriating local knowledge that later appeared in treatises such as those of the missionaries Alonso de Ovalle (1603–1651), José de Acosta (1540–1600), or Bernabé Cobo (1582–1657) in Peru.[15]

JESUIT HISTORIES: A SCIENCE OF DESCRIPTION

The writing of natural and moral history in the New World antedated the Jesuits' arrival in Spanish America. Chroniclers, like Gonzalo Fernández de Oviedo, and members of mendicant orders, like the Franciscan Bernardino de Sahagún, conceived the description of plants, animals and geography as inseparable from inquiry into the precolonial history, religion, and governance of the Indies.[16] By the 1570s, when the Jesuits arrived in Spanish America, such inquiries proved hard to conduct, however. The king and his entourage had begun to consider that recording the Indians' past amounted to recording a period in which they governed themselves.[17] This explains why several important texts written by Jesuits at this time were either lost, or not published until the nineteenth century. In fact, the majority of published works appeared in the Old World, some written remotely by European Jesuits, others published by Jesuits who, like Acosta or Ovalle, returned to Europe bringing a manuscript with them. But not all Jesuits were able to publish. The majority of those who could were fathers, that is to say Jesuits who had completed all their vows. During their college training, although the curriculum did not explicitly include natural and moral history, Jesuits in training read classical texts, especially Pliny and the Aristotelian corpus, and studied naturalia as a way to celebrate the glory of God.[18] This encomiastic function of natural and moral history was especially true for exotica, and is exemplified in the well-known work of Juan Eusebio Nieremberg (1595–1658) or Athanasius Kircher (1602–1680).[19] In contrast to these examples, José de Acosta's work provided the Jesuits with a more practical way to write natural and moral history which influenced many authors, both within and outside of the Society of Jesus, including Alonso de Ovalle, Bernabé Cobo, Johann Steinhoeffer, Sigismund Aperger, Pedro Montenegro, Marcos Villodas, and José Sánchez Labrador. But even when published, these later works were not really addressed to a European public. Rather, they served as tools for advancing the goals of the missions in the Americas, and were distributed in the form of copies, printed or handwritten, that circulated alongside other kinds of texts, such as recipe leaflets.[20]

JESUIT RECIPES AND PHARMACIES

In an urban context, the writing of such natural histories was linked to the presence of a Jesuit pharmacy.[21] After the foundation of the first Jesuit pharmacy in Rome in the mid-1550s by Baldassare Torres, a former physician, countless colleges equipped themselves with an apothecary shop. Their original purpose was the care of Jesuits themselves, and the diffusion of remedies outside a college was limited to charitable treatment of the poor. But the Jesuits soon began to participate in the medical marketplace. In addition to recipes and memoranda for internal use, Jesuit "apothecaries" or "nurses" might write instructions for their clients. The majority of these were not fathers, but brothers: Jesuits who did not complete their vows, but committed to the Society's apostolic project by undertaking its practical activities. Placed in charge of dispensaries, they needed to master the basics of the Galenic pharmacopoeia, to know how to use certain genres of text like recipe books or medical treatises, and to be able to write instructions for the use of remedies.

Among the archives where such recipes survive today, these college libraries are of particular interest. That of the Colegio Imperial de Madrid, now held at the Real Academia de la Historia, includes several recipes using remedies from the Americas or the Philippines.[22] A text devoted to contrayerva, for instance, lists all the venomous bites cured by this counterpoison, as well as other diseases it was supposed to treat—fevers, typhus, measles, pox, kidney pains, urinary difficulties, melancholy—ending with the claim that it was a "unique remedy for every disease if used properly." Such a generalization was not surprising, as this was a promotional text serving to market the American remedy. But this text also provided practical instructions for preparing an infusion of contrayerva, and for selecting the right sort, which was that coming from New Spain.[23] Other similar texts were compiled in a collection entitled "Papeles de Medicinas," which included recipes for various balms (copaiba, or "Peruvian Balm dissolved in Wine Spirit") and a bean from Guatemala, as well as a compound remedy called "Elixir salutis," which included American drugs like guaiacum wood.[24] All these texts that ended up in Madrid relied on a complex set of practices, including conversion, economic exploitation, college teaching, and the retail of drugs from apothecary shops. Pharmacy was at the crossroads of various aspects of the Jesuit presence. But the example also neatly demonstrates the complex interplay between the movement of texts, people, and things. To put it another way, circulation was both a means and an end for the Society's apostolic project and the personal ambitions of its members.

DEALING IN THINGS: A SCIENCE OF CIRCULATION

The peculiarity of the Jesuits, compared to other religious orders, was not so much their ability to study local drugs during their missions as their ability to bring information about them back to Europe and introduce new remedies there. Behind the texts mentioned in the previous section, there was a science of circulation.[25] In this section, I will show how the introduction of American drugs by the Jesuits took place within this broader system of exchange that was internal to the Society, but also tightly connected to non-Jesuit spheres. This will allow me to discuss the link between curiosities and drugs, gifts and commodities, with a particular emphasis on one aspect: the sample.

The Society's internal exchange system was materialized, in the first instance, by the exchange of letters. Alongside official correspondence (letters, missionary reports, catalogues of Jesuits, book accounts, necrologies, etc.) addressed to the upper echelons (superiors of the missions, rectors of the colleges, provincials, and the superior general), the Jesuits also exchanged letters among themselves, dealing with a wide range of topics—theology, literary novelties, historical events, commercial affairs, personal issues, etc. Such communications were encouraged by Ignatius of Loyola, for whom the information about the geography, the plants, and the animals of the Indies constituted a "sauce to the palate of harmless curiosity."[26] Nourishing the human appetite for curiosity was almost a Jesuit rule, for at least two reasons: first, spreading news from the Indies was held to contribute to their intellectual formation and spiritual and moral edification; second, this news could also make missionary work attractive, something apparent from the *Indipetae*, letters of request to go to the Indies.[27] Besides the exchange of letters, the display of curiosities like Athanasius Kircher's cabinet, installed at the Collegio Romano, allowed the Jesuits to gratify urban elites' desire for the exotic. Curiosity, defined as the will to know, use, or possess the world in all its variety, was the common tongue spoken by the Society. Information, objects, and people circulated within its framework.

If Jesuit networks of exchange relied on letters, they also relied on the circulation of an overlooked item: boxes. Books, manuscripts, correspondence, gifts: all of these kinds of objects were transmitted from the Americas to Europe in boxes. According their size, they might be described as a *caja* (regular box), *cajón* (large box) or *cajeta* (small box). Shipping inventories (*memoria*) and travel logs (*libros de viático*) offer precious details of their circulation. A significant part of these exchanges involved the movement of one or more members of the order called the *procuradores*, a Spanish term translatable into English as "procurator" or "deputy." Being a *procurador* meant acting as an agent on behalf of certain people or institutions, and representing these to others. Unsurprisingly, a lot of functions fell under this rubric, not only within a Jesuit context—other religious orders, such as the Franciscans, also had both procurators and civil authorities.[28] In the Jesuit case, two kinds of *procurador* are of interest here. The first was the *procurador de provincia* (provincial procurator), who was a Jesuit who, often accompanied by a fellow priest, represented his province at the order's congregations in Rome. The second kind of procurator—specific to the Spanish territories—was the *procurador general de las Indias* (general procurator of the Indies), of whom there were two types.[29] One represented the order before the Casa de la Contratación in Seville, but in practice worked at the Colegio of San Hermenegildo, organizing the reception, registration, and departure of missionaries, and seeking financing for the missions.[30] Another represented the order before the Council of the Indies in Madrid, negotiating the foundation of new missions, visas for missionaries and fiscal exemptions from the crown. Everything coming from the Indies theoretically passed through the hands of these two general procurators.

The combined action of these provincial and general procurators enhanced the effectiveness of the Jesuit exchange network. A good example is the inventory (*memoria*) of two *cajones* the Jesuits from Peru sent to "Castilla" in 1630, entrusting them to the merchant Pedro de Saldías. Inside the boxes were letters, silver, bezoar stones, and other things stored in "cajetas" and "cajetillas" (small and tiny boxes). The majority of the shipment was sent to Jesuits, especially prominent figures like Fabián López, general procurator of the Indies, and Juan de Lugo (1583–1660), professor at the Collegio Romano.[31] These boxes materialized the friendships and sometimes the patronage relations between Jesuits in Europe and America.

But the circulation of boxes raised a double problem: entrusting a box to a shipowner was expensive and also created the risk that it might be opened. To avoid such inconveniences, the Jesuits could rely on the provincial procurators traveling to Rome. Their tours were not only undertaken to represent their province at the congregation but also to recruit new missionaries. To do so, procurators had to show off the wonders of the Indies in gifts, medicines, and curiosities. Thus, on their return to America from Europe, they not only brought with them new missionaries, but also artifacts, books, and letters.[32] In Seville, the general procurator of the Indies contributed to the logistics of such operations. In 1666, for instance, Juan de Ribadeneyra, general procurator of the Indies, made an advance payment towards the travel costs of two Peruvian procurators, Father Felipe de Paz—who died during the trip—and Brother Alonso Gómez. He also assisted them with distributing the bezoar stone, vanilla, chocolate, and "cascarilla" in Spain or in Rome.[33]

This organization of exchange not only crossed the Atlantic but also followed European trade routes. In a letter dated January 29, 1636, Lugo asked his Sevillean correspondent Father Rafael Pereyra to source him some tacamahaca, an American resin: "A lay brother in this Collegio, whose pharmacy is the best in the world, according to the Pope, . . . has pressed me to have someone bring some tacamahaca for him. If a procurator should come to the Congregation, could Your Reverence do me the favor of asking him to endeavor to undertake this task, because this brother will pay him the whole cost, or in any commodity from Rome of any type, because he will know better than anyone how to find [tacamahaca]?"[34] This letter interestingly highlights the articulation between transatlantic and intra-European circuits. To obtain the rare drug, Lugo, as broker for the unnamed pharmacist of the Collegio Romano, was ready to pay either in money or in exchange for commodities from Rome.[35] The internal exchanges of the Society thus took the form of a non-monetary economy, allowing very different things to be exchanged: by giving a gift, one could receive the protection of important people like Juan de Lugo; by giving a balm or bezoar stone, one could receive European books, or information about diplomatic events.[36]

The procurators' journeys enabled European Jesuits to admire unseen artistic productions, to consult books and manuscripts that were rare if not forbidden in Europe, such as descriptions of Amerindian rites, and to receive new drugs and curiosities.[37] As early as the 1570s, they regularly carried balms, bezoar stones, vanilla, chocolate, and medicines with them.³⁸ Once they arrived in Europe, all these things would be dispatched along various channels, depending on patronage relations and friendship. The example of tobacco and chocolate is especially interesting in this regard.³⁹ Although the Jesuits were not the only actors spreading these products, they had access to high-quality sources, thanks to their presence in America. But at the same time, they also had to contend with regulations that limited the use of these two products to the clergy.⁴⁰ It was, then, through interpersonal correspondence that they could exchange tobacco, and more especially, chocolate (which, as a precaution, they referred to as "the medicine") with one another.⁴¹ Jesuits in Seville, Granada, or Madrid supplied their coreligionists around Europe.⁴² They also accompanied their packages with recipes in which cacao was mixed with other ingredients (annatto, cinnamon, chili, vanilla, amber, musk).⁴³ Father Benito de Sojo, born in Granada, the professor of theology at the College of Vilnius, distributed Russian artifacts to other colleges in Vienna, Warsaw, Rome, and Andalucía, and sought to spread chocolate in Eastern Europe. On July 29, 1631, he asked Juan Camacho, general procurator of the Indies, the correct dosage for preparing the drink: "how much chocolate, how much sugar, when the sugar should be added, how long it needs to cook, whether slowly or not, whether it needs to boil."⁴⁴ By creating a kind of informal market where drugs, curiosities, and texts could be exchanged, giving rise to forms of obligation, the Jesuits also created a space of inquiry into exotic products. Samples had a fundamental importance in this process, holding the status both of a curiosity, because of their rarity, and of an aid to experimentation, allowing something unknown to be tested for the first time, and, after refining preparation methods, accredited. The circulation of samples, gifts, and curiosities was thus fundamental to the process of conversion into a merchandise. To understand this circulation, we must also consider how exchanges within the Jesuit order made contact with non-Jesuit circles.

BETWEEN GIFT AND COMMODITY: THE SAMPLE

If the fact of being a Jesuit allowed comparatively easy access to this informal market, members of the Society also knew how to profit by opening trade to non-Jesuits. The first profit to be made from their science of circulation was political. In America as well in Europe, the added security afforded by the provincial procurators' role as couriers was used by individuals and institutions the Jesuits were happy to serve: the

Inquisition, the Council of the Indies, and powerful families needing to transfer letters or money.[45] In the same spirit, boxes could also be addressed to influential people outside the Society. The aforementioned inventory of the two boxes sent from Peru to Europe in 1630 contained, for instance, gifts and letters for the Secretary of the Indies and the Countess of Lemos.[46] In Seville and Madrid, the general procurators of the Indies constantly received such packages of curiosities, drugs, chocolate, or money for important persons.[47] Going from the Americas to Europe, provincial procurators did the same. In 1602–3, for instance, the founder and provincial of the Paraguay mission, Diego de Torres Bollo, traveled to Europe as procurator to the province of Peru. He distributed a number of American curiosities between Madrid, Milan, and Rome, and was able to recruit forty new missionaries. In Milan, where he hired Agostino Salumbrino—a man who would end up heading the pharmacy at San Pablo College, Lima—Torres Bollo gave some of the most precious items he had collected in America to the powerful cardinal Carlo Borromeo.[48] In 1646, again in Milan, Alonso de Ovalle, provincial procurator to Chile, gave precious American objects to another prominent member of the urban elite, the collector Manfredo Settala.[49] This generosity was supposed to secure the friendship of princes, bishops, counselors of the Indies or members of the Curia Romana who, when the time came, could repay these debts, for example by granting fiscal exemptions to Jesuits, or visas to their missionaries.[50]

Besides brokering political relations, the circulation of artifacts, natural objects, and images from the New World could also yield economic profit for the Jesuits. Because of canonical restrictions imposed on clerical involvement in trade, they had had to negotiate the sale of remedies with papal and urban authorities, justifying it on the grounds of healthcare.[51] While their drug retailing activity was often contested—especially by secular apothecaries—it was also a reality, documented, for instance, in the account books of college pharmacies.[52] In fact, Jesuit involvement in the medicine business was not unique. From the Middle Ages onward, many pharmacies based in different religious houses actively sold a range of both common medicines and more specialized remedies, which in some respects resembled what would later be called "secret remedies."[53] To promote these, clerical retailers distributed printed instructions with an address where they could be procured. The Jesuits' peculiarity lies in the part they played in the commodification of particular remedies. While they sold theriac or mithridate, they also became known for introducing certain simple remedies to Europe, something that becomes clear from letters and recipes they exchanged, more than from their pharmacies' account books. When Sojo asked about the preparation of chocolate in 1631, his intention was not just to make it for himself: he also thought the product would be "esteemed" in his city.[54] Jesuits thus certainly considered a commercial perspective in their exchanges about drugs.[55]

Herein lies the importance of the samples that Jesuits exchanged. Between curiosity and drug, gift and commodity, the sample is the key to understanding how exotic products became European goods. This process had occurred in the sixteenth century with other American remedies introduced by missionaries, soldiers, and merchants.[56] But it reached unprecedented dimensions with the Jesuits, who brought, or planned to bring, a wide range of remedies to Europe, thanks to their effective internal exchange system.[57] In order to explore this articulation further, the following section concerns what was perhaps the most important drug that the Jesuits contributed to introducing in Europe: Peruvian or Jesuits' bark.

A SCIENCE OF TESTING: THE INTRODUCTION OF JESUITS' BARK

The European "discovery" of Peruvian bark has been a subject of fascination since the remedy's introduction. A mural still visible at Santo Spirito Hospital, Rome, illustrates a widely circulated account first narrated by Genoese physician Sebastiano Bado in 1663.[58] The story recounts how the Countess of Chinchón, wife of the Viceroy of Peru, was "miraculously cured" from her fever by bark brought to the court by the *corregidor* of Loja. Later, Jesuit fathers brought the remedy to Rome, where, as depicted in the mural, Cardinal Juan de Lugo dispensed it to the sick at Santo Spirito.

This narrative was less about discovery than appropriation: the Loja tree became the fever-tree, Peruvian bark became "Jesuits' bark," an American remedy became a European medicine. Several studies have been devoted to this reddish and bitter bark.[59] A major feature of its early history is the involvement of the Jesuits in its introduction to Europe, but this has often been asserted rather than analyzed. In this section, I would like to reconstruct the introduction of the remedy by relating it to three aspects mentioned earlier: the politics of medical interaction and knowledge in the missions, the politics of circulation and commodification, and the politics of testing.

FIGURE 7.1. Episodes in the history of cinchona, from the Santo Spirito Hospital in Rome. Public domain.

TOP. The count of Chinchón receives the febrifuge from his native servant. https://wellcomecollection.org/works/cuz3x83r

MIDDLE. The countess of Chinchón takes cinchona. https://wellcomecollection.org/works/hkg6atd4

BOTTOM. Cardinal de Lugo orders the use of cinchona in the Hospital of Santo Spirito. https://wellcomecollection.org/works/d442mcpe

If Sebastiano Bado's account remained the official story of the discovery of cinchona for some decades, it was no longer being taken for granted by the 1730s.[60] Indeed, it contained numerous inconsistencies. Some related to the botanical identity of the tree bearing the bark: the Quechua name *quina quina* in fact belonged to another tree from the genus *Myroxylon*. Perhaps because its bark was also used for curing fever, Bado included several characteristics of *Myroxylon* in his description of the cinchona tree. Other inconsistencies concerned the countess's recovery: her "miraculous" cure is not confirmed by any contemporary source.[61]

A possible answer to these difficulties appears in a 1663 text by the Spanish physician Gaspar Caldera de Heredia (1591–16??), which claimed that the bark was not introduced to Seville by the Jesuits, but by Juan de Vega, the physician to the Count of Chinchón, Viceroy of Peru until 1641, on his return to Spain , in 1642–43.[62] But it was the Jesuits who, according to Caldera de Heredia, learned the secret of the bark from the Indians:

> At the end of this land, in this province of Quito, close to the Amazon river, some Indians come to earn money and are brought to a gold mine. . . . On the way, the Indians are forced to cross a river . . . so that the majority of them, on reaching the other bank, frozen and shaking, complain pitifully. And immediately, for their relief, they take the bark of a tree they know, powdered, ground and dissolved in hot water. . . . Seeing this, the fathers of the Society of Jesus . . . asked them from which tree they took the bark.[63]

Contrary to Bado's claims, the bark is here "discovered" not so much as a febrifuge used by the Indians, but as a remedy that reduced shivering. Likewise, while in Bado's account, as related by this Italian physician, the Indians initially refused to give their ancestral remedy to the Spaniards, in the Spanish account, the Indians offered it willingly. Such divergences reveal how the *leyenda negra* was inscribed in the history of drugs, forcing us to handle such accounts with care.[64] At least in part, however, there is reason to follow Caldera de Heredia's version, namely the link between the knowledge of the remedy and the exploitation of gold.

Although the place of origin of the Indians mentioned by Caldera de Heredia is unknown, the Loja region, especially the Maynas, was an important gold-mining area.[65] Documents held in Quito, Lima, and

Santiago de Chile show that some Jesuits involved in transmitting the bark between Ecuador, Peru, Spain, and Rome were also overseeing gold extraction.[66] The best evidence for this golden discovery of the bark is the project by Francesco Viva at the beginning of this chapter. In his 1689 letter, Viva emphasized that "close to our Missions are the Jívaros, a nation rebelling for 90 years, in those mountains where there is so much gold." Ultimately, as we have seen, Viva failed to reach Europe, and his mission was fulfilled by another Jesuit sent as provincial procurator. But he continued to sell the bark, using the proceeds to buy gunpowder.[67] Jesuits' bark, in other words, was a weapon in the military and spiritual war against "Jívaros," and to control gold.

COMMODIFYING THE BARK

What Viva's project shows is that Jesuits' bark was a large-scale commodity across the Atlantic by 1690. The date is particularly relevant, for it echoed the remedy's triumph in European courts during the 1680s, particularly in France, where it was taken up by King Louis XIV in 1686.[68] The following year, the monarch's ministers were placing large orders in Iberian ports.[69] In order to supply this increased demand, the elites of Loja overexploited the trees, and probably falsified the genuine bark by mixing it with others. In his 1689 letter, Viva claimed he was alone on the market, because too much bark, and of bad quality, had been sent the previous year.[70] In the last quarter of the seventeenth century, Loja's elite became convinced of the profitability of the plant, and massively increased its exploitation: it was in association with some of the members of this elite that Viva planned to harvest the fifty muleloads of cascarilla.[71]

How did this commodification take place, and what role did Jesuits play in it? Caldera de Heredia identified one Gabriel, or Graviel, de España, a master apothecary of Lima, as the first to sell the bark there.[72] Over ensuing years, Jesuits like Agostino Salumbrino and Claude Chicaut then spread the remedy within Peru.[73] But Caldera de Heredia's second route of transmission was to Seville, where Juan de Vega, physician to the Viceroy of Peru, introduced the remedy. After the Count of Chinchón completed his mandate as Viceroy in 1641 and returned to Spain, he caught a "tertian fever" in Seville, and the best doctors in town were called. Vega did not suggest using the bark, perhaps—according to Caldera de Heredia—because he saw no effects from it in Spain. Later, a woman from the count's family suffered a fever, and on this occasion,

Juan de Vega used the bark. It seemed effective at first, but on the fourth day the fever returned, leading Seville's medical milieu to be hesitant about the remedy's usefulness. However, such precautions were not followed by other people, especially "religious and vulgar physicians," who administered the remedy incorrectly, according to Caldera de Heredia.[74]

This testimony may allow us to reconstruct events in Seville around 1642 to some extent. First, the trial on the "woman of the Count's family"—who may have been the "Countess" invoked by Bado—shows how important initial trials, whether successful or unsuccessful, were in shaping the remedy's subsequent reputation. The defect attributed to the bark—its failure to prevent fever from recurring—would be a lasting feature of later experiences. Secondly, it is clear that part of the Sevillian medical community regarded Juan de Vega as the introducer of the remedy, something confirmed in account books of the Hospital de Cinco Llagas from 1643 onward, and the Hospital de Amor de Dios.[75] It seems therefore that the new remedy began to be marketed in Seville around 1642–43, in connection with the return to Spain of the Count of Chinchón and Juan de Vega. From this point of view, the Jesuits' main contribution lay in spreading the remedy beyond the port. Both the travels of procurators in Spain and Italy, and the transmission of boxes, were crucial to this process. Father Bartolomé Tafur (1589–1665) may have brought back the remedy as early as 1643.[76] Brother Pedro Salina, Alonso de Ovalle, and others offered the bark to people both inside and outside the Jesuit order, alongside chocolate, vanilla, bezoar, and other American curiosities.[77] This system was still in use by the 1690s, as Viva's letter reveals.[78] The Jesuits also participated in the increasing commodification of the bark.[79] A series of letters exchanged between Father Rafael Pereyra in Seville and Father Sebastián González in Madrid illustrates this wonderfully. In 1648, González discovered the effect of the "barks for quartan fevers" that Pereyra had sent him.[80] A few weeks later, he asked Pereyra to send him more, since it had produced "such a good effect." This request offers an insight into how demand created a commodity. González probably began by transmitting the bark as a gift or sample. But once many people wanted more, he had to find a way to secure his supply. Later in the letter, he asked Pereyra for more information about the marketing of the remedy: "Let me know whether it is sold in apothecaries' shops, and what it costs, so that when others ask for it we can direct them where to look for it, and also its name, so that they know what to ask for."[81] Here we see the very beginning of the commodification process. A product was used in a place even before it had a price or a name. At this stage, González did not intend to sell it himself: he was just trying to secure a supply for his Madrid connections. In a later letter, he thanked Pereyra for sending him more of the bark and urged him again to say how much it cost, which pharmacies sold it, and under what name.[82] In 1648, therefore, Madrid apothecaries did not know the bark, and it was via a Jesuit, well supplied by one coreligionist, that other people, probably members of social elites, came to know of it. For Pereyra in Seville, sending such samples was simultaneously a way to offer a gift to a friend, and to test the remedy's commercial potential. It is unclear from the remaining correspondence whether Gonzalez ended up selling the remedy, or how much it cost. But one thing seems certain: if Juan de Vega and local merchants supplied the Sevillian medical marketplace with the febrifuge, it was via the Jesuits that the Peruvian drug moved out of the port and began to be conceived of as merchandise. The Jesuits not only spread a material, they also spread a name, a price, and a recipe they designed, thanks to their science of testing samples.

JESUIT CHARITY: COMMODIFICATION AND EXPERIMENTATION

Commodification was thus a qualitative and polycentric process, in which the bark was accredited in several places and by several actors. Evidence of this may be seen in information as basic as dosage. While Juan de Vega used one dram of powdered bark in his febrifuge in Seville, in Rome, Juan de Lugo and the apothecary Pietro Pucciarini used two drams. Both were trialing the bark in hospitals.[83] When Lugo was depicted generously giving the bark to the sick at Santo Spirito, he was also using the bodies of the pilgrims as subjects of experiment.[84] The bark was not the only new remedy tested in hospitals: guaiacum wood, in the sixteenth century, and ipecacuanha, at the end of the seventeenth, followed a similar pattern. The involvement of charitable institutions, especially religious hospitals, in such experiments is an old tradition.[85] The distinctiveness of the Jesuits' approach, especially regarding the bark, lay in experimenting at almost every stage of the process, from the Loja region to Lugo's trials in Rome. The tests conducted by the Jesuits had several goals: to confirm their method, improve it if possible, and, in response to good results, install the remedy among the habits of a hospital or a town. In order to help with the trials, Pucciarini circulated a method that, according to him, produced the same good effects everywhere:

Take two drams of [the bark] and run them through a sieve; and three hours before the fever comes, put it in an infusion in a glass of very strong white wine. And when the chill begins, or when you feel the slightest symptoms, take the whole of the prepared infusion and put the patient to bed.

Be warned that it will be possible to give the said bark in the aforementioned manner in the tertian fever, when it has stopped for several days.

The continued experience has freed almost all those who have taken it, having first purged the body well, and not taking any kind of medicine for four days afterwards, but be warned not to give it except with the permission of the Physicians, so that they may judge whether it is the time to take it.[86]

In later sources, this text became known as the *Schedula Romana*. It was more than a recipe: it also supplied indications of how to take the remedy that derived directly from the Roman trials. Inevitably, with the circulation of Pucciarini's text and the multiplication of tests, changes appeared. For instance, a Spanish version of the recipe preserved in Madrid contains slight additions concerning the diet the patient should follow, as well as other important information like the following: "[Of] this wood or bark of the Tree of the Indies that was brought from Rome, Brother Marin says that it is established that in the Indies where it grows, it calms quartan [fevers] among all native peoples, and from there they distribute it to many regions for that purpose. And, having been brought to Rome by some among us, it removed many people's fevers, and they brought it to Spain and particularly to Valladolid."[87] Thus the Jesuits of Valladolid only came to know the remedy after it became known in Rome. Commodification and testing, in other words, were nonlinear processes. Both inside and outside the Jesuit network, the trajectory of the bark did not necessarily follow a single path from Loja to Lima, Lima to Panama, Panama to Seville, Seville to Madrid, Madrid to Rome.

Outside Spain, the Jesuit method of administering the bark was transcribed in several works, including a manuscript book of recipes written in Toulouse, and books published in Genoa, Delft, and Copenhagen.[88] Some of these borrowings, however, were not used to confirm the virtues of the bark, but rather to contest them. During the 1650s, the Jesuit bark raised an intense controversy within the European medical milieu. The history of this quarrel is well known, and the subject of several studies.[89] It began in Brussels in 1652, when the physician Jean-Jacques Chifflet (1588–1660) trialed the bark on his patron, Archduke Leopold Wilhelm of Austria (1614–1662), the governor of the Spanish Low Countries. Although he followed the Jesuit method, the physician was unable to prevent the archduke from relapsing. He was commissioned to write a treatise against the drug's use, which appeared in 1653.

In subsequent years, several authors either defended or attacked the remedy. Among them were Jesuits like Honoré de Fabri (writing under the pseudonym Antimus Conygius), and physicians close to the Society, like Sebastiano Bado. Early on, this controversy essentially divided Italians, who tended to praise the remedy, from Flemish, who tended to discredit it. Their controversy was largely framed in the language of humoral theory. The bark's detractors argued that the drug's bitterness proved that it was very hot, and therefore could not cure a hot disease like tertian fever. Moreover, the bark could not operate against the fever, since it did not expel any humors. These two reasons explained why the bark could not prevent the return of fever, although it did bring some relief.

Alongside these theoretical considerations, the bark was also attacked by some physicians on account of its Jesuit origins. This was especially the case for Guy Patin (1601–1672), dean of the Paris Faculty of medicine, who distributed Chifflet's book to his correspondents as early as 1653.[90] While Patin's hostility to the Jesuits may relate to his sympathy for Protestantism or, more likely, Jansenism, his refusal of the bark was also linked to his aversion for the apothecaries and the "modernists" who praised any kind of novelty.[91] Indirectly, this critique helps to explain what the Jesuits did with the bark. Designing a safe method of administration was a way to generate trust in an unknown substance: the recipe had to procure them receipts. How much the bark actually enriched the Society is hard to determine without further analysis of its pharmacies' account books. But this case study at least shows that the Jesuits were more open to novelties than many other actors in the medical marketplace, an openness linked to the system they developed for the production of texts about the lands where they proselytized, and the circulation of curiosities. At the forefront of one of the main fields of innovation in early modern pharmacy—the arrival of non-European drugs—the Jesuits were able to impose themselves as merchants and designers of recipes to such an extent that the new drug came to be named after them. If the Jesuit origin of the remedy also led some to mistrust it, this suspicion had other grounds, principally fraud and the adulteration of the remedy during its transatlantic journey. Thanks to their network of exchange, the Jesuits could partly avoid such inconveniences. So the discrediting of the bark in the 1650s and 1660s was also linked to the fact that, during this period, other actors besides the Jesuits were selling it too. But during the 1670s, when the bark began to be accredited once more as the best febrifuge, another recipe, quite different from the Jesuit one, was designed, thanks to trials conducted by a young apothecary named Robert Talbor in Essex. This is another chapter of the European appropriation of Peruvian bark, a chapter that, in the 1670s and 1680s, involved not only the courts but also the state apparatus.

THE EXTENT OF JESUIT INTERMEDIATIONS

The history of Peruvian bark reflects the chronology of Jesuit intermediation in the drug trade. From the last quarter of the seventeenth century onward, the transmission of American remedies increasingly eluded the Jesuits. If they continued to be the informers and assistants of choice for naturalists sent on expeditions during the eighteenth century, the networks through which texts, samples, and curiosities circulated were now in the hands of merchants and officers, in particular. These last did more or less the same thing as the Jesuits: they learned about local remedies, wrote reports, passed them on to associates, and sometimes tried to sell them in European markets.[92] In a way, the logic is the same: it lies in the articulation between the science of description, the science of circulation and the science of experimentation. The eighteenth century was first and foremost characterized by a substitution of actors, and it would be interesting to see to what extent the Jesuits' organization formed the model for the kind of imperialism pursued by states during the eighteenth century in matters of exotic drugs, and also in other aspects of colonialism and trade.[93] Conversely, this development also means that the specificity of the exotic drug trade in the seventeenth century resides partly in the missionary intermediations of which the Jesuits were such an important example.

8

MASTERS OF THE EXOTIC?

The Stocklists of Parisian Grocers and Apothecaries, 1650–1730

E. C. Spary

Analysis by E. C. Spary, drawing upon research contributions from Samir Boumediene, Laia Portet i Codina, and Justin Rivest

For several decades, probate inventories have commanded historical attention as entry points into the history of early modern consumption.[1] Jon Stobart has published extensively on the grocery trade in England, while Anne McCants's invaluable study of 318 probate inventories provides a Dutch case study of the uptake of coffee, tea, and chocolate consumption in Amsterdam households.[2] The analysis of probate inventories has featured prominently in French historiography, but foods and drugs have largely been excluded.[3] In 1700s France, there was a legal requirement for the assets of deceased individuals to be inventoried in full and valued by an expert. This was because, across much of France, the default legal position on inheritance was that assets be divided equally between all heirs.[4] In this chapter, we present an analysis of 28 probate inventories of Parisian grocers and apothecaries, one of the main groups supplying medicinal simples in the city, where grocers and apothecaries were legally conjoint. Numbering among the "Six Corps," the six main city corporations, they had the capacity to sway domestic policymakers. High demand and high prices for their drugs meant that some of their number became very affluent.[5]

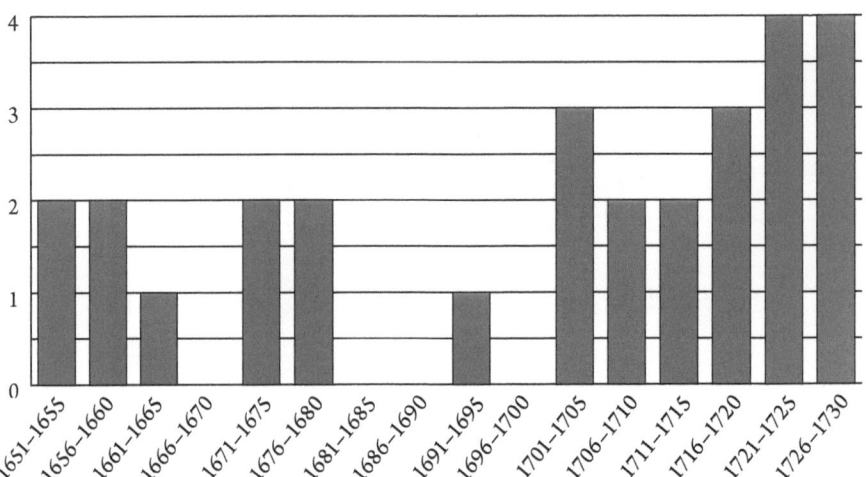

FIGURE 8.1. Dates of stocklists of Parisian grocers and apothecaries used in the study. Based on stocklists at the Minutier Central, Archives nationales de France, Paris. Credit: E. C. Spary.

We were able to identify 53 probate inventories produced between 1653 and 1730 for grocers and apothecaries.[6] Of these, just 28 contained detailed stocklists. As far as possible, we sought to capture a snapshot of "normal trade" for a specified period, so the inventories selected for inclusion were limited to those drawn up between 1650 and 1730. It is this sample—representing approximately 3 percent of the membership of these two communities—which forms the basis of my analysis. These years mark significant transformations in European consumption of exotic plant materials.[7] Coffee, tea, and chocolate became quotidian adjuncts to polite self-presentation.[8] Numerous medicinal drugs, including cinchona, ipecacuanha, and ginseng, were added to the French medicinal arsenal, while other arts using drugs, such as dyeing, were rapidly expanding in response to new techniques and textiles arriving in Europe.[9] Perhaps most importantly, the decades around 1700 marked the emergence and consolidation of the first French empire, raising questions about how wider patterns of globalization might have affected Parisians' relationship with the exotic.[10] Our survey includes no inventory dating from the period between 1680 and 1700, even though these two decades were critical in terms of growing crown support for drug entrepreneurs and urban uptake of several new drugs. Uncovering further inventories may alter the conclusions in what follows, therefore.

THE MERCHANTS

The individuals for whom we possess detailed stocklists number 22 grocers and 6 apothecaries, generally described as "marchand épicier" or "marchand apothicaire-épicier." This imbalance reflects the numerical superiority of the grocers' community over their apothecary counterparts. Not all in our sample belonged to Paris's guild of merchant apothecaries and grocers: César Lybault (IAD 1655) was a personal apothecary to the Prince de Condé, thus outside the guild system, while Vivant Carré (Inv 1729), whose stock was (exceptionally) inventoried at his marriage rather than his death, was based in Versailles, not Paris.

The working relationship between apothecaries and grocers, for all their legal union, was occasionally fraught by extended legal tussles, as one side or another sought to lay exclusive claim to particular practices or materials. In 1698, a law was passed preventing grocers from putting *albarelli*, the traditional pottery jars used to store drugs, on display in their shops. In 1701, a short-lived push was made to draw a legal line between medicinal and non-medicinal, when the apothecaries were ordered to draw up a list of the "compositions and preparations which are in use in ancient and modern medicine and pharmacy, and the grocers of those that do not serve in medicine at all," after which the medical faculty was to adjudicate over both lists.[11] The resulting "memoir or catalogue of the chymical remedies which the apothecaries of Paris ought to prepare, sell and retail to the exclusion of the merchant grocers" embraced a range of remedies, such as waters, oils, essences, salts, tinctures, and electuaries, that involved the processing of exotic simples. But the apothecaries also laid claim to jalap, benzoin, guaiacum, and other exotic simples, whose inclusion was successfully contested by the grocers, on the grounds that they were unprocessed plant materials. Grocers also continued to lobby, with some success, for the right to sell Galenic remedies like catholicon and theriac. In short, commercial boundaries over drugs within the medical marketplace remained porous throughout our period.[12] While grocers gradually ceded ground to the apothecaries as medical merchants, throughout this period they still remained key intermediaries, brokering access to exotic plant materials by Parisian consumers—in fact, they predominated over apothecaries in this role. Nicolas Delamare, the metropolis's chief of police, explained that "The grocers' trade is no less delicate and important to health than the apothecaries': if the latter compose remedies, it is the former who supply the greater part of the drugs and ingredients that enter these compounds. It is they who obtain [these materials] from the furthest lands, and sell them: there are few apothecaries who conduct, or even could conduct, such distant trade and travels."[13] If grocers were long-distance vendors of materia medica, indeed the only one of this dyad of merchant communities with the capital to purchase supplies of these costly substances, it seems curious that they have largely been omitted from historical studies of the early modern French medical marketplace. The grocers and apothecaries are one of the few Parisian guilds with a surviving archive that documents their relationships with authorities, rival healers, and one another.[14] Debates between the two communities touched on important questions about what purpose each trade served, and by extension also what purpose their merchandise served.

Until at least the end of the eighteenth century, the boundary between food and drugs remained fluid, especially given that the center of medical treatment remained the home.[15] This is one reason why apothecaries were rarely successful in attempts to challenge the grocers' stake in the medical marketplace. They retreated to safer ground instead, defending their exclusive right to display *albarelli* in their shops, and specializing in the preparation and sale of chymical remedies. Both grocers and apothecaries were invested in the medical marketplace, but divided over the ownership of particular processing techniques. In a series of *grotesqueries* printed by the L'Armessin family in the later seventeenth century, which represent tradespeople through their utensils and wares, the grocer's upper torso, shown in figure 8.2, is a *droguier*, or drugs cabinet. An alembic hat is the main difference between the apothecary and his grocer cousin, suggesting that chymistry defines the former trade. From the drugs listed on their bodies, it seems that L'Armessin associated the grocer more with comestibles—spices, sugar, and so on—and the apothecary more with medicinal drugs, or perhaps even more, with especially *potent* substances, such as the viper, held to be toxic in all its parts. But in practice, probate inventories show, someone called a "grocer" was just as likely to stock the simples accorded by L'Armessin to the apothecary figure, such as scammony, aloe, or myrrh, as someone called an "apothecary." It is rather the technically complex drugs that are commonly absent from the grocer's shelves—syrups, oils, elixirs, waters, unguents, and so on. However, in terms of what was stocked,

FIGURE 8.2. Nicolas de L'Armessin, "Habit d'Espicier" (Paris, ca. 1700). Collection Michel Hennin. Estampes relatives à l'Histoire de France. Tome 74, Pièce 6595. Credit: Bibliothèque Nationale de France.

FIGURE 8.3. Nicolas de L'Armessin, "Habit d'Apoticaire" (Paris, ca. 1700). Collection Michel Hennin. Estampes relatives à l'Histoire de France. Tome 74, Pièce 6569. Credit: Bibliothèque Nationale de France.

I argue below, even this distinction on the basis of technical expertise was more a matter of degree than a hard boundary. Cases like that of Joseph Seconds, who was described as "grocer-apothecary" in his 1724 IAD, were not uncommon.[16] It was probably not until the formal separation in 1777 that these two groups were fully disentangled, both in law and in commercial practice.

SUPPLIERS TO THE CITY

Several lists of guild members cover the period from 1692 to 1721.[17] Using these, we were able to identify a total of 886 Parisian grocers and apothecaries, representing some 618 families (inferred from the possession of identical surnames). Most entries listed an address, allowing us to map the density of grocer and apothecary shops in different streets. Contemporary estimates put the total population of Paris at around 425,000 at this time, meaning that there was approximately one apothecary-grocer's shop for every 500 city residents.[18]

Shops clustered in particular along main routes and at entry points such as the city gates. These locations had the double advantage of ready access to goods supplies and a clientele of travelers. The principal concentration of shops was however in the city itself, notably along two axes: the rue Saint-Denis, running north-south, and the rue Saint-Honoré, running east-west. By the 1710s, the rue Saint-Honoré was emerging as a residential street of choice for the nobility within Paris, and thus the heartland of an affluent city clientele.[19] A cluster of shops flourished around the Abbaye de Saint-Germain, another area frequented by elites, both French and foreign.[20] This, like the rue Saint-Denis, the Left Bank, and Montmartre, was a focal area for many arts and trades. The rue Saint-Denis connected the medieval market of Les Halles at the very heart of the city to the wider world. The concentration of grocers' and apothecaries' shops was statistically below average, however, on the side streets of larger roads. Evidently, grocers and apothecaries preferred to set up shop in high-traffic and affluent spaces within the city. Outside the city walls, an area of particular concentration was the Faubourg Saint-Antoine, popular with travelers coming into the city, and a center of hospitals and health spas in the later seventeenth century.[21]

SELLERS AND USERS

If we understand Parisian grocers as key suppliers of exotic drugs to the city's population, what drugs did they supply? We identified a

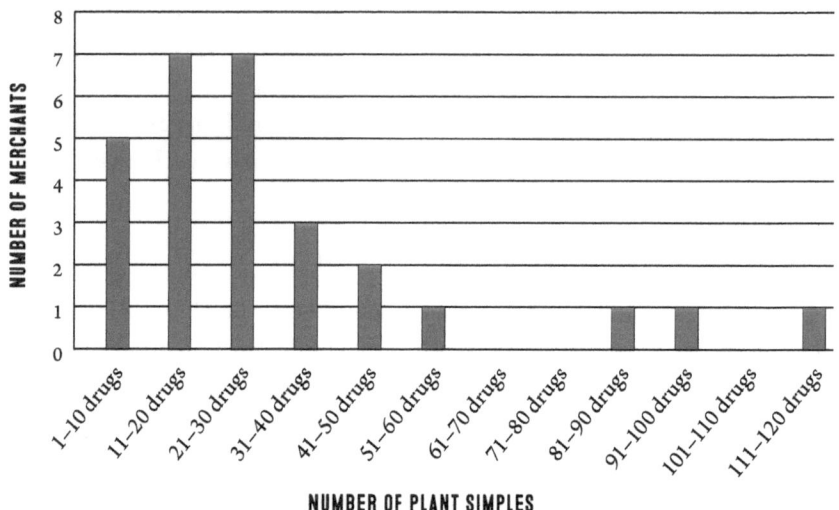

FIGURE 8.4. Graph showing numbers of plant simples stocked by grocers and apothecaries in Paris and Versailles, 1650–1730. Based on stocklists at the Minutier Central, Archives nationales de France, Paris. Credit: E. C. Spary.

total of 182 different plant simples in the stocklists of the 28 grocers and apothecaries studied. This was a broad range, if still some way off the ca. 240 different plant simples listed in the Parisian grocer and apothecary Pierre Pomet's famous book *Histoire generale des drogues* (1694). But it was far from being the case that each merchant stocked all, or even most, of these many different drugs. In fact, there was wide variation between shops: the smallest number of different plant simples stocked was just 2, and the largest was 113, with a mean of just under 31.[22] As figure 8.4 shows, the majority of merchants clustered in the 1–50 range; higher figures are outliers.

The stocklists can be compared with other kinds of sources that include a broad range of plant simples. One baseline is the plants mentioned in printed or manuscript recipe collections, which potentially shed light on what drugs medical clients actually sought out. Compiled over the same time span, the 550-odd recipes of the Oratorian house in the rue Saint-Honoré, preserved as Bibliothèque nationale de France (henceforth BnF), Ms Français 24251, mentioned 254 different plant simples as ingredients, comparable with the total referenced in Pomet's book. However, a comparison of the origins of drugs featuring in the probate inventories with those appearing in the Oratorian recipe collection

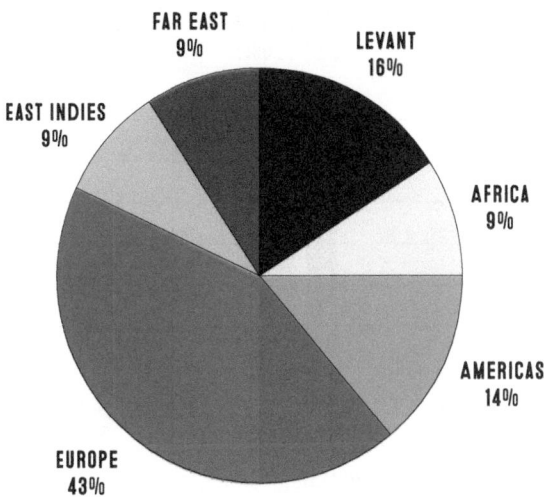

FIGURE 8.5. Chart showing the place of origin of plant simples in the stock of grocers and apothecaries in Paris and Versailles, 1650–1730. Based on information from the Minutier Central, Archives nationales de France, Paris. Credit: E. C. Spary.

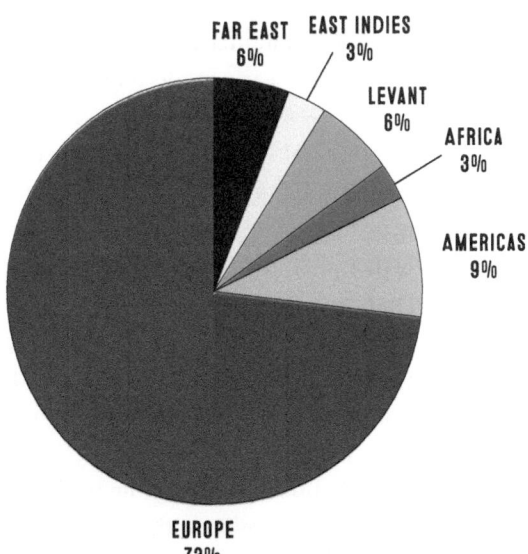

FIGURE 8.6. Chart showing the origin of unprocessed plant simples mentioned in the recipe collection of the Oratoire du Louvre. Derived from Ms. Fr. 24251, Bibliothèque Richelieu, Bibliothèque nationale de France, Paris. Credit: E. C. Spary.

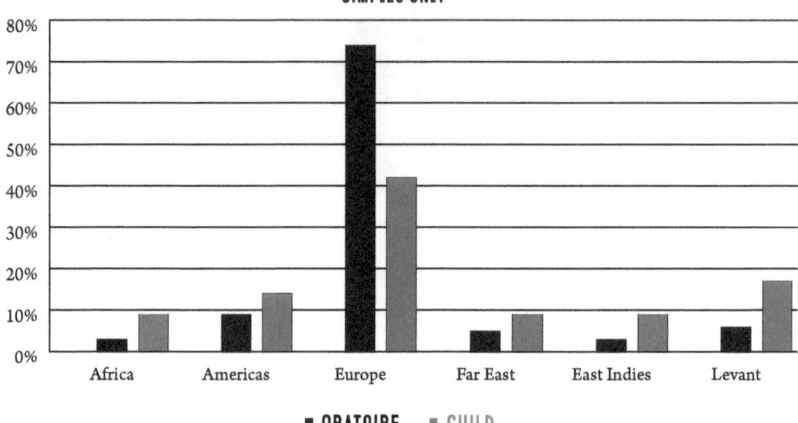

FIGURE 8.7. Comparison graph of the origin of plant simples in the stocklists of Paris and Versailles grocers and apothecaries, with that in the recipe collection of the Oratoire du Louvre. Based on information from the Minutier Central, Archives nationales de France, Paris, and Ms. Fr. 24251, Bibliothèque Richelieu, Bibliothèque nationale de France, Paris. Credit: E. C. Spary.

suggests some salient differences.[23] In the following analysis, geographical spaces are not defined in modern terms, but according to the trade routes and regions familiar to early modern Europeans, as follows:

- Africa: all of continental Africa including most of northern Africa and Ethiopia (but not Egypt, which was widely classified as Levantine, including by Pomet)
- Americas: North and South America, though most came from the Caribbean islands and Iberian empires
- East Indies: Indian subcontinent and areas south of the Himalayas
- Europe: all of continental Europe as far east as Athens
- Far East: Siam, Batavia, China, the Philippines, Moluccas, Cochinchina, Korea, Japan
- Levant: Eastern Mediterranean, Ottoman Empire, Persia, Syria, Arabia Felix, Cairo, Alexandria—the Old World trade route network since classical times

As figures 8.5, 8.6 and 8.7 reveal, European plants played a significantly larger role in the priests' stock of medicinal knowledge than in

TABLE 8.1. TWENTY MOST COMMON PLANT SIMPLES IN THE STOCKLISTS OF PARISIAN MERCHANT GROCERS AND APOTHECARIES, 1653-1730, AND TWENTY MOST FREQUENTLY-REFERENCED PLANT SIMPLES IN THE RECIPES OF THE ORATORIAN HOUSE ON THE RUE DU LOUVRE, PARIS, CA. 1650S-1720S

Merchant probate inventories		Oratoire recipe collection	
Simple	Percent of stocklists	Simple	Percent of recipes
sugar	85%	sugar	15.93%
licorice	74%	rose	7.26%
pepper (black)	67%	cinnamon	6.73%
rhubarb	67%	senna	5.84%
manna	63%	bugloss	5.13%
cloves	63%	borage	5.49%
cinnamon	59%	rhubarb	4.07%
nutmeg	59%	chicory	4.78%
pepper (white)	59%	cloves	3.89%
senna	59%	lemon	4.25%
china root	52%	orange	3.36%
guaiacum	52%	agrimony	3.54%
gum (Arabic)	52%	blessed thistle	3.36%
almonds	48%	licorice	3.36%
ginger	48%	marshmallow	3.19%
sarsaparilla	48%	anise	2.48%
anise	44%	sage	3.19%

Source: Data from this table was gathered from the inventories listed in Table 8.2.

the material stock of grocers and apothecaries. This reflects grocers' self-presentation before administrators as merchants "often constrained . . . to carry out long journeys in foreign countries in order to obtain and sell the merchandise of their art, on which they risk themselves and their wealth."[24] Grocers thus explicitly laid claim to the foreign where drugs were concerned.

This is borne out by a comparison between the most common plant simples appearing in each source (see table 8.1). Some of the most frequently appearing plant simples in the Oratorians' recipes, such as

CATALOGUE
DES MARCHANDS
ESPICIERS,
ET DES MARCHANDS
APOTHICAIRES-ESPICIERS
DE CETTE VILLE · FAUXBOURGS
& Banlieuë de Paris.

Fait le 16. *d'Octobre* 1717.

A PARIS,
De l'Imprimerie de JEAN-BAPTISTE COIGNARD, Imprimeur ordinaire du Roy, ruë saint Jacques, à la Bible d'or.

MDCCXVII.

FIGURE 8.8. Front cover of *Catalogue des Marchands Espiciers, et des Marchands Apothicaires-Espiciers de cette Ville, Fauxbourgs & Banlieuë de Paris* (Paris: Jean-Baptiste Coignard, 1717). Credit: Bibliothèque nationale de France, classmark 4-FM-24943.

borage, bugloss, chicory, agrimony, or marshmallow, are absent from the list of drugs most commonly sold by grocers and apothecaries. Indeed, one or two, such as agrimony, do not appear in the merchants' stock at all, while others, like borage, appear only once over the entire period of nearly eighty years. Parisians must have had other sources of indigenous herbs. One hypothesis might be that French drug plants could be grown in city gardens, so that clients did not need to shop for these simples. But this assumption is flawed: not every European herb used in medicine could be cultivated in Paris, nor could city gardens produce sufficient quantities to supply nearly half a million residents. Pomet often indicated distant French towns or regions, like Languedoc, Touraine, or the Alps, as his source of European herbs, blurring the distinction between "exotic" and "indigenous"; and yet herbs indigenous to Europe played a much smaller role in the stocklists. Given that grocers possessed networks and resources that should have afforded them access to a wide range of European plants, why then did they stock so few indigenous herbs, if these were desired by consumers?

One reason may have been the intense rivalry between urban merchant communities. The small but significant community of herbalists had long specialized in the procurement and cultivation of living plants. Under Henri IV, the Robin family introduced several North American plants to French gardens.[25] In a discussion of "French plants" such as betony, chamomile, artemisia, lemon balm, and hyssop, Pomet observed that "in Paris we do not sell these sorts of Plants, on account of the Herbalists." Within the city, he explained, apothecaries could not readily walk into a field and pick the simples they required, but rather depended on herbalists for their supply, "who often give them one in place of the other."[26] Herbalists, too, thus acted as medical gatekeepers over which plants could play which therapeutic roles. Few grocer-apothecaries were in the fortunate situation of Pierre Boulduc, keeper of the apothecaries' physic garden, who not only had knowledge of many healing plants, but also access to a space where these could be grown. Different merchant groups thus "owned" specific plant techniques, from cultivation, procurement, processing, preservation, and compounding to distillation.

By the later seventeenth century, the most prominent Parisian herbalist was Mathurin Sibille. When Paris's royal academy of sciences embarked upon a vast project of chymical analysis of plants, it was Sibille who supplied much of their experimental material.[27] The academicians'

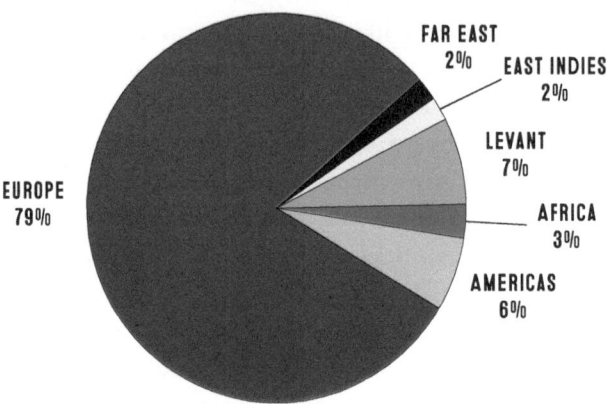

FIGURE 8.9. Chart showing the place of origin of plant simples analyzed in the laboratory notebooks of the Académie Royale des Sciences, Paris, 1669–1698. Derived from Nouvelles Acquisitions Françaises 5147, Bibliothèque nationale de France, Paris. Credit: E. C. Spary.

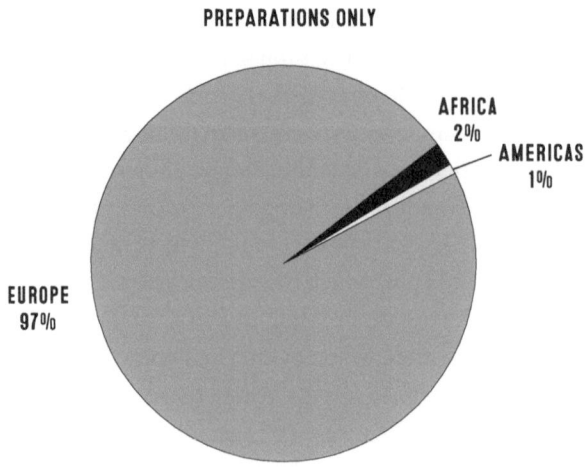

FIGURE 8.10. Chart showing the origin of plants named in compound remedies in the stock of grocers and apothecaries in Paris and Versailles, 1650–1730. Based on information from the Minutier Central, Archives nationales de France, Paris. Credit: E. C. Spary.

laboratory notebooks show they often paid handsomely for European as well as non-European plants. Between 1667 and 1698, 240 different species were analyzed, predominantly European herbs supplied by Sibille and numerous unnamed women, or plants grown at the royal physic garden.[28] Grocers' and apothecaries' enhanced access to exotic plants is again thrown into relief by comparing the geographical origin of plants they stocked with that of the specimens analyzed by the academicians, especially if we consider that the latter had royal resources at their disposal.

Chymical experimenters thus procured plant materials from multiple sources. Some view into the otherwise invisible agency of herb women in the commercial heart of the city, Les Halles, comes from a 1680 police interview with one Jeanne Colignon, a witness in the famous Poisons Affair, who described her interactions with "a tall, dark man with pockmarks, whom she heard called La Chaboissière," and who distilled by day and by night in a house in the faubourg Saint-Antoine. Among his associates was a man "who called himself a doctor, and who was named Monsieur Delorme."[29] "[La Chaboissière] sent the . . . gardener to search for herbs to distil, and even sent her, Colignon, once to search in the bois de Vincennes, and around the Minims and towards Ménilmontant. . . . And besides this she saw many other kinds of herbs that she does not know, being brought to La Chaboissière during the period in question, and [said] that it was women who brought them from the *halle*, all of which herbs the said La Chaboissière distilled."[30] Any suggestion in Colignon's report of exciting herbal shenanigans going on beyond the purview of orthodox medical practice can quickly be dispelled. "La Chaboissière" was almost certainly Jean Poisson de la Chabossière, apothecary and valet to the king and queen's bedchamber, and therefore actively expected to supply drugs to the royal household; while Charles Delorme was the acme of respectability, a household physician to Henri IV and his successors. Dying in 1678, Delorme was highly regarded in scholarly circles, and extremely well-connected within the medical community.[31]

The plural terrain of expertise and commercial rivalries structuring the Paris medical marketplace may explain why indigenous herbs are so rare on the shelves of grocers and apothecaries. But chymical remedies allowed apothecaries to trade in indigenous plant materials they could not offer in unprocessed form, whether because these were the commercial territory of herbalists or because they were hard to preserve in the natural state. By "processed," I do not mean mechanical techniques such as harvesting, slicing, drying, and/or extraction, which (sometimes very far from Paris) converted plant commodities into the forms in which they reached European shops. Rather, I mean the further conversion and diversification of simples into a wide range of formats, vehicles, and commodities, such as tablets, pills, syrups, waters, elixirs, and so on. This necessitated recourse to technically demanding processes such as refining, distillation, purification, or extraction, practiced in city workshops. Capturing indigenous plants' virtues in such formats enabled apothecaries to sell them to customers or add them to remedies when making up a prescription, even in cases where they had no stock of a plant, or could not licitly supply it. If we include the plants *only* available on shop shelves in chymically processed form, the overall picture changes: of a total of 84 additional plant substances mentioned in chymical remedies, but not listed as simples in their own right, the overwhelming majority are indigenous plants, as figure 8.10 shows.[32]

In his overview of the Parisian medical marketplace in 1668, the royal physician Charles de Saint-Germain linked plant simples with the grocers and compound medicaments with the apothecaries.[33] As chymical medicine became fashionable in Paris from the 1670s onward, apothecaries extended this prerogative to seek hegemony over "chymical remedies"—as they were collectively termed in stocklists and books—drugs whose manufacture demanded complex technical skills, especially distillation.[34] In the shop of a well-supplied apothecary like François Rousseau (IAD 1705), processed plants predominated. For any given category of remedies, he offered numerous variants. While many grocers owned a syrup or three, Rousseau stocked no less than twenty, including apple, quince, wormwood, barberry, comfrey, capillary fern, coltsfoot, fumitory, and rose. Most grocers stocked fewer such products, preferring wares like dyestuffs, cheese, butter, tobacco, sugar, soda, starch, soap, candles, matches, and other household goods. Those plant materials they did sell in more "processed" forms, like sugar, licorice, and a variety of oils, were probably bought in that state.

Yet a distinction between grocers as long-distance experts and apothecaries as chymical experts proves less clear-cut than commentators sometimes made out and apothecaries asseverated. The grocer Étienne Regnault (IAD 1658), for example, possessed a wide range of syrups, made from European plants such as capillary fern, jujubes, limes, waterlily, chicory, apples, barberry, violets, and more. Another grocer, Charles Vignon (IAD 1702), sold chymical remedies such as polychrest salt and sweet mercury, while Joseph Seconds (IAD 1724), sold numerous processed drugs, including distilled waters of myrtle, orange flower, cinnamon, melissa, spoonwort, and "Divine water." The picture was the same for other categories of chymical remedies, such as oils, electuaries, confections, unguents, or waters.[35] In addition, numerous shops in Paris specialized in the manufacture of distilled and compound chymical remedies, which were sold both to private households and to merchants. The chymist Sébastien Matte La Faveur, for example, monopolized the supply of capillary syrup in Montpellier and Paris, made using ferns of that name from Europe and North America.[36] The possibility that Parisian grocers and apothecaries were buying in syrups, elixirs, and waters readymade from specialist workshops like this cannot be ruled out. However, the apothecaries' legal protectiveness over chymical processes strongly suggests that they expected to make an important part of their profits from chymical products, and the wide variety of such goods in stocklists suggests that the apothecary's home laboratory was the likely site of manufacture. In 1688, Pomet, though a grocer-druggist, was strong-armed into the apothecaries' guild, probably because his book itemized so many chymical remedies for sale in his shop.[37]

USING PRICE PER OUNCE TO ELICIT PRICING PRACTICES

Lawyers drawing up a grocer's or apothecary's probate inventory frequently called in guild experts to value the commercial stock. Accordingly, the process of valuing the drugs gives some sense of how far drug prices were subject to consensus within the city's community of specialist sellers, and potentially how they transformed over time. Merchants often sold different grades of a drug at different prices, attesting to stratification of the market in these substances. There might be multiple entries for a given drug in a stocklist at different prices. However, because valuers hardly ever discussed reasons for these variations, it is difficult to draw firm conclusions about the relationship between price and quality, as opposed to various other factors—such as wars, treaties, or changes in transport conditions—that might affect the global market in, or quality of, individual drugs.[38]

Although our modest sample size may prove unrepresentative when more stocklists become available, a useful working tool is price per ounce (henceforth PPO), which can be compared across time and

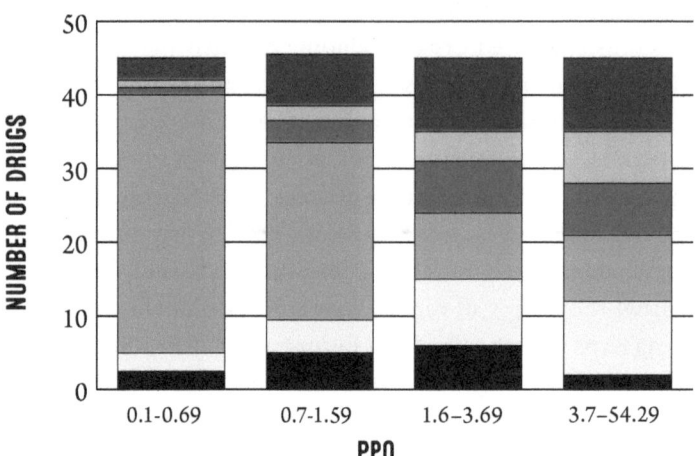

FIGURE 8.11. Price bands for plant simples from different parts of the world listed in the stock of grocers and apothecaries in Paris and Versailles, 1650–1730. Based on information from the Minutier Central, Archives nationales de France, Paris. Credit: E. C. Spary.

between stocklists, for all cases for which there is sufficient information. In analyzing prices, I took a pragmatic decision to exclude plant materials where quotidian uses predominated. Examples are dried or preserved table fruit (apples, lemons, oranges, figs, cherries), nuts (almonds, hazelnuts, walnuts, pine nuts), cereals (barley, wheat, rice), and major plantation commodities (tobacco, sugar). These plants commonly appeared in medicinal recipes, but in stocklists they were often itemized and valued separately from the substances that had specifically medicinal uses. Often they were stored in a separate space on the premises, or in different kinds of containers (e.g., barrels or bales, as opposed to drawers or *albarelli*). Such goods were generally stocked in bulk, unlike the smaller quantities of plant materials with a primarily medicinal use. However, all such decisions are attempts to generate particular kinds of meaning out of lists that sometimes drew no distinction between medicines, spices, preserves, herbs, artisanal raw materials and poisons. The prolixity of the shop is the omnipresent backdrop to each category decision taken.[39]

Breaking our entire set of data into quartiles divided according to PPO shows that the geographical origin of a drug affected its price. Overall, non-European plant simples, especially Levantine and

American drugs, were far more likely to figure in higher price brackets. This is an unsurprising finding, given the distance such commodities traveled. Our data do not suggest that drugs from the Americas served as cheap substitutes for Old World drugs, suggesting that an alternative explanation must underlie their entry into the pharmacopoeia. But other factors, such as availability and trade monopolies, also had a powerful effect upon the value of drugs. To take just one example, although the bark of guaiacum, a drug used to treat the pox, traveled all the way from South America, it numbered among the cheapest drugs available on the high street, because the medicinal material was a by-product of imports of the timber of the guaiacum tree, and thus effectively a waste product.[40]

The PPO suggests rough agreement among the merchant community over the market value of some drugs. For example, some simples varied only slightly in PPO over the period covered by the study. The PPO of anise varied between 0.38 *sous* and 1 *sou*; 7 out of a total of 11 entries priced it at either 0.38 or 0.5 *sous* per ounce. The PPO of aristolochia varied between 0.25 and 1.25 *sous*; 7 out of a total of 10 entries placed it between 0.25 and 0.44. Such consistency suggests that, where my analysis does disclose large and persistent shifts in one direction or another, these likely correlated with a change in the market value of the simple in question.

Such was the case for the PPO of clove and nutmeg, drugs monopolized by the Dutch East India Company (VOC). Price management by the VOC and tight control over the source ensured that, by 1700, the Dutch were able to keep the cost of these two drugs high and constant around Europe.[41] The stocklists bear the trace: the PPO of nutmeg rose from around 4.5–5.3 *sous* before 1700 to over 10 *sous* by 1730, effectively double what it had been in the seventeenth century. The PPO of clove, where a Dutch monopoly was established as early as 1656, was roughly similar to that of nutmeg, but varied little through the period.[42] In chronologically later inventories (Vedy 1716, Brechot 1720, Bonvallet 1722, Hébert 1723, Auger 1729), clove and nutmeg were often stocked as a mixture, again probably reflecting their monopoly status.[43] Significant trends are easier to identify in cases where there are numerous data points for a given drug over time. Extreme rarity meant that prices lacked a conventional reference point; for example, the value of an exceptionally rare drug like aloe-wood varied much more chaotically, between 2.19 and 8.13 *sous* per ounce for different merchants.

E. C. SPARY

TWO CASE STUDIES: THE BULK BUYER AND THE *CURIOSO*

a. The Bulk Buyer

Roche-Adrien Richard (IAD 1704), on the rue de la Verrerie, stocked just four different simples, the second-lowest number on the list. He was nonetheless far from the poorest merchant, possessing stock with a total value of 302 *livres tournois*, 3 *sous*, well above that of several others offering a more varied range of materia medica; in fact, when the merchants are ranked by total value of their stock of plant simples, he is exactly halfway down the list. It appears that Richard bought the few plant simples he did stock in bulk, perhaps acting as a supplier to other merchants nearby. For example, he owned 1974 pounds of cassia "grabot" or "garble," the term applied to the "dirt removed from spices, drugs, &c."[44] This was a cheap commodity, valued at the price of 6 *sous* (PPO), so Richard's clientèle was probably not affluent. He also owned 120 pounds of *grabot* of senna and 2 tons of licorice extract, both again inexpensive drugs. Such cases suggest that, even within the same guild, merchants could adopt widely differing market strategies when it came to plant simples.

Richard was not alone in stocking cheap variants of costly exotic drugs. In the stock of Jean-Baptiste de Fourcroy (IAD 1708) was to be found *grabot* or "dust" of black pepper; in that of the apothecary Jean-Baptiste Rotrou (IAD 1711) were coffee *épluchures*. "Éplucher," according to the Académie Française's dictionary for 1762, meant "to clean by separating out by hand the dirt and anything that is bad or spoiled," and *épluchures* were "the dirt that one removes."[45] *Épluchures* were around half the price of ordinary coffee. In other words, some exotic drugs were probably available even to less affluent city residents around 1700. The drugs sold in "garbled" versions fell into two main categories: purgatives (senna, cassia, agaric) and fashionable new drugs (coffee, cinchona); sweeteners such as sugar and licorice were also low-cost drugs. Evidently, purgation was a medical act that spanned the social scale, as were sweet tastes and fashionable drugs.

b. The Merchant-*Curioso*

At the opposite end of the spectrum from bulk buyers like Richard were those merchants who appear, from the wide range of diverse drugs they owned, to have interested themselves in simples as collectors, not just merchants. Valentina Pugliano has shown how sixteenth-century Italian apothecaries mingled with gentlemanly *curiosi* and used their shops as centers for the acquisition and transmission of natural knowledge, in part by assiduously collecting both specimens and information from their range of contacts worldwide, whether these were other merchants and agents, colonists or healers.[46] This pattern appears to hold good for the later French context too.[47] Pierre Pomet, for example, obtained his information on several American simples, including cochineal, mechoacan, vanilla, lacquer, and achiote from François Rousseau, a widely traveled Parisian candlemaker with ties to the Caribbean, with whom he struck up a correspondence in around 1693. This may have been the same individual whose stock, later inventoried in 1705, contained a total of 39 different simples.[48] Conversely, Pomet owed knowledge of cassia, camphor, coral, cuscuta, galbanon, cinnamon, and ipecacuanha to the Parisian botanical demonstrator and physician Joseph Pitton de Tournefort.[49] At this "curious" level of involvement, there appears to have been little difference between grocers and apothecaries. Among the largest numbers of different plant simples owned, we can single out individuals like the apothecary Jean-Baptiste Rotrou, whose wife Françoise Renard died in 1711. Rotrou's shop was on the rue Saint-Jacques de la Boucherie, and he had joined the guild on April 30, 1700. At the time his wife died, he stocked 95 different plant simples with a total value of £11,208.8s.3d, placing him at the top of all merchants studied. This was in addition to his vast range of chymical drugs, mentioned above. He also stands out for having large volumes of each drug on hand, measured in pounds rather than ounces: 62 pounds of galangal, 10 of turmeric, 5 of cinchona, 8 of opium, for example. In separate storerooms, he kept even larger volumes of the drugs available in the shop, such as a further 37 pounds of opium, valued at £203.6s.0d. Rotrou thus made large investments, buying in bulk and keeping a high stock value at hand in the shop, but he also had on hand an impressive range of different plant simples, the second-highest in our set of merchants. Another *curioso*, this time a grocer, was Charles Vignon, whose stock was inventoried in 1702. His shop carried 85 different simples, and, like Rotrou, he kept smaller volumes in his shop space, and a larger volume of certain drugs in a storage space. Sourcing so many different drugs would have required an extensive labor of correspondence and a wide range of informational resources, including people and books. Though Pomet was the only grocer to publish on his activity in this respect, inventories do disclose information about the books owned by individual merchants. Pierre Boulduc (IAD 1671), who ran the teaching garden of the apothecaries' corporation, owned many herbals, mostly Renaissance classics by Matthias de L'Obel, Caspar Bauhin, and Otto Brunfels, alongside books on distillation and pharmacy, a pattern repeated in other surviving booklists.

Sometimes individuals owned several drugs in smaller quantities, with little or no evidence of larger reserves elsewhere on the premises. Boulduc, a more modest retailer than Rotrou, is perhaps more typical in this respect. In the shop, he stocked his 23 different simples in precise weights of 4 or 8 ounces each, regardless of price, with few exceptions. But, like Rotrou, this apothecary was even more reliant upon his chymical and Paracelsian remedies, holding stock of 36 waters, 29 syrups, 19 oils, and so on.

The inventories afford insights not only into the use of space for housing stock, but also into the types of containers and furniture in which they were stored, as well as (in some cases) suggesting a shop layout from the order in which the inventory was drawn up. That of the grocer-apothecary Joseph Seconds (IAD 1724) was alphabetized, and other lists also show traces of an alphabetical arrangement for the drugs. Other stocklists suggest a shop order according to plant part, the same arrangement followed in Pomet's book: seeds, roots, woods, barks, leaves, flowers, fruits, gums, and juices. That is, knowledge of a simple revolved around the parts used in trade, a commodity view of exotics used to cure.

OLD AND NEW DRUGS

Finally, the probate inventories provide a valuable ancillary resource for exploring how and when new plant drugs entered trade, a phenomenon otherwise hard to document. Probate inventories, as earlier work has shown, are a valuable way to compare the sometimes exaggerated or mythologized reports of new drug assimilation found in both printed primary and secondary sources with actual practice in material and commercial culture.[50] When new plant materials entered European cultures around 1700, how important were shops like these in making them available? For each of these plants, a more detailed individual history could obviously be generated, drawing on a wider range of different sources; so here I will limit myself to exploring some specific trends disclosed by the stocklists.

Judgments about when particular materials entered, or alternatively dropped out of, trade are complicated by wide variation between individual stocklists. Fifty-two of the simples identified, or just under 30 percent, appeared in no more than one stocklist. By contrast, in the case of a simple like galangal, which appeared sporadically across the entire period surveyed (1658, 1665, 1703, 1705, 1711, 1729), interruptions to supply are more likely to have been the main factor in its availability on grocers' shelves. For certain other drugs, such as balm of Mecca, cardamom, or copahu, rarity was almost certainly the reason for low representation and wide price variation. Some drugs seem to vanish altogether at certain junctures: cumin, an import from Malta, appears in several stocklists up to 1702, but none thereafter. A similar pattern is evident in the cases of euphorbia, incense, tamarind, purging nut, stavesacre, turbith, and vomic nut. Did these disappearances reflect the effects of the War of the Spanish Succession (1701–14) on French merchants' access to Mediterranean drugs? Individual merchants might preserve stock for decades after acquisition, so advancing chronologically precise reasons for the decline of individual drugs is difficult.[51] Some changes identified in scholarship on other settings are not in evidence in these stocklists. China root, imported to France by Dutch and English merchants, and sarsaparilla, imported by Spanish merchants, were two of the most commonly stocked drugs (see table 8.1), but throughout the period, the American root supplemented the Far Eastern one, rather than supplanting it, suggesting a market for both drugs.[52] Sassafras, though less common than these two, made a regular appearance throughout the period, rather than increasing its presence, as Katrina Maydom, in her chapter in this volume, finds to be the case in England.

For a few drugs, we know the decade when they entered French consumption with certainty. This is the case for cinchona, coffee, ipecacuanha, and pareira brava. Cinchona was known to *curiosi* from the 1650s, thanks to Jesuit trade in the bark, as discussed by Samir Boumediene in his chapter; but royal patronage boosted it into a valuable commodity from 1679 onward, and prices skyrocketed. None of the inventories used dated from the 1680s, so the earliest incidence of the drug in our data is in 1692. However, it appears in almost 50 percent of all stocklists after that date. Ipecacuanha shot to fame a few years later, from 1689 onward, but only appears in our stocklists from 1705 onward, with just 4 appearances in total and a PPO hovering around 20 *sous*. Pareira brava, like ipecacuanha and cinchona, was used within the Portuguese medical world. In 1694, Pomet reported that it had only recently become known in France, after the French ambassador to Portugal brought some to court. Pareira brava was also brought to Paris by the botanist Tournefort after his voyage to Spain and Portugal in 1687–90.[53] In 1704, the physician Adrian Helvetius, renowned as the "discoverer" of ipecacuanha, described his experiments on *Pareira brava*, in an apparent attempt to start a fashion akin to the ipecacuanha trend, which had made him rich in the 1680s. He continued to push the drug in subsequent years.[54] The academy of sciences in Paris commissioned the physician Étienne François Geoffroy to conduct research into pareira brava in 1710. Only following this publicity did the drug began to appear in guild stocklists: specifically, in two probate inventories drawn up in 1724, those of the apothecary François Rousselot and the grocer Joseph Seconds.[55] The publication of favorable judgments by fashionable physicians evidently could generate demand for a drug.

Coffee makes a surprisingly late first entrance, in 1702, but thereafter stands out for its ubiquity, appearing in 55 percent of stocklists, for the large volumes in which it was usually stocked, and for its relatively high PPO, around 2 to 4 *sous*, half that of cinchona and a quarter that of ipecacuanha. These features are consistent with a rapid trajectory from medicinal rarity to quotidian luxury. By contrast, tea first appears a lot earlier in Pierre Boulduc's stock, in 1671, but remains much rarer, with only four appearances in total, and always in small volumes: a few ounces per merchant, compared to tens of pounds for coffee.[56] The only exception here is Robert Bonvallet (IAD 1722), who stocked 43 pounds of tea, as well as 6 pounds of lower-priced tea garble. Interestingly, cacao appears just once in the entire data set, in Rotrou's inventory, and chocolate only twice. Also striking in our set of data is the conspicuous absence of certain substances that were quite well-known in European consuming cultures, such as vanilla, an important commodity in Spain.[57] Vanilla was certainly being traded in France, as we know from a 1692 royal edict allocating a monopoly on the drug to François Damanne, which explicitly contained clauses banning the involvement of other merchants in the sale and processing of these drugs: "all Coffee, both as beans and as powder, Tea, Sherbet, & Chocolate, in loaves, rolls, tablets, pastilles, in whatever shape it has been formed, together with the drugs of which it is composed, such as Cacao & Vanilla . . . we ban all persons from involving themselves in the composition, sale and retail ... of the said drugs and merchandise."[58]

Inspections conducted by the Paris police over several days in January 1692, soon after the edict had been passed, sought to ascertain which merchants held tea, coffee, chocolate, and vanilla, and were therefore legally bound to sell their stock to Damanne. Their reports confirm that grocers and apothecaries were not the specialists in these new drugs. Of thirty-three shops inspected, the police found just six stocking tea, mostly in very small quantities of a few ounces. But five of these were distillers, and one a mercer. Chocolate was also stocked by six merchants, again five distillers and the mercer. Nobody owned any vanilla or cacao. In 1692, coffee was the only one of these goods present in the stock of grocers, and even then, only 6 out of 20 possessed any coffee, compared to eight out of the twelve distillers inspected.[59] This snapshot, at the very moment when these drinks were in the process of becoming quotidian experiences, indicates that other urban suppliers were also playing a leading role in promoting new drugs, in some cases more so than the grocers and apothecaries.

Damanne's monopoly ended in 1693.[60] While coffee went on to play a more prominent role among the goods sold by grocers and apothecaries, vanilla remained absent from the stock of all merchants we surveyed over the entire period. The low profile of cacao, chocolate, and vanilla in grocers' and apothecaries' stock in the early part of the period is striking, given that chocolate was being written about and consumed throughout the period, especially at court, and also given its visibility in secondary studies. Perhaps trade in chocolate and cacao remained linked to privileged court merchants and entrepreneurs. Prior to the privilege briefly held by Damanne, a twenty-year monopoly had been granted to one David Challiou in 1660 for the sale of chocolate. Kinship networks spanning the Pyrenees also brought ready-made chocolate from Spain to France.[61] When the courtier Madame de Guise wanted to serve chocolate to her cousin, Philippe d'Orléans, the king's brother, in 1690, she ordered it from Jean de Herrera, a privileged court merchant. Likewise, she had to pay a large sum for a shipment of vanilla specially ordered from Florence in 1694.[62] The aberrant spelling "Banille" in the first edition of Antoine Furetière's dictionary of 1690 confirms the unfamiliarity of this drug to French readers.[63] Possibly vanilla was largely unavailable in its own right to Parisians at this time, although it certainly entered consumption as an ingredient of chocolate.

WHO OWNED DRUGS?

Probate inventories, while an imprecise tool for the study of the drugs trade, afford useful general insights into the trade of apothecaries and grocers. When stocklists are compared with other large sets of drugs used in healing and for experimentation, the boast of the grocers' and apothecaries' corporation to be "masters of the exotic" is borne out: they were indeed specialists in non-European drugs, though not *all* such drugs, as the cases of vanilla and cacao show. Just one or two merchants in our sample were the only vendors to stock a very broad range of different plant drugs. This leaves many questions about the relationship between supply and demand unanswered. Of one American plant drug, arara, which the sixteenth-century Flemish physician Carolus Clusius had identified as a fruit serving to soften the belly and cleanse ulcers, the famous Parisian apothecary Nicolas Lémery noted: "This fruit is very rare in Europe, & when you have any you keep it for curiosity."[64] If a grocer or apothecary owned a plant material that was so rare as to make him the only person in the city selling that substance, was its primary purpose cure, or curiosity? Did customers come asking for it, or did they depend on him to recommend it? How and where did he learn about and obtain his rarities? Evidently, knowledge and possession alone did not transform a plant simple into a commodity. On the other side, how did apothecaries and grocers know which simples were most heavily in demand, or determine what prices to ask for them? It is impossible to know how representative our small sample of merchants is of the corporation as a whole.

The early modern metropolis of Paris was a patchwork of groups contending to "own" specific drugs and the technical practices that formed them, from distillation and other highly skilled chymical operations to gardening or just sourcing these materials. Many constraints governed merchants' entitlement to sell, display, and work upon particular simples. A range of circumstances might facilitate or block a merchant's access to specific drugs, from international trade or warfare to local legislation and competition. Only further research into the complex chains of procurement, credit, and debit may shed light on the routes by which exotic plant materials traveled from their places of origin to the shelves of city shops, and on the diverse range of individuals involved in sourcing, selling, and processing these plant simples along their journey.

LISTS OF NAMES AND ADDRESSES OF GUILD MEMBERS

- *Roolle des Marchands Espiciers et des Marchands Apothicaires-Espiciers de cette Ville, Faux-bourgs et Banlieue de Paris. Fait le cinquiéme May 1671* (Paris: Jean Baptiste Coignard, 1671) BnF, Ms. Français 21738, fols. 202r–213v.
- Bibliothèque Interuniversitaire de Santé—Pôle Pharmacie, Paris (henceforth BIUS-P): Registre 110, untitled printed sheet, dated May 17, 1692.
- *Catalogue des Marchands Apothicaires-Epiciers, et des Marchands Epiciers de cette Ville, Fauxbourgs & Banlieuë de Paris. Fait le troisiéme jour d'Octobre 1716* (Paris: Jean-Baptiste Coignard, 1716). British Library, classmark D.B. 2/5 (7), document 88.
- *Catalogue des Marchands Espiciers, et des Marchands Apothicaires-Espiciers de cette Ville, Fauxbourgs & Banlieuë de Paris. Fait le 16. d'Octobre 1717* (Paris: Jean-Baptiste Coignard, 1717). BnF, classmark 4-FM-24943.
- BIUS-P: Registre 110, "Extrait du Rolle & Repartition des droits qui doiuent se Leuer sur tous les Marchands apoticaires Epiciers de La Ville et fauxbourgs de Paris en Execution de L'arrest du Conseil d'Estat du Roy du 14 Nouembre 1718 pour L'anné Mil Sept cent Vingt huit."
- *Catalogue des Marchands Espiciers, et des Marchands Apothicaires-Espiciers de cette Ville, Fauxbourgs & Banlieuë de Paris. Fait le 30. Juin 1721* (Paris: Jean-Baptiste Coignard, 1721). BnF, classmark V–13191.

TABLE 8.2. INVENTORIES ANALYZED

	AN Minutier Central Call No.	Name and Occupation	Date
IAD 1653	Étude LXXXVIII, 152	Thomas Goré, merchant grocer	August 27, 1653
IAD 1655	Étude XX, 308	Cesar Lybault, apothecary to the Prince de Condé and merchant grocer	May 15, 1655
IAD 1658	Étude C, 250	Anne de Lassus, wife of Étienne Regnault, apothecary grocer	March 6, 1658
IAD 1660	Étude LXXXVIII, 174	Michel Faverel, apothecary	April 22, 1660
IAD 1665	Étude XCVII, 25	Barthélémy Boisseau, merchant grocer	June 3, 1665
IAD 1671	Étude XXXV, 431	Pierre Boulduc, apothecary	March 21, 1671
IAD 1675	Étude XLIV, 55	Pierre Brusle, merchant grocer	May 27, 1675
IAD 1679	Étude I, 173	Gabriel Papin, merchant grocer	September 27, 1679
IAD 1692	Étude I, 196	Jean Binois, merchant grocer	December 1, 1692
IAD 1702	Étude I, 221	Charles Vignon, merchant grocer	November 22, 1702
IAD 1703	Étude XLI, 323	François Niceron, merchant grocer	November 20, 1703
IAD 1704	Étude I, 225	Roche-Adrien Richard, master grocer	August 18, 1704
IAD 1705	Étude CXIII, 211	François Rousseau, merchant apothecary grocer	May 9, 1705
IAD 1707	Étude I, 231	Martin Coras, merchant grocer	April 1, 1707
IAD 1708	Étude LIV, 699	Marie Valentin, wife of Jean-Baptiste de Fourcroy, merchant grocer	October 2, 1708
IAD 1711	Étude II, 368	Françoise Renard, wife of Jean-Baptiste Rotrou, merchant apothecary and grocer	May 18, 1711
IAD 1713	Étude II, 376	Mathieu-Michel Martin, merchant grocer	December 22, 1713
IAD 1716	Étude II, 386	François Vedy, merchant grocer	September 25, 1716
IAD 1720a	Étude XXIX, 348	Louis Brechot, merchant grocer	April 13, 1720
IAD 1720b	Étude II, 407	Elisabeth Jorand, wife of Étienne-Dominique Théveneau, merchant grocer-wax chandler	November 7, 1720
IAD 1722	Étude LXX, 271	Robert Bonvallet, merchant grocer	July 27, 1722
IAD 1723	Étude I, 311	Louis Hébert, merchant grocer	February 3, 1723
IAD 1724a	Étude XII, 386	Marie-Claude Aprin, wife of Joseph Seconds, merchant grocer apothecary	July 3, 1724
IAD 1724b	Étude XXIX, 376	Geneviève Baudrier, wife of Nicolas François Rousselot, merchant grocer apothecary	August 12, 1724
IAD 1728	Étude I, 338	Florentin-Jean Couvreur, merchant grocer	October 12, 1728
IAD 1729	Étude I, 343	Henri-François Auger, merchant grocer	July 7, 1729
Inv 1729	Étude II, 435	Vivant Carré, merchant grocer, stocklist attached to marriage contract	August 15, 1729
IAD 1730	Étude XXIX, 396	Jean Noël Leclerc, merchant grocer	March 24, 1730

Source: Minutier Central (records of the Paris *notaires*), Archives nationales de France.

9
CONSUMING THE EXOTIC IN EIGHTEENTH-CENTURY SWITZERLAND

Laurent Garcin's "Maduran Pills"

Alexandra Cook

Circa 1700, it would have been hard to imagine a more internationally traded or renowned item of exotic materia medica than bezoar stones.[1] Portable, nonperishable and often worth more than their weight in gold, they had long been sourced and used medicinally across Asia and the New World. Widely desired as the ultimate transcultural commodity among materia medica, bezoars exemplify—perhaps more than any other remedy of the period—the mobility and adaptability of early modern exotic drugs. Their value was such that bezoars not only had use-value as medicines, but also *exchange*-value: they could be and indeed were traded *as if* they were gold. Bezoars performed this dual function in securing the secret remedy—Maduran pills—whose acquisition and European reception is the focus of this chapter.

FIGURE 9.1. Seventeenth-century man-made bezoar stone from Goa, most likely manufactured according to the Jesuits' proprietary recipe. Kunsthistorisches Museum, Vienna. Courtesy of user Vassil, Wikimedia Commons.

While the purported properties of bezoar stones as antidotes and panaceas were already known to Ibero-Arab physicians and Crusaders, it was the Portuguese who first marketed bezoars to a European clientele. Stones obtained from Asian mountain goats, porcupines (*Lapis malacensis, Pietra de porcospino,* or *Pedro do porco espinho,*) and monkeys were collectively known as *Lapis bezoar orientalis,* with porcupine and monkey stones commanding the highest prices; monkey bezoar (*Bezoar simiæ*, imported from Macassar) was reputed to be the best of all.² Treatises by Galen (131–ca. 201 CE), Nicolas Monardes (ca. 1512–1588), Garcia Da Orta (1501?–1568?), and Caspar Bauhin (1560–1624) all popularized bezoars' origins, characteristics, and uses. Bezoars acquired an outsize reputation, not only for their medicinal properties—especially as poison antidotes—but also as status symbols for the rich and famous.³ Their popularity with the well-heeled spread to the less affluent, who could rent bezoars from apothecaries.⁴ Man-made bezoar substitutes and counterfeits circulated widely, leading to a landmark 1603 decision by the English Court of Exchequer that proclaimed the principle of *caveat emptor,* "buyer beware."⁵ In 1691, Portugal granted the Society of Jesus a monopoly to produce its proprietary "Goa stone," a purportedly efficacious substitute, comprising ground gemstones, pearls, musk, true bezoar, and other costly ingredients. Unlike the Jesuits' more reputable Goa stones, many fakes in circulation were held to be either useless or toxic.

The high value of bezoars in early modern South Asia (see figure 9.1) holds the key to the hitherto untold story of the Maduran pills. In late 1722, the closely guarded secret of these pills left Ceylon in the baggage of Laurent Garcin (ca. 1681–1751). Garcin was a Dutch East India Company (VOC) ship's surgeon keenly interested in Persian, Chinese, and "Brahmin" medicine.⁶ Over twenty years later, having settled in the Prussian principality of Neuchâtel (today a Swiss canton), Garcin revealed his remedy's existence in print:

> I gave the name *Maduran* to these pills because I learned their composition on the Island of Ceylon from a Brahmin of the *Madura* Coast, belonging to the Dutch [East India] Company. I found myself at the end of the year 1722 at Colombo, capital of this Island, where a Priest of the Gentiles had stayed for several days. I knew the benefits of these Pills ... from the salutary effects I witnessed in the work of other priests at Cochin, a Dutch City on the Malabar Coast. The new praise I heard about them at Colombo, while visiting Mr. Rumpf, Governor of this Island and of the Madura Coast, confirmed my impression of their excellent virtues. Mr. Rumpf wanted to encourage my desire to learn their composition. I owe the discovery of this Secret to the Authority and Generosity of the Governor. A beautiful gift of various kinds of the best Bezoars that he made to the Brahmin, and the promise that in communicating their Composi-

tion to me alone, it would be neither revealed nor used in these countries, because I was to depart in a few days, and to distance myself [from Colombo] with great Voyages; all this persuaded him to communicate the formula to me. I was not mistaken about the effects of these pills, and have always employed them with success in comparable cases.[7]

Just what was this secret remedy, sufficiently precious to merit such a magnificent reciprocal gift of not just any bezoars, but "various *kinds* of the *best* Bezoars," particularly monkey bezoars, as a footnote to the original text indicated?[8] The present chapter is a case study of the Maduran pills, the secret remedy Garcin received from Isaac Rumpf, the Governor of Dutch Ceylon, as a result of a gift exchange with a Maduran Brahmin. It explores how this remedy traveled from South Asia to Europe, and addresses the following questions: What kind of remedy was the preparation that Garcin marketed as Maduran pills? Was it considered a specific against a defined disease, or was it a panacea, effective against many illnesses? Was it taken to have merely local or more broadly global application? How, where, and by whom were its ingredients prepared? What was the role of go-betweens, informants, or collaborators in South Asia? Why did Garcin, a European surgeon (and later physician), insert himself into the drugs trade in the first place? Finally, this study questions the presumed appeal of exotic drugs to the early modern European public. Following other researchers who have pursued this line of investigation, I point to the skepticism of European patients toward an unfamiliar "Brahmin" drug, due to its exotic ingredients and pagan associations: Was a drug used on foreign bodies in distant, foreign climes really suitable for conditions that afflicted Europeans in Europe?[9]

Even though the information currently available about Garcin's Maduran pills is largely drawn from European sources, this does not mean that non-European actors and their pharmaceutical-medical traditions are to be considered subordinate to those of Europe. On the contrary, these actors and traditions should be understood—as they were by the protagonists themselves—as integral to the story of the Maduran pills: in this case, the sources testify unequivocally to a hitherto unrecognized encounter between European and South Asian Siddha medicine. The story of Garcin and the Maduran pills shows us that an indigenous remedy connected with a non-Christian religious tradition could travel to Europe and find success in the medical marketplace.

FIGURE 9.2. The exchange of bezoars for the Maduran pills may have occurred in one of the buildings depicted here. Engraving on paper by Dirk Jongman of the Governor's homes in Colombo, "Vier gezichten op gebouwen in Colombo" (Four views of buildings in Colombo), in François Valentijn, *Oud en nieuw Oost-Indiën*, 5 vols. in 8° (Dordrecht and Amsterdam: Joannes van Braam, Gerard Onder de Linden, 1724–26). Courtesy of *Atlas of Mutual Heritage* and the Koninklijke Bibliotheek, Dutch National Library. Wikimedia Commons.

This suggests that, for all their differences, the traditions, practices, and ingredients of South Asian medicine were still compatible with their European analogues—in the sense of being capable of interfacing with them—in the early eighteenth century. Common practices such as gift exchange, as well as compatible traditions such as alchemy, and commonly held values of secrecy, all opened the way for such exchanges, even if the indigenous informants and precise modes of transmission have sometimes been obscured or written out of the historical record.

These commonalities are perhaps clearest in the nuances of transcultural practices of gift exchange, which played a key role in Governor Rumpf's—and by extension Garcin's—receipt of the Maduran pills. As

Anne Goldgar has demonstrated, it was not only the Trobriand Islanders whose society functioned as a gift economy; the European Republic of Letters likewise operated on the principle that receiving a gift entails a reciprocal obligation that could not be reduced to a cash transaction.[10] Note in this regard that Garcin does not *purchase* the Brahmin's secret remedy; nor, presumably, does the Brahmin *sell* it either to Garcin or to the governor. The governor mediated the transaction as well as financing it. Governor Rumpf's "generosity" would have been indispensable, since a ship's surgeon probably could not have afforded a gift of the "best bezoars," as acknowledged in effect by Garcin himself: "I owe the discovery of this Secret to the *Authority* and *Generosity* of the Governor."

Given that Garcin's description is the sole available account of the Colombo exchange, we can only speculate as to the details. But we can be certain that social interactions, perhaps even of a ritualized nature, would have been involved: one *presents* the gift (Garcin speaks of a "Présent," rather than "Cadeau," for example) following local custom and notions of rank or caste. During the "several days" the Brahmin spent in Colombo, he seems to have been treated with the utmost respect, as a highly regarded member of local society. In the Colombo case, the three participants occupied different social ranks, with Garcin probably the most socially inferior participant; indeed, he may not even have attended the negotiation for, or exchange of, bezoars for the secret remedy. And presumably the exchange was only finalized once the Brahmin was satisfied (see figure 9.2).

Garcin's account of the exchange thus highlights indigenous agency in co-constructing knowledge.[11] According to this paradigm of knowledge transfer, local people acted unconstrainedly in producing, translating, and exchanging various kinds of information in South Asian "middle grounds."[12] The Colombo exchange evidently conforms to this paradigm of knowledge exchange, for there is no hint of duress, striking a contrast with recent work on slave remedies, which has posited a coercive model of European knowledge acquisition.[13] Most important, Garcin's report of the exchange did not diminish the Brahmin's status in the transaction, even though Garcin failed to report the Brahmin's name or other details about him.

This exchange also highlights the importance of secrecy in early modern knowledge accumulation and circulation; both sides of the exchange are concerned with keeping the remedy secret, each for his own reasons. While the Brahmin's reasons can only remain a matter of speculation, he may have represented a traditional lineage of Tamil healers, for whom closely guarded secret remedies were, and still are, the norm.[14] Garcin's desire to keep the remedy secret had both reputational and commercial motives that I explore further below. And secret the Maduran pills would remain, at least with respect to their manner of preparation, until the end of Garcin's days.

While much work on knowledge circulation examines exchanges between *Europe* and other world regions, Geoffrey Gunn focuses instead on intra-Asian exchanges, striving to "capture the fluidity and ambiguity of boundaries, both physical and cultural" within Asia itself, as a globally connected "world region."[15] Within this Asian "world region," Garcin and the Brahmin promoted "overlapping communities of knowledge" constituted by "unsuspected linkages."[16] These "linkages" are, by their nature, difficult to reconstruct. Such "unsuspected linkages" most certainly shaped Tamil Ayurveda, that is, Siddha medicine, the likely origin of the Maduran pills. Siddha medicine evolved at the South Asian cultural crossroads, where many traditions met and interacted, including Hindu Ayurveda and Unani (the local term for Ionian or Greek) medicine as translated and transformed by Arab practitioners. Despite sharing many features with Ayurveda, Siddha medicine's "determining difference from Ayurveda is its anchorage in Tamil culture and in esoteric alchemy and tantrism."[17] Indeed, within lineages of traditional Siddha practitioners, medicinal formulas, prepared by arduous procedures of long duration and incorporating heavy metals such as arsenic and mercury, are jealously guarded as secret knowledge. Siddha medicine shared this alchemical pharmacy with early modern European iatrochemical medicine.[18]

However, Siddha medicine's distinctive alchemical medicines and esoteric practices descend from even more geographically and culturally remote sources in ancient China. As the noted Indian historian of medicine B. V. Subbarayappa observes, "the seed-ideas of this system ... were in the nature of an implantation from the Chinese-culture area."[19] While the movement of knowledge within geographical-cultural areas can often be traced or at least hypothesized, it is difficult to explain this Chinese Taoist influence. Joseph Alter argues that certain historical figures may have acted as cross-cultural vectors of esoteric health practices.[20] Consistent with Gunn's model of Asia as an interconnected region, these were not one-way trajectories of exchange. Ayurveda likewise influenced Chinese medicine, via the spread of Buddhism from South

Asia into China, early in the Common Era.[21] The prevalence and importance of such intercultural exchanges of ideas, practices, and materials renders ethnocentric categories such as "Indian," "Chinese," or "Arabic" medicine meaningless.[22]

Like the common practice of gift-giving, the crosscurrents of cultural exchange in early modern Asian medicine call into question the presumption that any purported "great divergence" divided major medical cultures such as European, Siddha, Chinese, or Ayurveda.[23] Not only did all of these medical cultures feature an alchemical component, but they also strongly relied on herbal remedies, even if theoretical differences might raise questions about specific cases of transcultural drug adoption. In Asia, Europeans like Garcin were thus entering an existing marketplace—one of both physical drugs, and also ideas and practices—in which the Maduran Brahmin would have been involved, and which had already existed for centuries.

But how had Garcin come to be in the East Indies in the first place, and where had he acquired his interests in non-European plants and drugs? Dropping down from the scale of entire traditions to that of biography, it is to these questions that we now turn.

A NAME WITHOUT A FACE

Laurent Garcin has sunk into obscurity, with many biographical details remaining unclear and no portrait being extant. The first published account of Garcin's life, which serves as the source for most others, is by Philippe-Sirice Bridel (1757–1845), the "doyen" of Swiss letters, and a friend of Garcin's son, Jean-Laurent Garcin (1733–1781).[24] Concerning Garcin's education, we only have Bridel's assertion that when Garcin was "of age," i.e., about fourteen years old (the traditional age of apprenticeship), he was sent to the United Provinces of the Netherlands to learn surgery and medicine.[25] It seems most likely the young Garcin was sent to the Dutch province of Zeeland, given his sustained, lifelong connections with members of the French-speaking (predominantly Huguenot refugee) community known as as "Walloons" (walonnes).[26]

Zeeland, then still an island, had a much higher profile than it does today, with the towns of Vlissingen and Middelburg flourishing due to the West and East Indies trade. Accordingly, surgeons interested in joining the the VOC trained in both towns. Given the likelihood that he trained in Zeeland, Garcin may have benefited from the innovative medical education offered to apprentice surgeons in Middelburg, where

FIGURE 9.3. The Salvadors' natural history cabinet, kept in their Barcelona pharmacy at the intersection of Ample and Fusteria streets, is likely where Garcin met Joan Salvador i Riera. Photo-montage from the Salvadors' natural history cabinet. Credit: Botanic Institute of Barcelona (CSIC-CMCNB). Reproduction authorized. Photos by Gerard Morada, arranged by Felix Yeung Shing Hay.

in 1720 he was engaged as a ship's surgeon by the Company's Zeeland chamber.[27]

Of Garcin's career prior to his employment by the VOC, we know only that he served for sixteen years as surgeon to a Dutch regiment deployed in Flanders, Portugal, and Spain.[28] Due to the dearth of surviving military records dating from the time Garcin joined the army, it has proven impossible to determine in which regiment he served. We may surmise, however, that he served with one of the Dutch Army's four Huguenot regiments formed to fight Louis XIV.[29] The War of the Spanish

Succession brought Garcin to Barcelona, where he became acquainted with the noted botanist and apothecary Joan Salvador i Riera (1683–1726).[30] During learned gatherings in the Salvadors' apothecary shop, Garcin would have experienced Barcelona's rich *savant* culture, and perused the Salvadors' magnificent natural history cabinet, preserved today at the Institute of Botany in Barcelona (see figure 9.3).[31]

While Garcin comments that he left for the Indies at an age (about forty) when others were returning, the high mortality rates among the VOC ship's surgeons meant that 50 percent of them could expect never to return home.[32] Moreover, even if a surgeon survived the outbound trip to the East Indies, he often succumbed to disease within a year of reaching Batavia, the endpoint of the outbound journey, or he might succumb on his return voyage.[33] Not surprisingly, therefore, only a handful of the 10,000 ship's surgeons employed by the VOC over its two-hundred-year history contributed anything to the sciences.[34]

Garcin departed Middelburg as chief surgeon of the *Oudenaarde* on May 19, 1720, and on January 19, 1721, he arrived in Batavia (modern-day Jakarta, Indonesia), the VOC's headquarters in Asia. Subsequent to his arrival at Batavia, where he would have received his work assignment from the company's chief medical officer, Garcin spent most of his time in the company's intra-Asian trade. Letters of introduction from the eminent Leiden professor Herman Boerhaave (1668–1738) to the Governor General of the Indies, Hendrick Zwardecroon (1667–1728), may have helped Garcin to secure his preferred assignment.[35] In this post, he was to collect plants throughout maritime South and Southeast Asia from 1720 to 1729, and would later remark, "[This posting] furnished me with an ability to make many observations in *natural history, pharmacy* and *medicine*, attaching myself in particular to everything that could perfect my practice of the last."[36] Departing the East Indies for good on November 1, 1728, Garcin boarded the East Indiaman *Valkenisse*, bound for Middelburg, arriving at his port of origin eight months later, on June 26, 1729.[37]

From his collecting activities while in Asia, Garcin constituted a herbarium of considerable scientific value.[38] This collection comprised Asian and South African plants that he collected for medical and botanical purposes. Its extent is currently unknown, and the number of specimens is impossible to calculate. Garcin's annotations—if there were any—are not extant. Prior to 1740, Garcin reportedly gave some or all of this "beautiful Collection of dried Plants" to the Amsterdam botany professor Johannes Burman (1707–1780).[39] Burman's son Nicolaas (1734–1793) used these specimens, together with those of other collectors, to compile *Flora Indica* (1768), an important treatise on Caribbean, South Asian, and Southeast Asian flora. Garcin's collection, contributing more than ninety-seven specimens to *Flora Indica*, constituted 25 percent of the total specimens consulted for the work.[40]

A significant number of Garcin's specimens are dispersed in the general and pre-Linnaean herbaria of the Conservatoire et jardins botaniques in Geneva, Switzerland. The latter comprises roughly 30,000 specimens, including those from the Burmans' collections. Unfortunately, it is only possible to reconstruct Garcin's herbarium to a very limited degree, although his unique signature, the red wax he used to attach the dried plant to the paper, is a useful guide to tracing his specimens.[41] Garcin's collecting is distinguished by its geographic range, as demonstrated by ninety-seven specimens acknowledged in the younger Burman's *Flora Indica*. Twenty-one of those reported in *Flora Indica* appear to have been collected in Persia (Iran), five in Surat (Gujarat state, India), thirteen in Ceylon (Sri Lanka), eleven on the Malabar Coast (Kerala state and Tamil Nadu state, India) and seventeen in Java (Indonesia). Another twenty-five are labeled as being from "India," which in the parlance of the time could refer to anywhere from modern-day Iran to Southeast Asia, or beyond. Garcin's prospecting introduced at least eight new plant species to modern European botany, while fifty-six specimens attributed to Garcin in *Flora Indica* are traditionally recognized for their medicinal properties; these include *Salvadora persica* L., a natural dentifrice, and *Andrographis paniculata* (Burm. f.) Nees, which is effective against dysentery and in prophylaxis against the common cold.[42] In the words of George Staples and Fernand Jacquemoud, Garcin was one of the "pioneers of the botanical exploration of the continental as well as insular areas of Asia."[43]

Garcin's hunt for medicinal plants complemented his rejection of popular iatrochemical remedies, such as mercury and arsenic:[44] "What we have discovered from plants, relative to Medicine, is so to speak, merely the preliminary sketch of the Science. I departed from Europe with the prejudices of Europe, but my voyages have convinced me that the true Therapeutics reside in plants. This truth will make itself evident sooner or later; I see it even now advancing toward us, such that the science will be stripped of its old jargon, and returned to its true principles."[45]

Upon his return to Europe in 1729, Garcin proceeded in short order to convert his surgeon's credentials into an MD taken at the University of Reims in July 1731, marry in Geneva in September 1731, and acquire naturalization in the principality of Neuchâtel in 1732.[46] Garcin likewise took concerted steps to make himself known locally and internationally in the Republic of Letters as a credible witness of natural phenomena, and trustworthy source of secret exotic remedies, publishing findings from his travels in the *Philosophical Transactions* and *Journal helvétique*.[47] These writings communicated knowledge transfers both global and European, ranging from treatises on various Asian plants in *Philosophical Transactions* to meteorological reports from the Netherlands.[48]

As a returning Swiss, Garcin was hardly unique in having a stint in the Indies behind him. It was not unusual for men from the Swiss cantons and allied territories (Geneva, Neuchâtel, St. Gallen) to find work in prosperous Protestant states such as the United Provinces.[49] Garcin was, it transpires, only one prominent Neuchâtel "bourgeois" who had overcome distance, cultural differences and many hazards to seek a livelihood in Asia, and even more remarkably, given the poor odds, to return and live out a long life at home. In 1732, Garcin either met or renewed contact with Jean Georges Bosset (1688–1772), a former Batavian merchant and watchmaker who retired to Neuchâtel in 1729 to enjoy his immense fortune as a *notable*. Since Bosset was living in Batavia in the 1720s, while Garcin was serving in the Dutch East India Company's intra-Asian trade fleet, the two men may have already been connected. Their close relationship is hinted at in the archives, where we learn that Bosset acted as godfather at the baptism of Garcin's first child, in July 1732. A few months later, Garcin was reported to be residing on Bosset's Neuchâtel estate, La Rochette, although on what terms is unknown.[50] Perhaps Bosset offered Garcin quarters out of friendship, and/or in return for medical services. The wealthy and influential Bosset was potentially a powerful patron for Garcin, and may have supported later ventures such as the Maduran pills.[51]

Even though Garcin had founded a family in Neuchâtel and achieved significant international recognition, he returned to the Low Countries around 1736 and apparently practiced medicine in Hulst until 1739, given his report of prescribing the Maduran pills there.[52] Garcin's sustained personal and professional connections in the United Provinces enabled him to profit from encounters with prominent Dutch *savants* such as Boerhaave, Johannes Burman, and Pieter van

Musschenbroek (1692–1761). While nowhere documented, Garcin may have met an up-and-coming Swedish naturalist named Carolus Linnaeus (1707–1778), who spent the years between 1735 and 1739 in the Netherlands. An acquaintance between Garcin and Linnaeus, perhaps mediated by their mutual acquaintance, Burman, is certainly conceivable, given that Linnaeus lodged with Burman, who acted as a mentor to the Swede.[53] Such an acquaintance might explain not only why Garcin introduced the Linnaean sexual system of classification to Neuchâtel, but also why Linnaeus named the genus *Garcinia* for Garcin, and *Salvadora* for Joan Salvador, on Garcin's recommendation.[54]

Upon finally settling in Neuchâtel in 1739, Garcin was warmly welcomed by the principality's scientific community. Jean-Antoine d'Ivernois (1703–65) exclaimed: "Mr. Garcin, our mutual friend, Member of the Royal Society of Sciences of England and Correspondent of that of Paris, has bestowed a new luster on Swiss botanists by settling among them," while Louis Bourguet (1678–1742) lauded Garcin for giving his herbarium to Johannes Burman, for his many and varied contributions to the revised *Dictionnaire universel de commerce*, and for his knowledge of the Indies.[55] Bourguet, d'Ivernois, and Garcin all wrote for the Neuchâtel *Journal helvétique*, an internationally recognized organ of the Republic of Letters. Garcin brought to the principality not only his knowledge of the Indies, but also, as noted above, the latest innovations from the Dutch Republic: "Doctor Garcin . . . has brought back from his voyages *a new system of botany* that is *very different* from the one [of Tournefort] we use today. The author is a certain Linnaeus—a Swede."[56] These words signaled a new era in Neuchâtel's scientific history, initiated by none other than Laurent Garcin.

MARKETING AN EXOTIC PANACEA

According to Garcin, all ills of the body derived from disturbance of the skin and the digestion.[57] He believed these disturbances to arise from a failure to follow the six "non-naturals" of ancient Greek medicine, such as sufficient sleep, exercise, or nutrition; similar concepts were likewise prominent in South Asian medicine.[58] In addition to regimen recommendations, Garcin offered patients the Maduran pills, which he claimed to have tested on multiple subjects on shipboard and in Europe. He alleged that they comprised "three simple Drugs, gentle and admirably beneficial to the Stomach and other parts involved in chylification. Their effect is imperceptible . . . and follows Nature's intentions."[59] In addition to promoting this remedy, Garcin endorsed the expertise of its creators: "We know from History and travelers' accounts that the Brahmins who are the Priests of the Gentiles of Malabar, are excellent Physicians, and that they have a particular talent for discovering by experience the best Remedies that their country, fertile in Plants and Aromatics, can furnish them.... They are certainly well versed in curing illnesses of the Body. They possess Secrets of great Antiquity, and they always search in a large Book that contains everything that time has taught them to value. They call this book *Manhaningatnam* and they regard it as very precious."[60] However, it is difficult to gain a clear view of the precise contributions of South Asian physicians to Garcin's medical knowledge. Like most Europeans engaged in collecting indigenous knowledge, he does not identify informants, collaborators, or go-betweens by name, so their contributions remain anonymous and ill-defined.[61] An additional lacuna is the disappearance of the Konkani Brahmin medical treatise(s) known collectively to both Van Rheede and Garcin as *Manhaningattnam*.[62]

Concerning the commercial side of Garcin's Maduran pills venture, ships' surgeons routinely traded privately in the Indies, having space allotted on board to carry goods for that purpose. There was also considerable illegal trade (*morshandel*), that VOC regulations strictly prohibited, albeit in vain.[63] Garcin was probably, therefore, no stranger to commerce before setting up as a physician and publicizing the Maduran pills, "in order to capitulate to Solicitations by my Friends to distribute [the pills] to the Public, for the benefit of those who might need them."[64] These "friends" might have included influential supporters such as Bosset, Gagnebin, and D'Ivernois. Besides the Maduran pills, Garcin also advertised his services as a natural history specimen collector for an international clientele; citing his Royal Society affiliation, among other credentials, he offered his expertise in rocks and minerals to the up-and-coming British naturalist Emanuel Mendes da Costa (1717–91), for example.[65]

Garcin's business acumen becomes apparent in the network he established to distribute the Maduran pills in major Swiss towns. In addition to selling the pills himself in Neuchâtel, he had outlets in Bern, Geneva, Lausanne, and Morges, where the remedy was sold by merchants such as Cyprien-Louis Levade (1707–83), a Lausanne surgeon and master apothecary, the Bern bookseller Niklaus Emanuel Haller, and Garcin's relatives in Geneva and Morges. A notice in the April 1745 *Journal helvétique* notified readers of additional points of sale in Vevey, Yverdon,

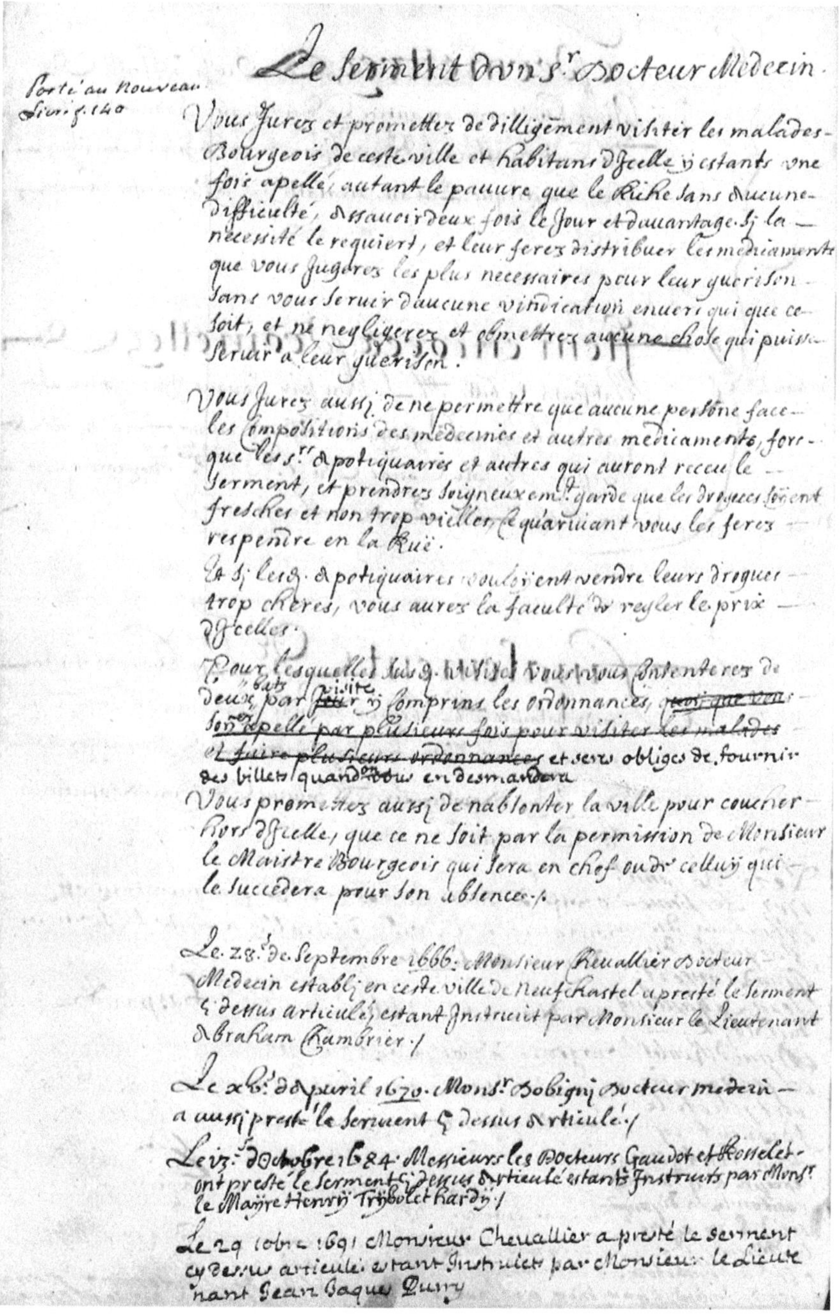

Porté au nouveau
Livre f. 140

Le Serment d'un S.r Docteur Médecin.

Vous Jurez et promettez de dilligemment visiter les malades Bourgeois de ceste ville et habitans d'icelle y estants une fois apellé, autant le pauvre que le Riche sans aucune difficulté, à savoir deux fois le jour et d'avantage si la necessité le requiert, et leur ferez distribuer les médicaments que vous jugerez les plus necessaires pour leur guerison, sans vous servir d'aucune vindication envers qui que ce soit, et ne negligerez et ometterez aucune chose qui puisse servir à leur guerison.

Vous Jurez aussi de ne permettre que aucune persone face les impositions des médecines et autres médicaments, fors que les S.rs Apotiquaires et autres qui auront receu le Serment, et prendrez soigneusement garde que les drogues soient fresches et non trop vieilles, le quariduant vous les ferez respendre en la Rue.

Et si lesd. Apotiquaires vouloient vendre leurs drogues trop cheres, vous aurez la faculté de regler le prix d'icelles.

Pour lesquelles d.s visites vous vous contenterez de deux batz par visite, comprins les ordonnances, soit vous soit apellé par plusieurs fois pour visiter les malades et faire plusieurs ordonnances et serez obligés de fournir des billets quand vous en demandera.

Vous promettez aussi de n'ablenter la ville pour coucher hors d'icelle, que ce ne soit par la permission de Monsieur le Maistre Bourgeois qui sera en chef ou d. celluy qui le succedera pour son absence.

Le 28.e de Septembre 1666. Monsieur Chevallier Docteur Médecin establj en ceste ville de Neufchastel a presté le Serment cy dessus articulé estant Instruict par Monsieur le Lieutenant Abraham Chambrier.

Le xb.e d'apvril 1670. Mons.r Bobignj Docteur medecin a aussj presté le Serment cy dessus articulé.

Le vij.e d'octobre 1684. Messieurs les Docteurs Gaudot et Besselet ont presté le serment cy dessus articulé estants Instruicts par Mons.r le Majre Henrÿ Tribolet hardy.

Le 29 octobre 1691 Monsieur Chevallier a presté le serment cy dessus articulé estant Instruict par Monsieur le Lieutenant Jean Jaques Qurry.

FIGURE 9.4. Neuchâtel physicians took this oath, requiring them to ensure medications were suited to the patient's condition, were not overpriced and were prepared by sworn apothecaries with fresh ingredients. Physician's oath (Serment du médecin), Neuchâtel, ca. 1666. "Livre où sont contenus & rapportés les serments que se prêtent pour le service de la Ville, reliés en 1739 et copiés au nouveaux en 1748," class mark AVN B 101.09.01.01, 37v. Courtesy of the Archives de la Ville, Neuchâtel, Switzerland.

and Basel. In Vevey, as at Lausanne, the pills were sold by an apothecary, a "M. Grenier" (probably Jean Abraham Grenier, 1704–81).[66]

Sale of Garcin's proprietary remedy was facilitated by the weak regulation of the drugs trade in the Swiss canton of Vaud and the principality of Neuchâtel. While Vaud barely regulated the sale of medicines, Neuchâtel physicians were required to regulate the prices and freshness of drugs, as well as respect restrictions on the sale of arsenic and other "poisons."[67] These limited rules left Garcin at liberty to advertise and sell his proprietary drug in both Neuchâtel and Vaud (see figure 9.4).

Among all of Garcin's points of sale, Basel was unique in subjecting the Maduran pills to approval by the university's medical faculty.[68] Unlike other Swiss cantons and their allies, this canton had regulated the ethics of pharmaceutical practice since the fourteenth century, in line with the legal code of Emperor Frederick II (1194–1250).[69] Notable among Frederick's innovations was the strict separation of medical from pharmaceutical professions, entailing a prohibition on physicians selling medications. Although the pills received the Basel medical faculty's approval for sale, Garcin understood that he needed to make his novel remedy appear safe and credible, so he offered patients reassurances and proofs of the pills' safety and efficacy.

First and foremost, the Maduran pills came with a kind of blanket guarantee, based on Garcin's qualifications as a physician with long experience and superlative international credentials. Writing as an experienced medical professional, Garcin enumerated the ailments the Maduran pills could treat: heaviness, weakness, excitability, paleness, women's complaints, melancholy, "depraved hunger," "cravings," "vapors," scurvy, hypochondria, suppressed appetite, and fevers, among others. Second, the Maduran pills came with a purity guarantee, being certified to contain absolutely no mercury or aloe: "The Analysis that I have carried out has assured me of the reliability of their Composition."[70] This emphasis on the formula's purity may have been of particular concern to Garcin, an opponent of chemical remedies in general, since the recipe for the pills originated in the homeland of Siddha medicine, which makes extensive use of mercury and other heavy metals.[71] Third, the *Journal helvétique*, Garcin's chosen publicity medium, acted as an additional quality indicator because leading *savants* numbered among its primary collaborators. Publication in the *Journal* implied that trusted authorities deemed the pills a safe and worthwhile treatment option. Finally, the Maduran pills' safety and value were attested to by *experiments* in a variety of places and contexts: on shipboard, in ports, and in the malaria-ridden southern Netherlands, where "we saw the most beneficial effects in 1737, 1738 and 1739 in the Low Countries, especially at *Hulst* and its environs, the regions where these sorts of Fevers are quite frequent and very stubborn."[72]

Yet, for all these quality assurances, not all patients were satisfied with the Maduran pills. One set of concerns centered on the pills' efficacy. Some patients reported that the pills did not help them very much. Garcin disposes of this concern with the observation that the pills' effects are proportional to the severity of the malady, and that the causes of illness and differences in temperament (an expression of a person's dominant humor in Galenic medicine) could contribute to the etiology of a disease. Those who did not find the Maduran pills effective were invited to contact him directly to identify the appropriate lifestyle changes to accomplish a cure. Garcin's policy of keeping the pills' ingredients and method of preparation secret also aroused concern, as reflected in "questions" from "several persons who are reluctant to use a Remedy if they don't know its composition."[73] In general, patients had every reason to question product purity and authenticity, given not only rampant adulteration, but also confusion around the identity of key drug ingredients, such as the esteemed febrifuge cinchona: barks from completely different genera were sold on the European market as the one true cure for intermittent fevers.[74]

But the public's concerns extended beyond the identity and purity of ingredients to "the *origins* of these Pills. People think that because they come from Indian *Priests* called Brahmins, they might contain *repugnant Ingredients*."[75] This comment is very telling about the seemingly contradictory attitudes of a clientele that questioned the Maduran pills' ingredients, while at the same time consuming all manner of substances from the "Indies," such as tea and spices, whose precise origins could not be readily examined. Yet these sentiments appear less contradictory when the term "priests" is highlighted. As Benjamin Breen notes, "associations with African, Asian and Amerindian religious practice continued to cast a shadow on drugs as a category of goods throughout the eighteenth century and beyond."[76] Garcin was not free of such sentiments himself, declaring that the Brahmins' "false religion blinded them to the true Medicine for the Soul," that is to say, the Christian gospel. Yet he remained committed to prescribing and disseminating a Brahmin treatment for the physical body alone.[77]

REVEALING SOMETHING, BUT NOT EVERYTHING

Garcin had a ready response to skeptical members of the public: "To destroy the Idea [*Pour detruire cette Idée*]" that the Maduran pills might contain something repulsive, he wrote, he was prepared to reveal the pills' ingredients: a move which, he says, causes him "no difficulty [*aucune peine*]." Why this is so quickly becomes apparent: The two principal ingredients, those with the "principal virtue [*principale vertu*]," are not available anywhere in Europe, while the third, although widely available, "requires particular preparation [*exige une preparation particulière*]."[78] In other words, even if patients, apothecaries, or other physicians knew the pills' ingredients, they would not be able to procure, much less process, them. What were these ingredients, and what was their relationship to their likely source, Siddha medicine?

"Sindoc" Bark

Siddha medicine uses ingredients not only from the inorganic realm of minerals and metals, but also from the animal and plant realms. Plant ingredients for Siddha drugs include tree barks of *Cinnamomum*, which would seem to be the genus of "sindoc."[79] Yet Garcin's "sindoc" is difficult to identify with any precision, since the tree known in Bahasa Indonesia as *sintok* (*Cinnamomum javanicum* Blume) is indigenous to the Indonesian archipelago, not to Ceylon, where Garcin claims "sindoc" is native. The difficulty is compounded by *Cinnamomum* species numbering in the hundreds, and being very similar to one another.[80] Garcin's "sindoc" could therefore have been any of a dozen or so Ceylonese *Cinnamomum* species known today, such as *C. malabatrum* (Lam.) C. Presl (= *C. malabatrum* [Burm. f.] Presl) or *C. tamala* (Buch.-Ham.) T. Nees & Eberm., *Folium Indicum*, already exported to the Mediterranean in antiquity. Given the difficulty in distinguishing *Cinnamomum* species, Garcin's "sindoc" could have even referred to more than one species.

Sea Urchin Shell

Shell ingredients play a prominent role in Siddha preparations such as *chunnam*, which entail the shell being "purified *as per classical literature* by using fuller's earth and limestone," which are "triturated with lemon juice."[81] The Maduran pills likewise contained a shell ingredient, "Digiti Marini," or sea urchin shell spines; citing Rumphius, Garcin calls these

"*Echinometra digitata Indica*," and credits them with stimulating menstruation, relieving children's fevers, treating intestinal ailments, and eradicating worms.[82]

Iron Filings

In Siddha medicine, iron denotes air, one of the five fundamental elements and a key ingredient in many preparations. Many classic Siddha formulas contain metals such as iron, copper, gold, lead, or zinc, prepared by "repeated heating of sheets of metal and plunging them into various vegetable juices and decoctions."[83] A "simple" or "complex" preparation method may be employed; in the latter, the minerals or metals are triturated with plant juices or distillates known as *pugai neer*. The resulting paste is dried, and heated to a very high temperature. Once cool, it is ground into a fine powder that is purportedly no longer toxic.[84] Without indulging in too much speculation as to Garcin's recipe, modern Siddha recipes for iron purgations certainly attest to such practices: "Pulverized powders of Lilium root (280 gm) and Alexandrian laurel roots (280 gm) were mixed with iron (35 gm). . . . The mixture [was kept] in a new mud pot closed with proper mud lid. This pot was heated followed by the addition of fermented old rice water for day and night. In between the heating old rice water was added to compensate [for] the loss o[f] heat. Finally . . . the pot [was cooled] and purified and detoxified iron was collected."[85] The processed metal can then be used to compound classic Siddha *chunnam*, "a fine white colour[ed] powder made up of minerals and metals treat[ed] with herbal juice. The word 'Chunn' refers to fine and white. The finished product is tested by using turmeric powder. When Chunnam is mixed with turmeric powder and water, the mixture turns red. This indicates the alkaline pH of Chunnam."[86]

That Garcin employed iron prepared in accordance with Tamil practice may seem curious, given his virtually blanket condemnation of mineral and metal treatments popular both in Tamil and European medicine of the time: "Time and experience have shown to those who are perceptive in Physic, that these substances more often produce bad, rather than good, effects. Today, Chemical Treatment in Medicine fortunately diminishes every day. Nature gives us Remedies *infinitely better than any Chemical Preparations*. In fact, what good can one expect from Remedies drawn from Minerals that, *if one excepts Iron*, agree so little with the nature of the Human body because they are more compact, more active and often more corrosive, in a word, more alien to our Body, than the Plants that we use." Iron, he therefore held, "is the only [metal] that is the Friend of Man."[87]

Garcin may have sourced his iron filings ("Limaille de fer") close to home, but the pills' unmistakable Siddha origin is revealed by the iron's preparation "with Fire & with the cold bath, and in a *vegetable juice*, in a manner *completely unknown in Europe*."[88] Given this preparation method, the Maduran pills may well have been a *chunnam* formula. That we cannot readily pinpoint this particular formula in any written source may be due to the paucity of translations and the secrecy with which the esoteric traditions are guarded to this day.[89]

Contemporary Siddha authorities defend the purity and long shelf life of their preparations, just as Garcin did: He asserts "the ingredients" of his proprietary drug are purportedly "incorruptible," with an indefinite shelf life.[90] But to convince his patients, Garcin adds that he has both the "sindoc" bark and the shell shipped directly to him from the Indies, where he maintains contacts, although he does not reveal these contacts' identities.[91] He might have thereby limited a persistent problem of the drugs trade: "the multiplication of intermediaries," who might misidentify drug ingredients or deliberately deceive their customers.[92]

Compounding the Maduran pills would have been a long, tedious procedure lasting weeks, even months.[93] How, when, and where Garcin might have learned to do this is unknown. An apprenticeship in the complex and lengthy manufacturing of such medicines is hardly even hinted at in Garcin's writings. Would he have had sufficient time on land to pick up such techniques, given his surgeon's duties on board ship? We do not know. Yet Garcin's application of these techniques does tell us that he was probably in direct contact with local experts, even if he did not name them, whose role in making the Maduran pills available to European patients was indispensable.

Authenticity and purity of the Maduran pills are thus assured by being prepared by a physician who sources and prepares the original ingredients himself, according to their original formula. This would presumably raise the value of the pills over that of drugs of actual or purported exotic provenance whose purveyors lacked Garcin's direct experience and contacts in the Indies. It is this direct involvement, not only in procuring drug ingredients, but also in processing them, that sets Garcin apart; he was unusual in uniting the knowledge of exotic pharmacopeia with the practice of preparing them himself. Other European actors, such as the German Pietist missionaries at Tranquebar, were working in the Tamil medical space at the same time as Garcin. While they did much to introduce Tamil medical knowledge to Europe, preserving palm-leaf manuscripts at their headquarters in Halle, there is no record of the Pietists having introduced Tamil drug manufacturing processes to Europe.[94]

As Garcin's 1747 revelations indicate, he hardly conceded everything he knew about his secret remedy, and continued to maintain control over two key factors: the *provenance* and *preparation* of its ingredients. In so doing, he undoubtedly had profit in mind. With no access to either the ingredients or the knowledge of how to process them, any possible competition was effectively stymied. Furthermore, an element of secrecy might serve to increase interest in the drug and its source, enhancing its marketability. But a concern for profit may only be part of the story. Drawing on Georg Simmel's work, Koen Vermeir and Daniel Margócsy argue for a "psychodynamics" of secrecy that explains the attraction of secrets whose content may even be "valueless": "Secrets evoke excitement and desire. Those who have a secret are under constant tension: they want to keep the secret, but they also want to indicate that they have a secret, to veil and unveil it at the same time."[95] The content of the secret alone, these authors argue, does not in such instances explain its attraction. Garcin seems to fit this model well, despite being a Fellow of the Royal Society, i.e., a *savant* who presumably subscribed to the Republic of Letters's altruism and the open access to knowledge that this implied.[96] Other prominent Royal Society Fellows, such as Robert Boyle, were similarly preoccupied with secrets, using them as currency for various kinds of exchange, while at the same time subscribing to the gentlemanly *savant*'s ethical code.[97]

Garcin's drug business could be described as vertically integrated, because he not only sold secret remedies, but also sourced their ingredients himself, and processed these using purportedly authentic "Brahmin" methods.[98] Such an integrated production process carried out by the prospector himself was, to the best of my knowledge, extremely rare. By and large, apothecaries, druggists, and grocers who processed and/or sold drugs did not themselves procure, much less discover, them in distant parts; rather, they obtained these goods from brokers, who in turn purchased them from trading companies.[99] Garcin's conduct is closer to that of patent remedy vendors, who often possessed one particular secret by which they made their fortune and reputation.[100]

Garcin appears to have successfully maintained control over the production of the Maduran pills until his death in April 1751, a death supposedly occasioned by a "chemistry" accident in which he burned his fingers, already paralyzed and lacking sensation from rheumatism or frostbite.[101] As he was unable to feel the injury until it was too late, gangrene set in, leading to his death. But in what sort of "chemistry" experiment was Garcin engaged? Was this "experiment" not really an experiment at all, but instead iron processing for the Maduran pills? If the latter hypothesis is true, it would mean that the pills (and Garcin's desire to market them) were also responsible for his tragic end. The secret of the Maduran pills apparently went with him to the grave, for from that point onward they disappear from view and from the written record.

While many exotic drugs and drug ingredients arrived in Europe over several centuries, the way they were prepared and applied did not always travel with them. It was even more rare for the person who sourced the drug material to learn and then import the preparation method to Europe directly and practice it himself. That after his return to Europe Garcin prepared drugs according to methods employed in Siddha medicine makes the Maduran pills stand out as a particularly unusual European appropriation of exotic pharmaceutical knowledge. Even the German Pietists who did so much to translate and preserve Tamil medical manuscripts did not, to our knowledge, import Siddha drug manufacturing processes to Europe.

Other aspects of the Maduran pills case likewise deserve notice, in the context of the exotic drugs being circulated, traded, and consumed in the early eighteenth century. One of these is secrecy, which appealed to both sides in the exchange of pills for bezoars with which this chapter began. Tamil physicians preserved their traditional proprietary drugs within their own family lineages. With a few exceptions, contemporary practitioners following Siddha medical traditions still maintain secrecy about their remedies.[102] Garcin, too, prized secrecy, probably for multiple reasons; despite the apparent pressure exerted by patients, he was only prepared to reveal the pills' ingredients, but not their preparation process, which he took with him to his grave. Patients' concerns expressed at the time foreshadow issues that dog the drug industry and the domain of intellectual property today: What do medicines contain, and how are they produced? For example, do vaccines—especially the new mRNA variety—cause harm? Should manufacturers be allowed to hold trade secrets that concern human health?

Finally, this case illustrates the crucial role of indigenous knowledge in the circulation of drugs, and their manufacture and application. In the gift exchange central to this study, traditional medical knowledge passes to a European, while a local medical expert receives in return precious materia medica worth at least their weight in gold. In a medical culture that prized secrecy, such knowledge was not normally sold or bartered; even exchanging the secret formula against a gift of equal value was probably unusual. So what is Garcin's attitude toward indigenous knowledge and collaboration with indigenous sources? The Brahmins, he consistently asserts, are excellent physicians; the cures they achieve are admirable. More generally, South Asian natural knowledge and materia medica circulated without due attribution by the Arabs who dominated the sea trade between Asia and the Mediterranean–that is to say, Garcin laments South Asians' erasure from contemporary accounts of knowledge circulation.[103] Garcin likewise praises Chinese and other indigenous doctors he encountered in the Indies, and in his works, he liberally cites indigenous plant names.[104] Yet, like the majority of his contemporaries, Garcin fails to provide any identifying details of his contacts, not even their given names. In some cases, such omissions may have been unavoidable. In the present case, Garcin may not even have directly encountered the "Brahmin," and perhaps had little, if any, information about the governor's honored guest. Unfortunately, as in so many other instances, this means that local informants who provided crucial knowledge remain faceless and anonymous.[105]

Why did Europeans of Garcin's time usually fail to describe the people who performed essential roles as informants, assistants, and the like? In response to this erasure or near-erasure of all of those who helped to make projects such as Garcin's possible, Daniel Rood proposes that it is time to "recast the global history of science as a coercive form of labor management."[106] I would argue that Garcin's acquisition of the Maduran pills does not fit this model, but it is nonetheless crucial to bear in mind how much Garcin and so many other Europeans profited from knowledge obtained from people rendered virtually faceless and nameless in the historical record. The story of the Maduran pills is but one poignant testimonial as to how knowledge of exotic substances and their uses passed from East to West around 1700, with only limited acknowledgement of indigenous agency. More than that, if even that, was not expected.

FIGURE 10.1. Mercurius in the storehouse. Michael Bernhard Valentini, *Museum Museorum, Oder Vollständige SchauBühne Aller Materialien und Specereyen*, 2nd ed. (Frankfurt am Main: Johann David Zunners Erben and Johann Adam Jungen, 1714), frontispiece. Image courtesy of the Hagströmer Library, Karolinska Institutet.

10

MERCURIUS IN THE STOREHOUSE

*The Exotic in the Here and Now
of Early Modern European Experience*

Hjalmar Fors

On the frontispiece of Michael Bernhard Valentini's *Museum Museorum* (1714) is an image of the Greco-Roman god Hermes/Mercurius, easily identifiable by his winged boots and cap.[1] Mercurius is surrounded by a wealth of symbols associated with his sphere of influence as the god of merchants and physicians. On his chest is the astrological/alchemical symbol of mercury. In his right hand is his herald's staff, the *caduceus* entwined by two serpents representing his status as the messenger of the gods. In his left hand is the message he conveys: a letter on which can be read the word "East-India" in German (*Ost-Indien*). Looking at his surroundings, we see that he is standing in a storehouse, surrounded by crates, sacks, and barrels. With regard to goods, all three kingdoms of nature are represented in the image. On the shelves at the back of the room sit labeled boxes, containing rare salts, metals and minerals, animal parts, and strangely shaped shells. The vegetable kingdom is present in even greater quantities. It is represented in the large sacks at Mercurius's feet, labeled as containing tree bark, flowers, roots, and seeds.

In a way, this image of Mercurius in a storehouse represents a microcosm of what early modern Europeans brought with them from the two Indies: vast numbers of objects from the plant realm used as spices and medicines, as well as a smaller number of objects from the mineral and animal realms. The latter, too, were used as medicine and food, but were, just as importantly, brought along to satisfy European curiosity and sense of wonder through their strangeness and novelty with regard to shape, color, and smell. Along with these goods was carried knowledge about the goods themselves, and about the foreign places and cultures from which they were taken or traded. This knowledge is represented by the letter that Mercury holds in his hand. Indeed, Valentini's book contained not only a copiously illustrated cabinet of curiosities in textual form, but also an addendum, in the form of German translations of key texts and reports about the East Indies, penned by learned Europeans who had traveled there.

But why let Mercurius—the god of physicians, merchants, and thieves—symbolize European overseas activities and expansion, rather than say, Mars, the god of war, Minerva, goddess of wisdom, or Jupiter, the king and ruler of gods? In fact, Mercurius is an excellent choice, a choice that tells us something important about how early modern Europeans conceived the world outside Europe, and what it was good for: the uses that Europeans actually had for exchanges with non-European spaces. Mercurius is a messenger, but also a traveler and translator: a deity who could make the unfamiliar familiar by imposing European ordering categories and set tropes.

Even more obvious is Mercurius's connection to trade, which brings us to the main proposition of this essay. In it, I suggest that the peoples of early modern Europe did not primarily conceive the world outside Europe in spatial terms, i.e., as an unknown place to conquer, understand or relate to, or a territory to be dominated and ruled. I suggest that they saw it as a more or less unknown foreign place of little consequence to them, except as a source of desirable trading goods. Following Edward Said's influential analysis of Orientalism and exoticism, many scholars have perceived European views of the exotic as a form of identity formation, proceeding from the binary pair of Self-Other. Here, I suggest that perhaps early modern Europeans simply did not care enough to formulate such an opposition. In place of an Orientalism of identity (broadly construed), I would suggest that early moderns' primary concern was to exoticize objects, rather than people or places. This *Orientalism of objects* is the subject of the present essay.[2]

Early modern categories were not our modern, twenty-first-century categories. We late-moderns tend to take pride in our ability to make sense of the world in terms of difference, whether these differences are expressed in terms of culture, geography, or climate.[3] With this approach to the Other comes a lot of other assumptions, as well as inherited cultural goods (such as fear of going native, or joy of acceptance by the Other; the possession of an exploratory cartographic gaze; a tendency to hierarchize cultures in various colonial schemes, or a dislike of this activity, etc.). Most of these viewpoints were developed, or at least elaborated, in the nineteenth and twentieth centuries.[4]

Early moderns, on the other hand, insofar as they engaged with the non-Christian/Jewish parts of the world at all, tended to do so as traders, looters, or treasure hunters, though there was also a decidedly high level of colonial domination and exploitation going on. I would argue that this early modern engagement did not, primarily, proceed from an articulation of difference. Rather, it proceeded from a desire for assimilation, grounded in the direct sensory appreciation of objects present in the here and now. We can perceive this as an engagement led by the devouring stomach, rather than by the evaluating gaze. Mind you, this type of engagement through assimilation had both purpose and intellectual aspects: it was a significant feature in the encounter between Galenic medicine and non European drugs and medical traditions.[5]

I will suggest the outlines of an early modern outlook, which also may be called a world-view, epistemology, or even ontology, depending on how far one wishes to take this argument. Whatever weight we assign this outlook as an historical or analytical entity, I would say that it has rarely been acknowledged by historians who study the relationship between Europe and the wider world of early modernity.[6] In the conclusion, I will also discuss how my position can be taken to be an elaboration of one of the key insights presented in Edward Said's highly influential *Orientalism* (1978).

A further aim of this essay is to remind historians of global interactions around trade and knowledge exchanges that the vast majority of early modern Europeans never once left their home country, and presumably cared little about what happened outside of its borders. If we focus our attention on them, rather on scholarly specialists and the world-traveling few, it becomes apparent that the world outside of Europe was first and foremost a place that produced unusual and curious objects, stories, and images that were consumed *at home*. Trade goods offered Europeans a way to relate to faraway places as something more than fairy-tale stories, and fixated meaning that would otherwise be unstable and fleeting.[7] This may explain why many early moderns could make do with surprisingly imprecise distinctions between residents of, and goods from, the Americas, Asia, and Africa. Distinctions of difference were imprecise because they were assigned little importance. Factors and relationships that seem highly important to late modern observers such as ourselves seem to have carried much less weight with early moderns. In the Americas, Europeans encountered wholly unknown peoples and customs. In the aftermath of the encounter, the epidemic spread of Old World diseases eradicated large parts of the original populations. Together with superior European weaponry and other factors, this catastrophe soon permitted Europeans to dominate, exploit, and commit large-scale genocide on Native American populations. India, on the other hand, had been a known entity and distant trading partner for Europeans since at least antiquity. During the early modern period, locals and Europeans were interacting on an a more or less equal basis, and up until the second half of the eighteenth century, Europeans were consistently thwarted in their attempts to gain a colonial foothold in the Indian interior.[8] Nevertheless, most early modern Europeans seem not to have cared much whether the word "Indian" referred to a resident of India proper, a Native American, or for that matter a person of European descent born in the Dutch East Indies. Thus, people to whom late moderns assign different categories were easily assimilated to a single category of the previously known. Furthermore, although they were given the label of "Indian," they did not lose the feature which was most important: that of being an exotic curiosity. Or, to use the words of William Shakespeare, commenting on the fad for exotic curiosities among his contemporaries, "What have we here? A man or a fish? Dead or alive? A fish. He smells like a fish, a very ancient and fish-like smell. . . . A strange fish! Were I in England now, as once I was, and had but this fish painted, not a holiday fool there but would give a piece of silver . . . When they will not give a doit to relieve a lame beggar, they will lay out ten to see a dead Indian."[9] There were, of course, several discourses concerned with how to understand and order the vast influx of novelties that threatened to overwhelm the minds and sensibilities of early modern Europeans. There was, for example, a substantial market for travel narratives, as well as the emergence of what has been termed "proto-ethnographic" texts. But there are reasons to be suspicious of the written word. It is possible to discern a disconnect between discourse in writing and European sensuous appreciation of objects. While production of the former was predominantly an elite endeavor, the latter took place in all parts of society, and at most social levels. The notion that customers and users did not rely entirely on descriptions and medical interpretations supplied by intermediaries and professionals is supported by several contributions to this volume. For example, the contribution by E. C. Spary notes that large quantities of medicinal simples were sold directly to customers by grocers, presumably to make home remedies.[10] Similarly, the contributions by Wouter Klein and Alexandra Cook suggest that the advertisement and sale of secret remedies could be a highly viable marketing strategy.

It is also important to remember that written texts can have several purposes and be aimed at several different audiences at once. In their influential study, *Wonders and the order of nature* (1998), Lorraine Daston and Katharine Park showed how medieval and early modern elite texts describing wonders tended to be integrated into or proceed from a moralizing discourse.[11] A similar point is made by Samir Boumediene in this volume. For our purposes here, let us note that such discourses did not necessarily appeal to a broader audience, which perhaps did not want to have its exotic objects of consumption delivered together with sermons and/or propaganda. Indeed, Boumediene notes that, although the Jesuit order was a major importer and exporter of certain exotic remedies, a majority of Jesuit works on these topics remained unpublished until the eighteenth century.[12]

Nevertheless, even in the printed medium, interest in material *things* often tended to take center stage. Books which are nowadays characterized as belonging to different genres, such as travel narratives, works in natural history, pharmaceutical texts, and books on materials, often discussed and showcased, side by side, objects that were unique or inaccessible, objects that could be purchased at great cost, and objects (such as spice) that could be picked up at a local apothecary shop. In this way, unknown worlds were made intelligible and accessible, and imagination and wonder, but also a sense of comprehension and control, were connected to European overseas trade and economic exploitation.[13] Valentini's *Museum Museorum* is an excellent example of this. Assembling information and images from a vast number of sources, Valentini created a representation of a cabinet of curiosities in the form of a hefty folio volume intended for a broad, German-speaking audience. Scholarly authors (whose ambitions were somewhat less overtly commercial) presented the world in similar ways. Willem Piso's and Georg Marggraf's *Historia Naturalis Brasiliae* (1648) serves as an example, with its numerous illustrations of South American animals, fish, plants, and stones (see fig. 10.2).[14] The bottom line is that objects from the three realms of nature—including also, as we have seen, remains of humans, as well as human-made artifacts—remained firmly embedded at the center of cultural attention, not least because to own and consume exotic spice, medicine, clothes, and objects was the height of fashion, and these practices were signifiers of power, wealth, and cultural refinement. The contrary was true too, as the individual, household, city, or state that had no or little access to beautiful or strange exotic things accordingly showcased poverty and insignificance.[15]

Consequently, it makes sense to assume that the world outside Europe was primarily seen as a repository of objects that could be funneled, or *assimilated*, into Europe through trade or conquest. Places were not represented in European minds as geographical locations ("spaces") far away, but as goods present in the here and now. If this proposition is correct, it follows that the early modern view of the world differed significantly from our late modern one. After several hundred years of expanding global trade, Europeans, or perhaps rather the global upper and middle classes, are now in a place where they expect that even their fresh vegetables should travel regularly between continents. Consequently, the thread of continuity has been severed with the sense of exoticism that early moderns associated with, say, the drinking of a cup of tea or the smell of cinnamon. Instead, affluent late moderns acquire a sense of the exotic from very different practices, whether these are grounded in fantasies about authentic cultures, consumption of images of foreign vacation spots, or the perusal of statistical or newspaper reports of faraway poverty, crime, oppression, and war. Unlike late moderns, the majority of early modern Europeans would have had very vague ideas about what was going on in faraway places, and what could be found there. Insofar as they cared to do so, they would probably have furnished these places with objects they had seen, and which they believed to come from those places. Simultaneously, they would have had rather precise knowledge about the spices they consumed and the silks they wore (or longed to wear), and vivid memories of any parrots or monkeys, or indeed deceased Indians, they might have paid to gawk at. Hence my suggestion that places were not primarily represented in European minds as geographical locations far away, but as objects and images, smells and tastes present in the here and now.

Let us consider the meaning of China. Today, the word "China" tends to be regarded as a place name. Early moderns, on the other hand, could use it as an adjective that described objects with certain

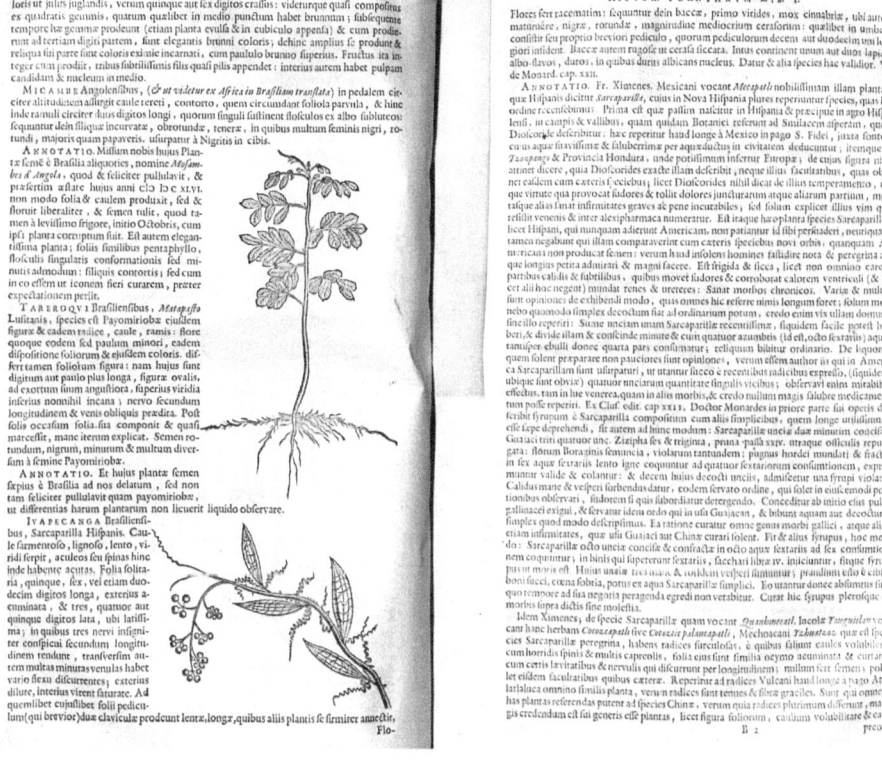

FIGURE 10.2. Illustration of American sarsaparilla with marginal annotation of the plant's Linnean name (possibly in the hand of Daniel Rolander). From Georg Marggraf, *Historia plantarum liber 1*, 10 (detail). From Willem Piso and Georg Marggraf, *Historia Natvralis Brasiliae* (Leiden: Franciscus Hackius and Amsterdam: Lud. Elzevirium, 1648). Image courtesy of the Hagströmer Library, Karolinska Institutet.

properties. In English, the word even made the transition from adjective to substantive, becoming the common English word for true porcelain. The reason, of course, is that until the early eighteenth century, Chinese manufacturers and traders held a global monopoly on this desirable product. The English name of porcelain reflects a conflation of, or at least a perceived similitude between, the product and its place of origin.[16] There is another, even more illuminating, example. Let us return again to Valentini's *Museum Museorum*, and its entry on "China-wurtzel," that is, China root. Valentini says that this root came from the East Indies, indeed from China, and claims that it had medical properties not dissimilar to those of the North American plant sarsaparilla.[17]

However, the substance known as "China-China," or fever-bark, did not come from China, or even from the East Indies, but from Peru.[18] Similar terminological entanglements can be found among several authors, for example, Georgus Francus in his *Lexicon Vegetabilium Usualium* (1672). Here, Francus claimed that "China" is a root, also called "Cina" and "Chinna," which can be found in the Orient, or in (the land of) "Sina" as well as in the Occident, that is, Peru and New Spain.[19] It may well be that the drugs these authors were discussing were *sarsaparilla china* (China root), *sarsaparilla aristolochiifolia* (american sarsaparilla), both of which are species within the genus *Smilax*, and the drug "Peruvian bark," derived from various species within the genus *Cinchona*, native to South America, and from which the drug quinine would later be derived. But my point is this: When these authors talk about American botanicals as "China," it is not a mere matter of linguistic and etymological confusion, but serves an intelligible purpose. Peruvian bark is a good example. Although first known in the Netherlands as "Pulvis Indicus," "Cortes Indicus," "Cortex Americanus," or "Indian" or "American Bark," this drug would come to be called "China" in several European languages, including Italian, Danish, and Swedish. In the German lands, the names "Cortex China de China," "Cortex Chinensis," "Cortex Sinensis," "Gentiana Indica," and "Chinese bark" were names still in use in the early nineteenth century.[20] Traditionally, it has been assumed that the term was derived from the word "Quina" in the South American Quechua language. But there is another possible explanation, namely that the word "China" was not a reference to the drug's geographic origin, but a statement about its potency, aimed at potential customers.

There was a long European tradition of associating the East with particularly effective medicines, and the continent's pharmacies depended strongly on exotic botanicals. This was in part a consequence of European dependence on the classical pharmacopoeia, which had been codified in Roman times, during a previous era of global trade. Indeed, the medical tradition of Galenism, as well as the theriac of Andromachus, its most prized medicament, were initially created during Hellenic and Roman times. Both made use of a wide variety of substances that could only be obtained through extensive intercontinental trade (see the chapters by De Vos and Di Gennaro in this volume.) Some of these substances were simultaneously well known and unknown entities. They were known, insofar as they had a long history of European usage, and were well-known as to their effect, and the way they looked, smelled and tasted in the form in which they reached Europe. But they were also unknown, insofar as, throughout much of the early modern period, several of the traditional substances of the European pharmacy grew in unknown places, and were sourced from plants which had neither been described nor seen by Europeans.

Places such as India and China (or "Cathay"), were traditional spaces for projecting the strange and marvelous, and could serve as the places of origin for substances that could work miraculous cures.[21] In the 1560s, the ambassador Ogier Ghislain Busbecq quickly embarked upon this topic when reporting on his meeting with an unnamed "Turkish vagabond," who claimed to have traveled to China.

> Now let me tell you what I heard about . . . Cathay from a certain Turkish vagabond.—I thought I would ask if he had not brought back from his travels any curious kind of root, or fruit or pebble or what not? "Nothing whatever," he said, "except this little root that I carry about with me, and if I am knocked up with fatigue or cold, by chewing and swallowing a tiny morsel of it, I feel quite warmed and stimulated." And so saying, he gave it to me to taste, telling me to be careful to take but the smallest quantity. My doctor . . . tasted it, and got his mouth into a state of inflammation from its burning quality. He declared it to be regular wolfsbane.[22]

A contemporary account (ca. 1550) to that of Busbecq was written by Giovanni Battista Ramusio, a Venetian. Ramusio told of a meeting with Chaggi Memet, a Persian trader who had visited China. Memet described the high esteem in which the Chinese held various locally available drugs, among them a wonderfully beneficial root which he called "Mambroni Cini" and which he did not think was imported to Europe. He also described the use of a substance which clearly was tea, claiming that it could remove "fever, head-ache, stomach-ache, pain in the side or in the joints." This substance was so highly valued that the Chinese would gladly give "a sack of rhubarb for an ounce of Chiai Catai."[23]

Medicinal rhubarb was highly esteemed in Europe, due to its excellent purgative properties, and its origins in China, rather than some other faraway place such as Ethiopia, were just about to become more generally acknowledged.[24] Memet said that rhubarb grew in abundance in the Chinese region of Campion (Ganzhou/Zhangye).[25] There they did not use it for medicine, but instead as a component of incense, or as fuel, "whilst others give it to their sick horses, so little esteem have they for this root in those regions of Cathay."[26] Stories such as these sent a clear message to Europeans. Rhubarb's purgative properties were highly valued in the Galenic medical tradition, due to its emphasis on keeping the body open to allow disease to pass through it. Nevertheless, pervasive reports would have it that the Chinese had little regard for the drug: the reason being, of course, that they had even more powerful medicines in their arsenal. Matteo Ricci, in his account of the Jesuit mission to China (published 1615) also commented on the wealth of medical herbs available in China, among them "that famous remedy for many diseases, called Chinese Wood by the Portuguese and Sacred Wood by others."[27]

As is evident from these examples, the idea that the Chinese had access to wondrous medical substances unknown in the West was a powerful trope that was pervasive in both popular accounts and the reports of travelers. But for historians to try and ascertain precisely which medicines these authors were talking about, and from whence they "really" came is to miss the point.[28] This is to let Minerva, goddess of Learning, take precedence over Mercurius. In my view, the imported object did not represent its place of origin. Rather, the place of origin was imagined and shaped through the object.

This line of reasoning only makes sense if we acknowledge that the most significant property of these objects was their exoticism, and that *exotic* was primarily a sensory property. Only objects that were alien, wonderful, colorful, and strange came out of the fabled realm of Cathay.[29] From this, it followed that Cathay must be a place endowed with these properties. Add to this that distance lends enchantment, and that enchantment adds to the commercial value of goods, and it is easy to see that the East India trade could have real benefits from this specific brand of exoticism. Since China remained closed to almost all Europeans but the Jesuit order, it could continue to serve as a repository for wondrous, weird, and fantastic objects throughout the early modern period. In this sense, it performed a similar function in the European imagination to the fabled realms of gold reputedly hidden in the African and American interiors. But whereas the latter imaginary spaces carried the association of hosting primitive societies, ready to be conquered by enterprising Europeans, China and East Asia could be used as projection spaces from which came secrets and objects connected to civilization.

It may seem paradoxical that *not knowing* might fill an important cultural function. Yet for the discourse on marvelous Eastern medicines to work, too precise a knowledge of its object must be absent. A discourse of not knowing—an agnotology of exotic objects—to use a term coined by Robert Proctor, does not exclude but complements a discourse of knowing. Some groups produce value from providing knowledge, others from withholding it. A comparison can be made: Throughout early modernity, many highly popular patent remedies were sold by both traveling mountebanks and established practitioners (see Klein's and Cook's contributions to this volume.) Such medicaments, as a rule, were of secret composition. This thriving market coexisted with pharmacies, which sold strictly controlled and regulated medicinal simples and compositions. A space of uncertainty and wonder could be filled with meaning and commercially exploited alongside a space of (imagined) certainty.

Thus, this essay does not imply that early modern production of knowledge about the world outside of Europe, and of import substances, was unimportant or inconsequential. Clearly, a great number of early modern actors derived both meaning and profit from the production of knowledge about the world. The aim of this essay is more modest: to investigate a possible disjunct between scholarly, commercial, and medical discourses on the one hand, and consumption practices on the other. It is this disjunction, or empty space, which I believe can be explored using notions such as *agnotology* (see below) and experimenter's regress.

This latter key concept in the sociology of science points to the inherent circularity of knowledge claims.[30] By shifting from an object-centered perception of the world ("only fabulous objects come from the fabled realm of Cathay, hence Cathay is a fabled realm") to a spatial-geographical perception of the world, the loop of internal European references and correspondences was broken. Geographical knowledge and the connection of exotic objects to geographical sites inscribed on maps thus fulfill a function of furthering knowledge, but also of making non-European spaces available for exploitation. But something is also lost in this operation. It is time for the god Mercury to make his exit, and to be replaced by Jupiter and Mars: that is, imperial domination and subjugation of territories through war. Such activities allowed Europeans increased control of overseas production. Thus they permitted trade in exotic and rare substances to be replaced with bulk imports of mundane consumables. Untangling this change, we have also uncovered a tension which runs through the centuries-long story of Western expansion. We can perceive how an older European culture operating on the principle of assimilation through trade, plunder, and wonder was gradually replaced with a culture preoccupied with colonization and extraction, i.e., colonial-territorial domination.[31]

The period between roughly 1670 and 1740, which is the focus of this volume, is significant. During this period, previously rather modest imports grew to a torrent, and the consumption of exotic produce of all kinds penetrated deeply into wide social strata, as well as geographically into European interiors and peripheries (see E. C. Spary and Justin Rivest's introduction to this volume.) As a consequence, relative indifference may gradually have been replaced with a need to relate to a steady influx and presence of non-European objects. Perhaps it was this "need to relate" that eventually would be met by the eighteenth-century quest for order, and in particular by Linnean natural history. This elaborated powerful methods to handle the influx of the exotic in an orderly, and rationally comprehensible way, which furthermore made it possible to further and accelerate commodification without any threat to existing European worldviews and sensibilities.

The quest for new medicaments was a central driving force behind natural history prior to Linnaeus. It focused on medical and commercial utility. It connected new objects to previously known ones through assimilation, thus turning them into objects of use for European medicine. As observed by Kroupa, Boumediene, and De Vos in this volume, such an outlook both made the introduction of new drugs and the use of substitute drugs, *succedanea*, rather unproblematic. This was because the identity of a substance was of less importance than its effect on the patient. Natural history was thus ancillary to the Galenic medical system. Admittedly, it was also quite often rather difficult to establish a secure correspondence between an object (substance), its name, and its effect. It was this underlying conceptual fluidity which was challenged by Linnaeus. His system was not based on the notion of assimilation to the previously known based on similitude. It proceeded from the careful, scrupulous establishment of difference. Comparison is an effective method to establish identity. It does so by establishing a semantic field in which differences are produced. Thus, we can see how Linnean science created a watershed moment, with a clear before and after (see Cook's chapter in this volume). That this watershed moment would eventually create a deep ontological disjuncture is attested to by the fact that we late moderns still have such apparent difficulty in understanding the nature of natural history, and of Galenic pharmaceutical endeavors, prior to Linnaeus.[32]

And, as I have suggested above, an important aspect of this shift is the move from attention to similitude followed by assimilation, to the careful scrutiny of difference. The establishment of difference was then followed by hierarchical ordering. This mode of reasoning is of course what Edward Said, in his famous study, uncovered in Western ethnographical descriptions of the Orient. Although the ethnographers studied by Said proceeded from the Self-Other divide, and Linnaeus proceeded from pistils and stamens, their *modus operandi* were similar. It can thus be argued that the Orientalism scrutinized by Said, which would not begin to emerge properly as a cultural feature until the decades around the beginning of the 1800s, began in natural history, with the emergence of the Linnean conception of identity based on difference. This model of perceiving the world replaced an earlier conception of identity based on similitude followed by assimilation. And it is interesting to note that this epistemic shift happened in conjunction with the politico-economic shift from an extractive commercial imperialism to an imperialism focusing on territorial domination.

Colonial cartography and natural history reinforce the bond between ruler, territory, and the resources in that territory. They are forms of knowledge that create links that make it possible to take symbolic control over not just trading stations, but also the inland territories where the trading goods are produced. Here, too, is located the experimenter's regress. Because, although colonial empires amassed knowledge, that knowledge only partially managed to break the loop of internal references and correspondences within European culture. Some parts of the traditional fantasies about non-European knowledge and goods remain to this day. When all Western medicine fails, many still think that there may be "Oriental" acupuncturists, yogis, and healers who can perform miracle cures. And there is still a place in the Western cultural imagination for the concept of the *Fabulous Chinese Medical Substance*™. This place has been held by medicinal rhubarb, tea, even Peruvian bark, and is presently held by ginseng, which, in all likelihood, will soon be replaced by some other substance from the Chinese pharmacopoeia.

AN EXTRACTIVE ORIENTALISM OF OBJECTS

In *Orientalism* (1978), Edward Said freely admitted that the starting point of his study was the end of the eighteenth century. He studied the emergence of modern Orientalism as an ideology of an ascendant West that had the underlying aim of constructing "the Westerner" as rational and manly, and casting "the Oriental" as the irrational, feminine Other, thus facilitating colonial subjugation of *the Orient* through (mis-)representation.[33] Said's book remains a foundational text of postcolonial studies. It has had an immense influence, and has also engendered a healthy debate in the field of early modern history. It is nevertheless cumbersome, and probably not at all desirable, to transpose Said's theory wholesale to the period before about 1800.[34] But it is useful to adopt one of Said's basic premises: that Western representations of the Other were basically Eurocentric and unidirectional.[35] Drawing on recent work conducted in the history of material culture, this essay develops this idea. It suggests that early modern Europeans in Europe may have considered the projection of representations into external geographical and cultural space as a rather unimportant, inconsequential activity. Instead, it was for the most part objects that filled the symbolic role of representing both the Self and the Other. But while Said argued that the aim of Orientalism was (and is) to gain symbolic control of *the Orient* as part of a colonial system, early modern Orientalism worked under other constraints.[36] It was less concerned with *domination* than *extraction*, and had the ultimate aim of extracting and assimilating specific objects from non-European regions, and facilitating their transformation into European objects of use.[37] Hence European representation was not primarily focused on distant *spaces*, but rather on *objects* present in Europe. Given European preoccupation with physical objects, and comparative lack of interest in other aspects of non-Christian/Jewish cultures, what is studied here may be described as cases of *agnotology*, a term coined by Robert Proctor to signify the scholarly study of non-knowing, and the not-known, and defined by Londa Schiebinger as the study of "culturally induced ignorances" or "instances of the nontransfer of important bodies of knowledge."[38] Proctor distinguishes between three basic types of ignorance, "ignorance as a native state (or resource), ignorance as a lost realm (or selective choice), and ignorance as a deliberately engineered and strategic ploy (or active construct)."[39] Early modern non-knowing about the world had aspects of all three: (1) It existed due to lack of information about the world that would later become available. (2) It was a choice, thanks to a selective preoccupation with things rather than spaces. And (3) it was maintained by traders and others in order to bolster the value of traded goods, and keep such goods within a discourse of exoticism.

However, such latter-day interpretations serve as a distorting filter too. The early modern system of knowledge was, above all, an instance of itself. We should be careful about expressing outrage or irritation at the early moderns for not caring about, or seeing, the same things as we late moderns do. We should acknowledge that early moderns thought deeply in ways we do not readily understand, and that they sought to orient themselves in their world using the tools at their disposal. Or to put this differently: European understanding of non-European regions was conditioned and created by European priorities and pursuits and perceptions, by available knowledge and interpretative frameworks. From this, it follows that we should be very careful when we discuss encounters between Europeans and inhabitants of non-European regions. Of course, as historians we want to uncover historical processes that make sense from the point of view of our present concerns. And of course we should be aware of instances of colonial exploitation, or encounters laden with positive, liberal associations, such as cultural exchange, circulation of knowledge, or the establishment of contact zones. These things existed and happened.[40] But nevertheless, such things mattered little to most early modern Europeans in Europe: Their aims and interests were other, and should be acknowledged as such.

Within early modern Europe, foreign geographies, material environments, and cultures were of little consequence, except as sources of objects and knowledge useful to Europeans. Perception and understanding of the world outside of Europe proceeded from the handling of physical objects. From this, it follows that comprehension also included full-body perception, grounded in the sensory experience of smell, taste and touch, as well as sight and hearing. There was curiosity and interest in foreign customs and geography. There was a rage for travel accounts and maps. But, from the point of view of stationary Europeans, these discourses about what was going on in faraway places and what could be found there were overshadowed by something much more material and tangible: objects that could be eaten, drunk, smoked, handled, and gawked at.

NOTES

INTRODUCTION: THE TROUBLING EXOTIC

1. Wolfgang Schivelbusch, *Tastes of Paradise: A Social History of Spices, Stimulants, and Intoxicants* (Pantheon Books, 1992).

2. All citations come from *Correspondance complète et autres écrits de Guy Patin*, 4th ed., ed. Loïc Capron, May 2022, online at:

http://www.biusante.parisdescartes.fr/patin/ (accessed October 12, 2019). Letter 151, Guy Patin to Charles Spon, March 10, 1648. All translations by the authors.

3. *Correspondance*, letter 153, Guy Patin to Charles Spon, March 24, 1648.

4. *Correspondance*, letter 504: Guy Patin to Charles Spon, November 23, 1657.

5. *Correspondance*, Latin letter 87, Guy Patin to Johannes Antonides Vander Linden, June 22, 1657.

6. *Correspondance*, Latin letter 87, Guy Patin to Johannes Antonides Vander Linden, June 22, 1657.

7. *Correspondance*, letter 334, Guy Patin to Charles Spon, December 16, 1653.

8. *Correspondance*, Latin letter 87, Guy Patin to Johannes Antonides Vander Linden, June 22, 1657.

9. *Correspondance*, Latin letter 118, Guy Patin to Sebastian Scheffer, March 7, 1659.

10. A summary of medical objections to tea is in Philippe Hecquet, *Traité des Dispenses du Carême* (Paris: François Fournier and Frederic Leonard, 1709), 482–88. See, in particular, Simon Paulli, *Commentarius de Abusu Tabaci Americanorum veteri, et Herbæ Thee Asiaticorum in Europa Novo* (Argentorati: filii Simonis Paullii, 1665).

11. Markman Ellis, Richard Coulton, and Matthew Mauger, *Empire of Tea: The Asian Leaf That Conquered the World* (Reaktion, 2015); Woodruff D. Smith, *Consumption and the Making of Respectability, 1600–1800* (Routledge, 2002); Chris Nierstrasz,

"The Popularization of Tea: East India Companies, Private Traders, Smugglers and the Consumption of Tea in Western Europe, 1700–1760," in *Goods from the East, 1600–1800: Trading Eurasia*, ed. Maxine Berg, Felicia Gottmann, Hanna Hodacs, and Chris Nierstrasz (Palgrave Macmillan, 2015), 263–76.

12. Mary J. Dobson and Steven R. Meshnick, "The History of Antimalarial Drugs," in *Antimalarial Chemotherapy: Mechanisms of Action, Resistance, and New Directions in Drug Discovery*, ed. Philip J. Rosenthal (Springer, 2001), 15–27. Quinine was gradually replaced by various synthetic antimalarials from the 1920s through to the Second World War. For recent historical studies, see especially Matthew James Crawford, *The Andean Wonder Drug: Cinchona Bark and Imperial Science in the Spanish Atlantic, 1630–1800* (University of Pittsburgh Press, 2016); Samir Boumediene, *La Colonisation du savoir. Une histoire des plantes médicinales du "Nouveau Monde" (1492–1750)* (Les Éditions des mondes à faire, 2016).

13. Marcy Norton, *Sacred Gifts, Profane Pleasures: A History of Tobacco and Chocolate in the Atlantic World* (Cornell University Press, 2008), 141, 260.

14. Carla Nappi, "Surface Tension: Objectifying Ginseng in Chinese Early Modernity," in *Early Modern Things: Objects and Their Histories, 1500–1800*, ed. Paula Findlen, 2nd ed. (Routledge, 2021), 31–52, at 31.

15. Renata Ago, "Denaturalizing Things: A Comment," in *Early Modern Things*, 4 ed. Findlen, 433–38, at 34.

16. See for instance Londa Schiebinger's work on the peacock flower, widely known as an abortifacient in the Caribbean, but whose uses as such were not transmitted to Europe: *Plants and Empire: Colonial Bioprospecting in the Atlantic World* (Harvard University Press, 2004).

17. For a similar ambivalence to exotic novelty in the Dutch case, including over tea, see Thijs Weststeijn, "Unease with the Exotic: Ambiguous Responses to Chinese Material Culture in the Dutch Republic," in *Making Worlds: Global Invention in the Early Modern Period*, eds. Angela Vanhaelen and Bronwen Wilson (University of Toronto Press, 2022), 436–76.

18. Nancy G. Siraisi, *History, Medicine, and the Traditions of Renaissance Learning* (University of Michigan Press, 2007), chapter 7; Joseph M. Levine, *Between the Ancients and the Moderns: Baroque Culture in Restoration England* (Yale University Press, 1999); Anthony Grafton, April Shelford, and Nancy G. Siraisi, *New Worlds, Ancient Texts: The Power of Tradition and the Shock of Discovery* (Belknap Press, 1992).

19. Bibliothèque Interuniversitaire de Santé, ms 2007, f°. 59r.; Carolus Clusius (Charles de l'Écluse), *Exoticorum libri decem: Quibus Animalium, Plantarum, Aromatum aliorumque peregrinorum Fructuum historiæ describuntur* ([Leiden]: Raphelengius, 1605).

20. Dipesh Chakrabarty, *Provincializing Europe: Postcolonial Thought and Historical Difference* (Princeton University Press, 2000).

21. Calvert Watkins, ed., *The American Heritage Dictionary of Indo-European Roots*, 3rd ed. (Houghton Mifflin, 2011), 16, 17.

22. Antoine Furetière, *Dictionnaire universel François & Latin*, 3 vols. (Trevoux: Estienne Ganeau, 1704), vol. 2, art. "Exotique."

23. See, for example Exodus 30:23, Proverbs 7:17, Cant. 4:14, Revelations 18:13.

24. See Christoph Mauntel, "The T-O Diagram and Its Religious Connotations," in *Geography and Religious Knowledge in the Medieval World*, ed. Christoph Mauntel (De Gruyter, 2021), 57–82.

25. On the antiquity of European views of the Levant as a source of mystical and potent drugs, see especially Myerson, *The Desire for "Syria" in Medieval England* (Cambridge University Press, 2025), chapter 2; Schivelbusch, *Tastes of Paradise*; Jean-Louis Flandrin, "Assaisonnement, cuisine et diététique," in *Histoire de l'alimentation*, ed. Jean-Louis Flandrin and Massimo Montanari (Fayard, 1996), 491–509; Paul Freedman, *Out of the East: Spices and the Medieval Imagination* (Yale University Press, 2008), chapter 1. On concerns with civilization and civility in Europe, see especially Bruce Mazlish, *Civilization and Its Contents* (Stanford University Press, 2004).

26. For some examples, see Thabit J. Abdullah, *Merchants, Mamluks, and Murder: The Political Economy of Trade in Eighteenth-century Basra* (State University of New York Press, 2001), 72–80; Francesca Trivellato, *The Familiarity of Strangers: The Sephardic Diaspora, Livorno, and Cross-cultural Trade in the Early Modern Period* (Yale University Press, 2009), chapter 9; Peter Frankopan, *The Silk Roads: A New History of the World* (Bloomsbury, 2018), chapter 13; Sebouh Aslanian, *From the Indian Ocean to the Mediterranean: The Global Trade Networks of Armenian Merchants from New Julfa* (University of California Press, 2014).

27. On the Eastern location of the Garden of Eden and its connection to spices, drugs, etc., see Freedman, *Out of the East*, chapter 3. On the local/exotic dispute in early modern medicine, see especially Alix Cooper, *Inventing the Indigenous: Local Knowledge and Natural History in Early Modern Europe* (Cambridge University Press, 2007).

28. Edward Said, *Orientalism* (Pantheon Books, 1978).

29. The Portuguese and Venetians, already succeeding in Asiatic trade by 1600, were displaced by the Dutch by around 1650, with the English and French following them; the Spanish and Portuguese began colonizing the Americas first, followed by other European powers. Nonetheless, by 1700, all trading emporia and European colonies remained hotly disputed between European trading consortia. See Holden Furber, *Rival Empires of Trade in the Orient, 1600–1800* (University of Minnesota Press, 1976).

30. See Boumediene, *La Colonisation du savoir*; Pratik Chakrabarti, *Materials and Medicine: Trade, Conquest, and Therapeutics in the Eighteenth Century* (Manchester University Press, 2010).

31. Recently, see Berg et al., eds. *Goods From the East*; Zoltán Biedermann, Anne E. Gerritsen, and Giorgio Riello, eds., *Global Gifts: The Material Culture of Diplomacy in Early Modern Eurasia* (Cambridge University Press, 2018); Anne Gerritsen and Giorgio Riello, eds. *The Global Lives of Things* (Routledge, 2016); Jordan Goodman, Paul E. Lovejoy, and Andrew Sherratt, eds., *Consuming Habits: Global and Historical Perspectives on how Cultures Define Drugs* (Routledge, 2007); Bernd-Stefan Grewe and Karin Hofmeester, eds., *Luxury in Global Perspective: Objects and Practices, 1600–2000* (Cambridge University Press, 2017); Findlen, *Early Modern Things*.

32. Pierre Pomet, *Histoire générale des Drogues* (Paris: Jean-Baptiste Loyson and Augustin Pillon, 1694), part 1, 197.

33. Norton, *Sacred Gifts*; Londa Schiebinger, *Secret Cures of Slaves: People, Plants,*

and Medicine in the Eighteenth-Century Atlantic World (Stanford University Press, 2017).

34. Hector Roddan, "'Orientalism is a Partisan Book': Applying Edward Said's Insights to Early Modern Travel Writing," *History Compass* 14, no. 4 (2016): 168–88. Similar points are made by the contributors to *Re-Orienting the Renaissance: Cultural Exchanges with the East*, ed. Gerald L. MacLean (Palgrave Macmillan, 2005); on Said's neglect of the precolonial period, see MacLean, "Introduction," 7. See also Carina L. Johnson, *Cultural Hierarchy in Sixteenth-Century Europe: The Ottomans and Aztecs* (Cambridge University Press, 2011). We refer here specifically to changed attitudes to the Mughal, Ottoman, and Safavid Empires; European violence was already entrenched in the Americas and Southeast Asia by 1700.

35. Norton, *Sacred Gifts*, 108; see similarly Schiebinger, *Plants and Empire*.

36. For responses to tobacco in the Ottoman Empire, see James Grehan, "Smoking and 'Early Modern' Sociability: The Great Tobacco Debate in the Ottoman Middle East, Seventeenth to Eighteenth Centuries," *American Historical Review* 111, no. 5 (2006): 1352–77; and Aslıhan Gürbüzel, "From New Spain to Damascus: Ottoman Religious Authorities and the Making of Medical Knowledge on Tobacco," *Early Science and Medicine* 26, nos. 5–6 (2021): 561–81; for China, Carol Benedict, *Golden-Silk Smoke: A History of Tobacco in China, 1550–2010* (University of California Press, 2011).

37. Sidney W. Mintz, *Sweetness and Power: The Place of Sugar in Modern History* (Elisabeth Sifton Books and Viking, 1985).

38. For some case studies of the complex processes involved in uptake, see, e.g., Clare Griffin, "Disentangling Commodity Histories: *Pauane* and Sassafras in the Early Modern Global World," *Journal of Global History* 15, no. 1 (2020): 1–18; Matthew P. Romaniello, "True Rhubarb? Trading Eurasian Botanical and Medical Knowledge in the Eighteenth Century," *Journal of Global History* 11 (2016): 3–23; Benjamin Breen, "The Failed Globalization of Psychedelic Drugs in the Early Modern World," *Historical Journal* 65, no. 1 (2022), 12–29.

39. Mintz, *Sweetness and Power*; David Wootton, *Bad Medicine: Doctors Doing Harm Since Hippocrates* (Oxford University Press, 2007), e.g., 11, on cinchona; Bob Zebroski, *A Brief History of Pharmacy: Humanity's Search for Wellness* (Routledge, 2016), e.g., 97, on cinchona and ipecacuanha; Daniel Lord Smail, *On Deep History and the Brain* (University of California Press, 2008), 179–87.

40. On drugs as a "tracer" or "sampling device," see the introduction to Sergio Sismondo and Jeremy A. Greene, eds., *The Pharmaceutical Studies Reader* (Wiley-Blackwell, 2015).

41. Pablo F. Gómez, *The Experiential Caribbean: Creating Knowledge and Healing in the Early Modern Atlantic* (University of North Carolina Press, 2017); Ralph Bauer, *The Alchemy of Conquest: Science, Religion, and the Secrets of the New World* (University of Virginia Press, 2019).

42. E.g. Martha Baldwin, "The Snakestone Experiments: An Early Modern Medical Debate," *Isis*, 86, no. 3 (1995): 394–418.

43. Louise Hill Curth, "Medical Advertising in the Popular Press," *Pharmacy in History* 50, no. 1 (2008): 3–16, 5; Patrick Wallis, "Exotic Drugs and English Medicine:

England's Drug Trade, c. 1550–c. 1800," *Social History of Medicine* 25, no. 1 (2012): 20–46.

44. For the Dutch Republic, Anne E. McCants, "Exotic Goods, Popular Consumption, and the Standard of Living: Thinking About Globalization in the Early Modern World," *Journal of World History* 18, no. 4 (2007): 433–62; and Wouter Klein and Toine Pieters, "The Hidden History of a Famous Drug: Tracing the Medical and Public Acculturation of Peruvian Bark in Early Modern Western Europe (c. 1650–1720)," *Journal of the History of Medicine and Allied Sciences* 71, no. 4 (2016): 400–21. For France, see André Lespagnol, "Cargaisons et profits du commerce indien au début du XVIIIe siècle. Les opérations commerciales des Compagnies Malouines 1707–1720," *Annales de Bretagne et des pays de l'Ouest* 89, no. 3 (1982): 313–50. For Portugal, see Timothy D. Walker, "The Medicines Trade in the Portuguese Atlantic World: Acquisition and Dissemination of Healing Knowledge from Brazil (c. 1580–1800)," *Social History of Medicine* 26, no. 3 (2013): 403–31. For Spain, see Teresa Huguet-Termes, "New World Materia Medica in Spanish Renaissance Medicine: From Scholarly Reception to Practical Impact," *Medical History* 45, no. 3 (2001): 359–76.

45. Noble David Cook and Alexandra Parma Cook, "Afterword," in *Science in the Spanish and Portuguese Empires, 1500–1800*, ed. Daniela Bleichmar et al. (Stanford University Press, 2008), 311–24, at 320; Daniela Bleichmar, *Visible Empire: Botanical Expeditions and Visual Culture in the Hispanic Enlightenment* (University of Chicago Press, 2012), 29; Antonio Barrera Osorio, *Experiencing Nature: The Spanish American Empire and the Early Scientific Revolution* (University of Texas Press, 2006).

46. Crawford, *Andean Wonder Drug*, 6.

47. James E. McClellan III, "Scientific Institutions and the Organisation of Science," in *The Cambridge History of Science*, vol. 4, *Eighteenth-Century Science*, ed. Roy Porter (Cambridge University Press, 2003), 87–106, at 101–2; Alette Fleischer, "(Ex)-changing Knowledge and Nature at the Cape of Good Hope, Circa 1652-1700," in *The Dutch Trading Companies as Knowledge Networks*, ed. Siegfried Huigen, Jan L. de Jong, and Elmer Kolfin (Brill, 2010), 243–65.

48. Loïc Charles and Paul Cheney, "The Colonial Machine Dismantled: Knowledge and Empire in the French Atlantic," *Past and Present* 219, no. 1 (2013): 127–63. See also James E. McClellan III and François Regourd, *The Colonial Machine: French Science and Overseas Expansion in the Old Regime* (Brepols, 2011); and Philip P. Boucher, *France and the American Tropics to 1700: Tropics of Discontent?* (Johns Hopkins University Press, 2008).

49. On networks as an analytical category, see esp. the introduction to Paula Findlen, ed., *Empires of Knowledge: Scientific Networks in the Early Modern World*, 1st ed. (Routledge, 2019). On "connected histories," see Caroline Douki and Philippe Minard, "Histoire globale, histoires connectées : un changement d'échelle historiographique," *Revue d'histoire moderne et contemporaine* 54-4bis, no. 5 (2007): 7–21; and Sanjay Subrahmanyam, "Connected Histories: Notes Towards a Reconfiguration of Early Modern Eurasia," in *Beyond Binary Histories: Re-Imagining Eurasia to c.1830*, ed. Victor B. Lieberman (University of Michigan Press, 1999), 289–316.

50. Gilbert Chinard, *L'Amérique et le rêve exotique dans la littérature française au XVIIe et au XVIIIe siècle* (Droz, 1934); Anna Winterbottom, "Of the China Root: A

Case Study of the Early Modern Circulation of *Materia Medica*," *Social History of Medicine* 28, no. 1 (2015): 22–44; Cooper, *Inventing the Indigenous*, 27.

51. Matthew James Crawford and Joseph M. Gabriel, eds., *Drugs on the Page: Pharmacopoeias and Healing Knowledge in the Early Modern Atlantic World* (University of Pittsburgh Press, 2019).

52. Crawford, *Andean Wonder Drug*; Klein and Pieters, "Hidden History"; Boumediene, *La Colonisation du savoir*.

53. Klein and Pieters, "Hidden History."

54. Cristina Bellorini, *The World of Plants in Renaissance Tuscany: Medicine and Botany* (Routledge, 2016); Florike Egmond, *The World of Carolus Clusius: Natural History in the Making, 1550–1610* (Routledge, 2015); Leah Knight, *Of Books and Botany in Early Modern England: Sixteenth-Century Plants and Print Culture* (Routledge, 2016).

55. Steven J. Harris, "Long-Distance Corporations, Big Sciences, and the Geography of Knowledge," *Configurations* 6, no. 2 (1998): 269–304.

56. See notably Michelle DiMeo and Sara Pennell, *Reading and Writing Recipe Books, 1550–1800* (Manchester University Press, 2013); Elaine Leong, *Recipes and Everyday Knowledge: Medicine, Science, and the Household in Early Modern England* (University of Chicago Press, 2019); Wendy Wall, *Recipes for Thought: Food and Taste in the Early Modern English Kitchen* (University of Pennsylvania Press, 2015); Anne Stobart, *Household Medicine in Seventeenth-Century England* (Bloomsbury, 2016).

57. See for instance David Gentilcore, *Healers and Healing in Early Modern Italy* (Manchester University Press, 1998); Gianna Pomata, *Contracting a Cure: Patients, Healers, and the Law in Early Modern Bologna* (Johns Hopkins University Press, 1998); Margaret Pelling, *Medical Conflicts in Early Modern London: Patronage, Physicians, and Irregular Practitioners, 1550–1640* (Clarendon Press, 2003).

58. Winterbottom, "Of the China Root"; Kevin Siena, *Venereal Disease, Hospitals and the Urban Poor: London's "Foul Wards" 1600–1800* (University of Rochester Press, 2004), chapter 2.

59. According to Victor Segalen, "Exoticism . . . is nothing other than the notion of difference, the perception of Diversity, the knowledge that something is other than one's self." See Segalen, *Essay on Exoticism: An Aesthetics of Diversity* (Duke University Press, 2002), 19.

60. *Correspondance . . . de Guy Patin*, letter 156, Guy Patin to Charles Spon, May 29, 1648.

61. On the revival of Galenism, see Vivian Nutton, "Renaissance Galenism, 1540–1640: Flexibility Or an Increasing Irrelevance?," in *Brill's Companion to the Reception of Galen*, ed. Petros Bouras-Vallianatos and Barbara Zipser (Brill, 2019), 472–86. On pharmacology, see, e.g., Philip M. Teigen, "Taste and Quality in 15th- and 16th-Century Galenic Pharmacology," *Pharmacy in History* 29, no. 2 (1987): 60–68. On bezoar and theriac, see, respectively, the chapters by Cook and Di Gennaro Splendore, this volume.

62. For a summary of Patin's conflict with the Paris apothecaries, see Félix Larrieu, *Gui Patin, doyen de la Faculté de médecine de Paris : sa vie, son œuvre, sa thérapeutique (1601–1672)* (Alphonse Picard, 1889), 41–51.

63. Philippe Albou, "Guy Patin et son martyrologe de l'antimoine," *Histoire des sciences médicales* 50, no. 4 (2016): 455–66.

64. *Correspondance . . . de Guy Patin*, letter 441, Guy Patin to Claude II Belin, July 20, 1656; Laurence Brockliss and Colin Jones, *The Medical World of Early Modern France* (Clarendon Press, 1997), 282.

65. On which, see Philippe Albou, "Histoire des 'Œuvres charitables' de Philibert Guybert," *Histoire des sciences médicales* 32, no. 1 (1998): 11–26.

66. Harold J. Cook, "Good Advice and Little Medicine: The Professional Authority of Early Modern English Physicians," *Journal of British Studies* 33, no. 1 (1994): 1–31.

67. Brockliss and Jones, *Medical World*, 127–28, 137, 332–33.

68. On "the new" as a cultural category, see especially Colin Campbell, "The Desire for the New: Its Nature and Social Location as Presented in Theories of Fashion and Modern Consumerism," in R. Silverstone and E. Hirsch, eds., *Consuming Technologies: Media and Information in Domestic Spaces* (Routledge, 1992), 48–64.

69. See especially Schiebinger, *Plants and Empire*.

70. Sarah Easterby-Smith, "Recalcitrant Seeds: Material Culture and the Global History of Science," *Past and Present*, supplement 14 (2019): 215–42; also Christopher Parsons and Kathleen S. Murphy, "Ecosystems Under Sail: Specimen Transport in the Eighteenth-Century French and British Atlantics," *Early American Studies* 10, no. 3 (2012): 503–29.

71. Breen, "Failed Globalization." On assemblages, see, e.g., Stephen J. Collier, "Global Assemblages," in the special issue "Problematizing Global Knowledge," *Theory, Culture and Society*, 23, nos. 2–3 (2006): 399–401; and Martin Müller, "Assemblages and Actor-Networks: Rethinking Socio-Material Power, Politics, and Space," *Geography Compass*, 9, no. 1 (2015): 27–41. On similarities between the acculturation of things and people, see Igor Kopytoff, "The Cultural Biography of Things: Commoditization as Process," in *The Social Life of Things: Commodities in Cultural Perspective*, ed. Arjun Appadurai (Cambridge University Press, 2014), 64–92.

1. THE UNEXOTIC WORLD

1. Edward Said, *Orientalism* (Penguin, 2003); Benjamin Schmidt, "Inventing Exoticism: The Project of Dutch Geography and the Marketing of the World, Circa 1700," in Pamela Smith and Paula Findlen, eds., *Merchants and Marvels: Commerce, Science, and Art in Early Modern Europe* (Routledge, 2002), 347–69; Benjamin Schmidt, *Inventing Exoticism: Geography, Globalism, and Europe's Early Modern World* (University of Pennsylvania Press, 2015).

2. See for example José Rabasa, *Inventing America: Spanish Historiography and the Formation of Eurocentrism* (University of Oklahoma Press, 1993).

3. Xiaomei Chen, "Occidentalism as Counterdiscourse: 'He Shang' in Post-Mao China," *Critical Inquiry* 18, no. 4 (1992): 688. See also Rachael Hutchinson, "Occidentalism and Critique of Meiji: The West in the Returnee Stories of Nagai Kafū," *Japan Forum* 13, no. 2 (2001).

4. Schmidt, "Inventing Exoticism"; Schmidt, *Inventing Exoticism*.

5. Russian Federation, Russian State Archive of Ancient Documents, Moscow (henceforth RGADA), fond 143 Apothecary Chancery, op. 2, ed. khr. 1554. All translations are my own unless otherwise stated.

6. Rebecca Earle, *The Body of the Conquistador: Food, Race, and the Colonial Experience in Spanish America, 1492–1700* (Cambridge University Press, 2012).

7. Ann Stoler and Frederick Cooper, "Between Metropole and Colony: Rethinking a Research Agenda," in *Tensions of Empire: Colonial Cultures in a Bourgeois World*, ed. Frederick Cooper and Ann Stoler (University of California Press, 1997), 11.

8. See for example Mikhail Lotman and Boris Uspensky, "Binary Models in the Dynamics of Russian Culture (to the End of the Eighteen Century)," *Studies in Soviet Thought* 33 (1987): 376–80.

9. Loren R. Graham, *Science in Russia and the Soviet Union: A Short History* (Cambridge University Press, 1993).

10. For one of his more recent works, see R. A. Simonov, *Matematicheskaia i kalendarno-astronomicheskaia mysl' Drevnei Rusi* (Nauka, 2007).

11. Valerie Kivelson, *Cartographies of Tsardom: The Land and its Meanings in Seventeenth-Century Russia* (Cornell University Press, 2006); Valerie Kivelson, "Claiming Siberia: Colonial Possession and Property Holding in the Seventeenth and Early Eighteenth Centuries," in *Peopling the Russian Periphery: Borderland Colonization in Eurasian history*, ed. Nicholas Breyfogle, Abby Schrader, and Willard Sunderland (Routledge, 2007); Eve Levin, "The Administration of Western Medicine in Seventeenth-Century Russia," in *Modernizing Muscovy. Reform and Social Change in Seventeenth Century Russia*, ed. Jarmo Kotilaine and Marshall Poe (Routledge Curzon, 2004); Rachel Koroloff, "Travniki, Travniki, and Travniki: Herbals, Herbalists and Herbaria in Seventeenth- and Eighteenth-Century Russia," *Vivliofika: E-Journal of Eighteenth-Century Russian Studies* 6 (2018).

12. K. A. Bogdanov, *O krokodilakh v Rossii. Ocherki iz istorii zaimstvovanii i ekzotizmov* (Novoe literaturnoe obozrenie, 2006).

13. Dániel Margócsy, "The Camel's Head: Representing Unseen Animals in Sixteenth-Century Europe," *Nederlands Kunsthistorisch Jaarboek* 61 (2011).

14. Bogdanov, *O krokodilakh v Rossii*, 183.

15. See for example Samir Boumediene, *La colonisation du Savoir. Une histoire des plantes médicinales du 'Nouveau Monde' (1492–1750)* (Les Editions des Mondes à Faire, 2016), 149–52.

16. Bogdanov, *O krokodilakh v Rossii*, 10.

17. Nikolaos A. Chrissidis, *An Academy at the Court of the Tsars: Greek Scholars and Jesuit Education in Early Modern Russia* (Northern Illinois University Press, 2015).

18. E. A. Savel'eva, ed., *Katalog knig iz sobraniia Aptekarskogo prikaza* (Al'faret, 2006), 93–4.

19. Clare Griffin, "Bureaucracy and Knowledge Creation: The Apothecary Chancery," in *Information and Empire: Mechanisms of Communication in Russia, 1600–1850*, ed. Simon Franklin and Katherine Bowers (Open Book Publishers, 2017).

20. Bogdanov, *O krokodilakh v Rossii*, 8.

21. Koroloff, "Travniki, Travniki, and Travniki," 2018.

22. Schmidt, *Inventing Exoticism*.

23. Said, *Orientalism*; Chen, "Occidentalism as Counterdiscourse."

24. Marina Tolmacheva, "The Early Russian Exploration and Mapping of the

Chinese Frontier," *Cahiers du monde russe. Russie-Empire russe-Union soviétique et États indépendants* 41, no. 1 (2000).

25. Gregory Dmitrievich Afinogenov, "The Eye of the Tsar: Intelligence-Gathering and Geopolitics in Eighteenth-Century Eurasia" (PhD diss., Harvard University, 2015), 42.

26. Peter Perdue, "Boundaries, Maps, and Movement: The Chinese, Russian, and Mongolian Empires in Early Modern Eurasia," *International History Review* 20, no. 2 (1998): 267.

27. Erika Monahan, "Locating Rhubarb: Early Modernity's Relevant Obscurity," in *Early Modern Things. Objects and their Histories, 1500–1800*, ed. Paula Findlen (Routledge, 2013).

28. RGADA f. 143, op. 3, ed. khr. 319.

29. Schmidt, "Inventing Exoticism"; Valerie Kivelson, "'Between All Parts of the Universe': Russian Cosmographies and Imperial Strategies in Early Modern Siberia and Ukraine," *Imago Mundi* 60, no. 2 (2008).

30. N. N. Bolkhovitinov, *Rossiia otkryvaet Ameriky, 1732–1799* (Mezhdunarodnye otnosheniia, 1991).

31. Kivelson, "'Between All Parts of the Universe,'" 170. Translation Kivelson's.

32. Kivelson, "'Between All Parts of the Universe.'"

33. Kapil Raj, *Relocating Modern Science: Circulation and the Construction of Knowledge in South Asia and Europe, 1650–1900* (Palgrave Macmillan, 2007); Sanjay Subrahmanyam, "Connected Histories: Notes Towards a Reconfiguration of Early Modern Eurasia," *Modern Asian Studies* 31, no. 3 (1997).

34. Benjamin Braude, "The Sons of Noah and the Construction of Ethnic and Geographical Identities in the Medieval and Early Modern Periods," *William and Mary Quarterly* 54, no. 1 (1997); Surekha Davies, *Renaissance Ethnography and the Invention of the Human: New Worlds, Maps, and Monsters* (Cambridge University Press, 2016).

35. Dorothy Kim, "Introduction to Literature Compass Special Cluster: Critical Race and the Middle Ages," *Literature Compass* 16, nos. 9–10 (2019).

36. Geraldine Heng, *The Invention of Race in the European Middle Ages* (Cambridge University Press, 2018), 33–36; Davies, *Renaissance Ethnography*, 30–46.

37. Merrall L. Price, *Consuming Passions: The Uses of Cannibalism in Late Medieval and Early Modern Europe* (Routledge, 2004).

38. Davies, *Renaissance Ethnography*, 25–9.

39. Earle, *Body of the Conquistador*; Matthew P. Romaniello, "Humoral Bodies in Cold Climates," in *Russian History Through the Senses: From 1700 to the Present*, ed. Matthew P. Romaniello and Tricia Starks (Bloomsbury, 2016).

40. Kivelson, "Claiming Siberia."

41. Anthony Anemone, "The Monsters of Peter the Great: the Culture of the St. Petersburg Kunstkamera in the Eighteenth Century," *Slavic and East European Journal* 44, no. 4 (2000).

42. Heng, *Invention of Race*; Bruce S. Hall, *A History of Race in Muslim West Africa, 1600–1960* (Cambridge University Press, 2011).

43. Dieudonné Gnammankou, *Abraham Hanibal: l'aïeul noir de Pouchkine*

(Présence africaine, 1996); Hugh Barnes, *Gannibal: The Moor of Petersburg* (Profile Books, 2005).

44. John I. Edwards, "Looking for Abram Hannibal: Some Observations on the Supposed Portraits of Abram Petrovich Hannibal (1696–1781), the African Great-Grandfather of Aleksandr Pushkin," *Slavonica* 9, no. 1 (2003): 26.

45. Edwards, "Looking for Abram Hannibal."

46. Hall, *History of Race*, 69–103; Heng, *Invention of Race*, 43–5.

47. Heng, *Invention of Race*, 42.

48. Anne Lafont, "How Skin Color Became a Racial Marker: Art Historical Perspectives on Race," *Eighteenth-Century Studies* 51, no. 1 (2017).

49. RGADA f. 143, op. 2, ed. khr. 734.

50. Markman Ellis, *The Coffee House: A Cultural History* (Weidenfeld & Nicolson, 2011), 134.

51. RGADA f. 143, op. 2, ed. khr. 734.

52. Alan Mikhail, "The Heart's Desire: Gender, Urban Space and the Ottoman Coffee House," in *Ottoman Tulips, Ottoman Coffee: Leisure and Lifestyle in the Eighteenth Century*, ed. Dana Sajdi (I. B. Tauris, 2014).

53. Robert Collis, *The Petrine Instauration: Religion, Esotericism, and Science at the Court of Peter the Great, 1689–1725* (Brill, 2011), 359.

54. M. S. Meyer, "Russia and the Islamic World," *Russian Studies in History* 57, no. 2 (2018).

55. Nabil Matar, *Islam in Britain, 1558–1685* (Cambridge University Press, 1998), 115–19.

56. Bogdanov, *O krokodilakh v Rossii*; Audra Jo Yoder, "Tea Time in Romanov Russia: A Cultural History, 1616–1917" (PhD diss., University of North Carolina at Chapel Hill, 2016), 17.

57. Said, *Orientalism*, 3.

58. T. A. Oparina, *Inozemtsy v Rossii XVI-XVII vv.* (Progress-Traditsiia, 2007), 5–7.

59. Heng, *Invention of Race*, 27–31.

60. RGADA f. 143, op. 2, ed. khr. 1554.

61. On the Apothecary Chancery, see for example Levin, "The Administration of Western Medicine in Seventeenth-Century Russia"; and Sabine Dumschat, *Ausländische Mediziner im Moskauer Russland* (Franz Steiner, 2006). On Russian botany, see Koroloff, "Travniki, Travniki, and Travniki"; and A. B. Ippolitova, *Russkie rukopisnye travniki XVII-XVIII vekov. Issledovanie fol'klora i etnobotaniki* (Indrik, 2008).

62. RGADA f. 143, op. 2, ed. khr. 738.

63. Michael J. Schreffler, "Vespucci Rediscovers America: The Pictorial Rhetoric of Cannibalism in Early Modern Culture," *Art History* 28, no. 3 (2005).

64. Ângela Domingues, "The Portuguese Discoveries and Their Influence on European Medicine," *Workshop Plantas Medicinais e Fitoterapêuticas nos Trópicos. Instituto de Investigação Científica Tropical (IICT), Lisbon, Portugal* (Instituto de investigação científica tropical, Departamento de ciências humanas programa de desenvolvimento global, 2008): 6; Paula De Vos, "Methodological Challenges Involved in Compiling the Nahua Pharmacopeia," *History of Science* 55, no. 2 (2017): 226.

65. For a recent example of work on this, see Caroline Dodds-Pennock, "Women of Discord: Female Power in Aztec Thought," *Historical Journal* 61, no. 2 (2018).

66. Schmidt, "Inventing Exoticism."

67. Schmidt, "Inventing Exoticism," 364.

68. T. A. Isachenko, *Perevodnaia Moskovskaia knizhnost.' Mitropolichii i patriarshii skriptorii XV-XVII vv.* (Rossiiskaia gosudarstvennaia biblioteka, 2009), 135–53.

69. Published in V. M. Florinskii, *Russkie prostonarodnye travniki i lechebniki. Sobranie meditsinskikh rukopisei XVI i XVII stoletiia* (Tipografiia Imperatorskogo Universiteta, 1879), 2013–229.

70. For example, Florinskii, *Russkie prostonarodnye travniki i lechebniki*, 70.

71. Stephen Frederic Dale, *Indian Merchants and Eurasian Trade, 1600–1750* (Cambridge University Press, 2002); Scott Cameron Levi, *The Indian Diaspora in Central Asia and Its Trade, 1550–1900* (Brill, 2002); Audrey Burton, *The Bukharans: A Dynastic, Diplomatic, and Commercial History, 1550–1702* (Curzon Press, 1997); Sebouh Aslanian, *From the Indian Ocean to the Mediterranean: The Global Trade Networks of Armenian Merchants from New Julfa* (University of California Press, 2011).

72. Alix Cooper, *Inventing the Indigenous: Local Knowledge and Natural History in Early Modern Europe* (Cambridge University Press, 2007), 21.

73. Cooper, *Inventing the Indigenous*.

74. N. E. Mamonov, *Materialy dlia istorii meditsiny v Rossii*, 4 vols. (M. M. Stasiulevich, 1881), 2:228.

75. Janet Martin, *Treasure of the Land of Darkness: The Fur Trade and its Significance for Medieval Russia* (Cambridge University Press, 2004).

76. O. V. Belova and G. I. Kabakova, eds., *U istokov mira. Russkie etiologicheskie skazki i legendy* (Forum, Neolit, 2015).

77. Ippolitova, *Russkie rukopisnye travniki*.

78. Clare Griffin, "Russia and the Medical Drug Trade in the Seventeenth Century," *Social History of Medicine* 31, no. 1 (2016).

79. RGADA f. 143, op. 2, ed. khr. 1554.

80. Russian State Library, Moscow (RGB) f. 37 (Bol'shakova), No. 228, ll. 3v, 21v, 29r–31v; Russian National Library, St. Petersburg (RNB) koll. Titova, No. 3881, ll. 4–4ob.; *Florinova Ekonomiia, s nemetskago na rossiiskoi iazyk sokrashcheno perevedena i napechatana poveleniem eia Imperatorskago Velichestva Vsemilotsiveishiia Velikiia Gosudaryni Imperatritsy Anny Ioannovny Samoderzhitsy Vserossiiskia* (Imperatorskaia Akademiia Nauk, 1738), 288; *Florinova Ekonomiia v deviati knigakh sostoiashchaia; s nemetskago na rossiiskoi iazyk sokrashcheno Sergiem Volchkovym. Izdanie vtoroe* (Imperatorskaia Akademiia Nauk, 1760), 325.

81. RGADA f. 143, op. 3, ed. khr. 419.

2. GALENIC BODIES AND JESUIT BEANS

1. Raymond H. Thompson, ed., *A Jesuit Missionary in Eighteenth-Century Sonora: The Family Correspondence of Philipp Segesser*, trans. by Werner S. Zimmt and Robert E. Dahlquist (University of New Mexico Press, 2014), 71.

2. Pedro Chirino, *Relacion de las Islas Filipinas* (Rome: Estevan Paulino, 1604), 2.

3. Paula De Vos, "A Taste for Spices: Spanish Efforts at Spice Production in the

Philippines," *Mains'l Haul: A Journal of Pacific Maritime History* 41, no. 4 (2005): 33–42; Omri Bassewitch Frenkel, "Transplantation of Asian Spices in the Spanish Empire 1518–1640: Entrepreneurship, Empiricism, and the Crown," PhD diss., McGill University, 2017.

4. For example, Francisco Suárez de Ribera, *Escrutinio medico, o medicina experimentada* (Madrid: Francisco del Hierro, 1723), 124, 194, 243; Félix Palacios, *Palestra pharmaceutica, chymico-galenica* (Madrid: Juan de Sierra, 1725), 673–74; Samuel Dale, *Pharmacologia, seu manuductio ad materiam medicam* (Leiden: Johann Arnold Langerak, 1739), 357–58; Étienne Geoffroy Saint-Hilaire, *Tractatus de materia medica*, vol. 2 (Paris: Jean Desaint and Charles Saillant, 1741), 458–62; John Hill, *A History of the Materia Medica* (London: J. and J. Rivington, 1751), 506–7; Caspar Neumann, *Chymiae medicae dogmatico-experimentalis tomi secundi pars secunda* (Züllichau: Johann Jacob Dendeler, 1751), 288–94.

5. As Paula De Vos explores in chapter 6 of this volume, early modern Galenic pharmacopoeias were borne of cultural transfers that extended from antiquity to early modernity across the Indo-Mediterranean region and beyond. See also Paula De Vos, *Compound Remedies: Galenic Pharmacy from the Ancient Mediterranean to New Spain* (University of Pittsburgh Press, 2020).

6. For example, Antonio Barrera-Osorio, "Local Herbs, Global Medicines: Commerce, Knowledge and Commodities in Spanish America," in *Merchants and Marvels: Commerce, Science, and Art in Early Modern Europe*, ed. Pamela H. Smith and Paula Findlen (Routledge, 2002), 163–81; Londa Schiebinger, *Plants and Empire: Colonial Bioprospecting in the Atlantic World* (Harvard University Press, 2004); Pratik Chakrabarti, "Medical Marketplaces Beyond the West: Bazaar Medicine, Trade, and the English Establishment in Eighteenth-Century India," in *Medicine and the Market in England and Its Colonies, c.1450–c.1850*, ed. Mark S. R. Jenner and Patrick Wallis (Palgrave Macmillan, 2007), 196–215.

7. See also Samir Boumediene, *La Colonisation du savoir: Une histoire des plantes médicinales du "Nouveau Monde" (1492–1750)* (Les Éditions des mondes à faire, 2016); Sebastian Kroupa, "Missionary Remedies," in *Early Modern Medicine: An Introduction to Source Analysis*, ed. Olivia Weisser (Routledge, 2024), 167–89.

8. For Kamel, see Sebastian Kroupa, "Georg Joseph Kamel (1661–1706): A Jesuit Pharmacist at the Frontiers of Colonial Empires" (PhD diss., University of Cambridge, 2019).

9. Rebecca Earle, *The Body of the Conquistador: Food, Race, and the Colonial Experience in Spanish America, 1492–1700* (Cambridge University Press, 2012); Linda A. Newson, *Making Medicines in Early Colonial Lima, Peru: Apothecaries, Science, and Society* (Brill, 2017); De Vos, *Compound Remedies*, esp. chapter 4.

10. The Protomedicato was an official board of medicine, charged with training, examining, and supervising medical practitioners in the Spanish realms.

11. Earle, *Body of the Conquistador*.

12. Gregorio Saldarriaga, *Alimentación e identidades en el Nuevo Reino de Granada, siglos XVI y XVII* (Editorial Universidad del Rosario, 2011), 109–28; Earle, *Body of the Conquistador*, 54–83.

13. Miguel de Asúa, *Science in the Vanished Arcadia: Knowledge of Nature in the*

Jesuit Missions of Paraguay and Río de La Plata (Brill, 2014), 134, 136; Gianamar Giovannetti-Singh, "Galenizing the New World: Joseph-François Lafitau's 'Galenization' of Canadian Ginseng, ca. 1716–1724," *Notes and Records of the Royal Society of London* 75, no. 1 (2020): 59–72; Kroupa, "Missionary Remedies."

14. Georg Joseph Kamel, "De igasur, seu nuce vomica legitima Serapionis," *Philosophical Transactions of the Royal Society* 21, no. 250 (1699): 88–94.

15. Stephanie J. Mawson, *Incomplete Conquests: The Limits of Spanish Empire in the Seventeenth-Century Philippines* (Cornell University Press, 2023).

16. Dennis O. Flynn and Arturo Giráldez, "China and the Spanish Empire," *Journal of Iberian and Latin American Economic History* 14, no. 2 (1996): 309–38.

17. Ostwald Sales Colín, "Las actividades médicas en las Filipinas durante la primera mitad del siglo XVII," *Perspectivas Latinoamericanas* 2 (2005): 167–86; Mercedes G. Planta, *Traditional Medicine in the Colonial Philippines: 16th to the 19th Century* (University of the Philippines Press, 2017); Antonio García-Abásolo, "Médicos, enfermos y enfermedades en Manila (siglos XVI–XVIII)," in *Convivencia y conflicto en la frontera oriental de la Monarquía Hispánica*, ed. Martha María Machado López (Silex Ediciones, 2021), 215–80; Juan Carlos González Balderas, "The Medicine Monopoly Contract in Manila: A Perspective on Medical Provision and the Circulation of Medicine in the Early Eighteenth Century," *Crossroads* 21 (2023): 167–97.

18. The Spanish Inquisition defined *hechicería* as a form of witchcraft that encompassed various magical and superstitious practices, but, unlike the more serious charge of *brujería*, *hechicería* did not imply an explicit pact with the devil. See Stephanie J. Mawson, "Folk Magic in the Philippines, 1611–39," *Journal of Southeast Asian Studies* 54, no. 2 (2023): 220–44; David Max Findley, "Of Two-Tailed Lizards: Spells, Folk-Knowledge, and Navigating Manila, 1620–1650," *Journal of Social History* 56, no. 2 (2022): 294–325. See also Domingo Fernández Navarrete, *Tratados historicos, politicos, ethicos y religiosos de la Monarchia de China* (Madrid: Juan Garcia Infançon, 1676), 323–24; Juan Francisco de San Antonio, *Chronicas de la apostolica Provincia de San Gregorio de religiosos descalzos de N.S.P.S. Francisco en las Islas Philipinas, China, Japon, &c.*, vol. 1 (Manila: Juan de Sotillo, 1738), 41; Juan de Medina, *Historia de los sucesos de la Orden de Nuestro gran Padre San Agustín, de estas Islas Filipinas* (Manila: Chofré y Compañía, 1893), 103.

19. Findley, "Of Two-Tailed Lizards," 308.

20. For example, Luis de Jesus, *Historia general de los religiosos descalzos del orden de San Augustin*, vol. 2 (Madrid: Lucas Antonio de Bedmar, 1681), 29–30; San Antonio, *Chronicas*, 149–57; Juan José Delgado, *Historia general sacro-profana: política y natural de las islas del poniente llamadas Filipinas*, in *Biblioteca Histórica Filipina*, vol. 1, ed. Pablo Pastells (Manila: Imp. de El Eco de Filipinas de D. Juan Atayde, 1892), 367–70. This work was written between 1750 and 1754, but remained in manuscript until 1892.

21. Hernando de los Ríos Coronel, "Información sobre el Hospital Real de Manila," Manila, April 15, 1594, Archivo General de Indias (henceforth AGI), Filipinas 59, no. 31, fol. 1v.

22. Francisco Guerra, *El hospital en Hispanoamérica y Filipinas 1492–1898* (Ministerio de Sanidad y Consumo, 1994).

23. José Pardo-Tomás, "Conversion Medicine: Communication and Circulation

of Knowledge in the Franciscan Convent and College of Tlatelolco, 1527–1577," *Quaderni storici* 48, no. 1 (2013): 21–42; Gabriela Ramos, "Indian Hospitals and Government in the Colonial Andes," *Medical History* 57, no. 2 (2013): 186–205.

24. Fausto Cruzat y Góngora to Council of the Indies, Manila, December 16, 1690, AGI, Filipinas 14, rollo 3, no. 30, fols. 1r–v.

25. Miguel Sánchez, "Traslado de documentos relativos a la entrada de los Franciscanos en la administración del Hospital Real de Manila," Manila, May 20, 1685, AGI, Filipinas 83, no. 16, bloque 3, fol. 2r.

26. Joaquín Ramírez, "Traslado de la representación hecha por José Joaquín Ramírez, capellán mayor del hospital real de los españoles de Manila, al gobernador de Filipinas, sobre las necesidades materiales de dicho hospital," Manila, April 15, 1711, AGI, Filipinas 297, no. 1, bloque 2, fol. 6v.

27. Cruzat y Góngora to Council of the Indies, December 16, 1690, AGI, Filipinas 14, rollo 3, no. 30, fol. 1r.

28. Miguel García Serrano to Council of the Indies, Manila, July 31, 1622, AGI, Filipinas 74, no. 90, fols. 584r–628v; Guerra, *El hospital*, 535–78.

29. Carlos López Beltrán, "Hippocratic Bodies: Temperament and Castas in Spanish America (1570–1820)," *Journal of Spanish Cultural Studies* 8, no. 2 (2007): 253–89; Ann Twinam, *Purchasing Whiteness: Pardos, Mulattos, and the Quest for Social Mobility in the Spanish Indies* (Stanford University Press, 2015); Stephanie J. Mawson, "Philippine *Indios* in the Service of Empire: Indigenous Soldiers and Contingent Loyalty, 1600–1700," *Ethnohistory* 63, no. 2 (2016): 381–413.

30. For example, AGI, Filipinas 83, no. 16.

31. Ramírez, "Traslado."

32. These reforms are discussed in detail in González Balderas, "Medicine Monopoly Contract."

33. Sales Colín, "Las actividades médicas," 175.

34. Cruzat y Góngora to Council of the Indies, Manila, June 16, 1699, AGI, Filipinas 17, rollo 1, no. 32, fol. 1r. Around the year 1700, one peso was minted to roughly 25 g of fine silver; see William Graham Sumner, "The Spanish Dollar and the Colonial Shilling," *American Historical Review* 3, no. 4 (1898): 613–14.

35. Cruzat y Góngora to Council of the Indies, Manila, May 30, 1701, AGI, Filipinas 124, no. 15, fol. 6r–6v.

36. "Testimonio no. 2 de lo ejecutado en cumplimiento de la cédula de 6 de julio de 1714," Manila, February 15, 1719, AGI, Filipinas 132, no. 23, bloque 5, fol. 7r.

37. Instituciones coloniales, Gobierno virreinal, Reales cédulas originales y duplicadas, Reales cédulas duplicadas, vol. D49, expediente 224, Archivo General de la Nación, Mexico. I am most grateful to Steph Mawson for pointing me to this source. See Juan Villasana Haggard, *Handbook for Translators of Spanish Historical Documents* (University of Texas, 1941), 72, 79, 81.

38. "Testimonio no. 40 de los autos que precedieron de pedimiento de Miguel de la Torre," Manila, July 30, 1718, AGI, Filipinas 132, no. 23, bloque 9, fols. 23r–41r.

39. "Testimonio no. 40," fols. 41v–47v.

40. Royal College of Apothecaries of Valencia, *Officina medicamentorum, et methodus recte eadem componendi, cum variis scholiis, & aliis quam plurimis, ipsi operi*

necessariis (Valencia: Juan Chrysostomo Garriz, 1601), and Jerónimo de la Fuente Piérola, *Tyrocinio pharmacopeo methodo medico, y chimico, en el qual se contienen los canones de Ioanes Mesue, Damasceno* (Madrid: Diego Diaz de la Carrera, 1660). See Paula De Vos, "'The Prince of Medicine:' Yuhanna Ibn Masawaih and the Foundations of the Western Pharmaceutical Tradition," *Isis* 104, no. 4 (2013): 667–712.

41. "Testimonio no. 40," fols. 9v–10v, 47v–49v.

42. Charles Davis and María Luz López Terrada, "Protomedicato y farmacia en Castilla a finales del siglo XVI: Edición crítica del 'Catálogo de las cosas que los boticarios han de tener en sus boticas' de Andrés Zamudio de Alfaro, Protomédico General (1592–1599)," *Asclepio* 62, no. 2 (2010): 579–626.

43. "Testimonio no. 40," fol. 10r.

44. Newson, *Making Medicines*, 2.

45. De Vos, *Compound Remedies*.

46. Ramírez, "Traslado," fol. 7r.

47. Francisco Fernández Toribio to Council of the Indies, Manila, June 18, 1721, AGI, Filipinas 171, no. 20, fols. 1r–1v.

48. "Petición de Miguel de la Torre de boletas o botica," Manila, ca. 1739. AGI, Filipinas 197, no. 11, fols. 1r–2v.

49. Gaspar de San Agustín, "Carta a un amigo suyo dándole cuenta del natural y genio de los Indios de estas islas Filipinas," in *The Philippine Islands, 1493–1898*, vol. 40, ed. Emma Helen Blair and James Alexander Robertson (Arthur H. Clark Company, 1906), 194–95.

50. Ríos Coronel, "Información," fol. 2v.

51. Georg Joseph Kamel, "Certificación médica de Jorge Camel, médico y boticario del Colegio de San Ignacio de la Compañía de Jesús de Manila, sobre la falta de salud de Juan de Ozaeta," Manila, June 27, 1690, AGI, Filipinas 163, no. 24, bloque 3, fol. 1r.

52. Karl A. F. Fischer, "Catalogus generalis provinciae Bohemiae et Silesiae Societatis Iesu, 1540–1772" (1978), Archivum Romanum Societatis Iesu (henceforth ARSI), Bohemia 208, 69.

53. Matthias Tanner to the Father General of the Society of Jesus, January 25, 1687, ARSI, Bohemia 4 II, fol. 421r.

54. Pedro Murillo Velarde, *Historia de la Provincia de Philipinas de la Compañia de Jesus* (Manila: Imprenta de la Compañía de Jesus, por Nicolas de la Cruz Bagay, 1749), fol. 393v.

55. Murillo Velarde, *Historia*, fol. 394r.

56. Georg Joseph Kamel to Šimon Boruhradský, Manila, June 25, 1691, Moravský zemský archiv (Brno, Czechia), G11 571, fol. 55v.

57. For Kamel's correspondence with English scholars, see Sebestian Kroupa, "Ex Epistulis Philippinensibus: Georg Joseph Kamel SJ (1661–1706) and His Correspondence Network," *Centaurus* 57, no. 4 (2015): 229–59.

58. Probably *Cymbopogon citratus*, called "tanglad" by Indigenous Filipino communities; see Elmer D. Merrill, *A Dictionary of the Plant Names of the Philippine Islands* (Manila: Bureau of Public Printing, 1903), 110.

59. Georg Joseph Kamel, "Historia stirpium insula Luzonis et reliquarum

Philippinarum," in *Historiae plantarum tomus tertius*, ed. John Ray (London: Samuel Smith and Benjamin Walford, 1704), 28–29.

60. For example, John Gerard, *The Herball, or, Generall Historie of Plantes* (London: John Norton, 1597), 39–40.

61. Alain Touwaide, "*Quid pro Quo*: Revisiting the Practice of Substitution in Ancient Pharmacy," in *Herbs and Healers from the Ancient Mediterranean through the Medieval West*, ed. Anne van Arsdall and Timothy Graham (Ashgate, 2012), 19–61; Samir Boumediene and Valentina Pugliano, "The Substitute Route: Exotic Remedies, Medical Innovation and the Market for Substitutes in the 16th Century," *Revue d'histoire moderne et contemporaine* 66, no. 3 (2019): 24–54.

62. Kamel, "Historia stirpium," 9.

63. Andrés Serrano, "Memorial sobre limosna de medicamentos a jesuitas," November 8, 1706, AGI, Filipinas 94, no. 29, bloque 2.

64. The Procurador's responsibility was to manage the finances of the mission. Luis de Morales, "Petición del jesuíta Luis de Morales sobre medicinas," Madrid, April 1, 1686, AGI, Filipinas 28, no. 146, fol. 1259r; "Consulta sobre prorrogar medicinas a jesuitas de Filipinas," Madrid, April 1, 1686, AGI, Filipinas 3, no. 153, fol. 1r.

65. Murillo Velarde, *Historia*, fol. 393v.

66. Kroupa, "Missionary Remedies."

67. See Boumediene in this volume; and Boumediene, *La Colonisation du savoir*; Sabine Anagnostou, "Jesuits in Spanish America: Contributions to the Exploration of the American *Materia Medica*," *Pharmacy in History* 47, no. 1 (2005): 3–17; Matthew James Crawford, *The Andean Wonder Drug: Cinchona Bark and Imperial Science in the Spanish Atlantic, 1630–1800* (University of Pittsburgh Press, 2016).

68. Renée Gicklhorn, *Missionsapotheker: Deutsche Pharmazeuten im Lateinamerika des 17. und 18. Jahrhunderts* (Wissenschaftliche Verlagsgesellschaft MBH, 1973), 34.

69. Félix Garzón Maceda, *La medicina en Córdoba: Apuntes para su historia*, vol. 2 (Talleres Rodríguez Giles, 1917), 147–54. Cited in Asúa, *Science in the Vanished Arcadia*, 109–11.

70. Merrill, *Dictionary*, 67.

71. Nicholas Griffiths and Fernando Cervantes, eds., *Spiritual Encounters: Interactions Between Christianity and Native Religions in Colonial America* (University of Birmingham Press, 1999); Andrés I. Prieto, *Missionary Scientists: Jesuit Science in Spanish South America, 1570–1810* (Vanderbilt University Press, 2011), 48–89.

72. Kamel, "De igasur," 90: "casualiter hanc nucem secum habentem expertus fuit: qua occasione primum Hispanis innotuit igasur virtus, & efficacia."

73. Kathleen S. Murphy, "Translating the Vernacular: Indigenous and African Knowledge in the Eighteenth-Century British Atlantic," *Atlantic Studies* 8, no. 1 (2011)" 29–48; Christopher M. Parsons, "The Natural History of Colonial Science: Joseph-François Lafitau's Discovery of Ginseng and Its Afterlives," *William and Mary Quarterly* 73, no. 1 (2016): 37–72.

74. Joan-Pau Rubiés, "Were Early Modern Europeans Racist?," in *Ideas of "Race" in the History of the Humanities*, ed. Amos Morris-Reich and Dirk Rupnow (Palgrave Macmillan, 2017), 33–87.

75. Joannes Serapion, *De simplicibus medicinis*, ed. Otto Brunfels (Strasbourg: Georgius Ulricher, 1531), 115.

76. Kamel, "De igasur," 90–91.

77. Kamel, "De igasur," 90.

78. Kamel, "De igasur," 91.

79. For the introduction of the St. Ignatius bean in Europe, see Kroupa, "Georg Joseph Kamel," 176–79. I also examine this process in my draft monograph, *Drugs on the Move: Georg Joseph Kamel SJ and the (Un)Making of Cross-Cultural Knowledge in Southeast Asia, c.1650–1750*.

80. Cf. Dániel Margócsy, "'Refer to Folio and Number': Encyclopedias, the Exchange of Curiosities, and Practices of Identification Before Linnaeus," *Journal of the History of Ideas* 71, no. 1 (2010): 63–89; Staffan Müller-Wille, "Nature as a Marketplace: The Political Economy of Linnaean Botany," *History of Political Economy* 35, no. 5 (2003): 154–72.

81. De Vos, *Compound Remedies*, 11.

82. Prior to the arrival of the Spanish, Indigenous Filipino communities practiced distillation to produce palm liqueur and other products, having probably adopted the technology from Malayo-Muslim or Chinese migrants. See Paulina Machuca, *El vino de cocos en la Nueva España: Historia de una transculturación en el siglo XVII* (Colegio de Michoacán, 2018).

83. Nicolás Monardes, *Historia medicinal de las cosas que se traen de nuestras Indias Occidentales* (Seville: Alonso Escrivano, 1574), 3v, 9r, 12v, 18v, 24v, 28v, 51r.

84. Davis and López Terrada, "Protomedicato y farmacia," 594–5, 604.

85. José Pardo-Tomás and María Luz López Terrada, *Las primeras noticias sobre plantas americanas en las relaciones de viajes y crónicas de Indias, 1493–1553* (Consejo Superior de Investigaciones Científicas, 1993).

86. José Pardo-Tomás, "Pluralismo médico y medicina de la conversión: Fray Agustín Farfán y los agustinos en Nueva España, 1533–1610," *Hispania* 74, no. 248 (2014), 760.

87. Harold J. Cook, "Markets and Cultures: Medical Specifics and the Reconfiguration of the Body in Early Modern Europe," *Transactions of the Royal Historical Society* 21 (2011): 123–45.

88. Kamel, "Historia stirpium," 12.6, 21.6, 30.1, 30.2, 33.3, 39.17, 39.18, 39.23, 39.24, 54, 62.7, 62.9, 69, 77.4, 79bis.3, 82.10, 83.12, 87.1, 92.6.

89. Georg Joseph Kamel to James Petiver, Manila, October 15, 1704, British Library, Sloane MS 3321, fol. 151r.

90. Kamel, "Historia stirpium," 39.23–4, 60.10.

91. Kamel, "Historia stirpium," 39.17, 83.12, 85.4. Cuahmochitl is probably *Pithecellobium dulce*; see Hermes G. Gutierrez, *An Illustrated Manual of Philippine Materia Medica*, vol. 1 (National Research Council of the Philippines, 1980), 86–87. Tangal is probably *Ceriops tagal*, while ananapla is probably *Albizia procera* (Merrill, *Dictionary*, 17, 135).

92. John A. Parrota, *Pithecellobium dulce* (USDA Forest Service, Southern Forest Experiment Station, Institute of Tropical Forestry, 1991).

93. For example, Marcy Norton, *Sacred Gifts, Profane Pleasures: A History of Tobacco and Chocolate in the Atlantic World* (Cornell University Press, 2008); Markman

Ellis, Richard Coulton, and Matthew Mauger, *Empire of Tea: The Asian Leaf That Conquered the World* (Reaktion Books, 2015).

94. Alfred W. Crosby, *The Columbian Exchange: Biological and Cultural Consequences of 1492* (Greenwood, 1972).

95. For one exception, see Machuca, *El vino de cocos*.

3. THERIAC AS A DOMESTICATION TECHNOLOGY

The "Exporting Theriac" section builds on ideas first discussed in Di Gennaro Splendore, *The State Drug: Theriac, Pharmacy, and Politics in Early Modern Italy* (Harvard University Press, 2025), 160–61, 201–13.

1. Throughout this essay I refer to "exotic" as an emic category. Until the late seventeenth century, "exotic" was a term almost exclusively used in natural history. See Benjamin Schmidt, *Inventing Exoticism: Geography, Globalism, and Europe's Early Modern World* (University of Pennsylvania Press, 2015), 338n22. With reference to the early modern period, I use "Europe" and "Europeans" in a geographical sense.

2. In addition to other chapters in this volume, see Antonio Barrera-Osorio, "Local Herbs, Global Medicines. Commerce, Knowledge, and Commodities in Spanish America," in *Merchants & Marvels: Commerce, Science, and Art in Early Modern Europe*, ed. Pamela H. Smith, and Paula Findlen (Routledge, 2002); Anna E. Winterbottom, "Of the China Root: A Case Study of the Early Modern Circulation of Materia Medica," *Social History of Medicine* 28, no. 1 (2015): 22–44; Marcy Norton, *Sacred Gifts Profane, Pleasures: A History of Tobacco and Chocolate in the Atlantic World* (Cornell University Press, 2008); Matthew J. Crawford, *The Andean Wonder Drug: Cinchona Bark and Imperial Science in the Spanish Atlantic, 1630–1800* (University of Pittsburgh Press, 2016); Zachary Dorner, *Merchants of Medicines: The Commerce and Coercion of Health in Britain's Long Eighteenth Century* (University of Chicago Press, 2020).

3. On theriac see, Gilbert Watson, *Theriac and Mithridatium: A Study in Therapeutics* (Wellcome Historical Medical Library, 1966); and Di Gennaro Splendore, *State Drug*.

4. Symphorien Champier and Paul Dorveaux, eds., *Appothiquaires et Pharmacopoles par Symphorien Champier* (Paris: H. Welter, 1894; first published in Lyon, 1532), 34.

5. On theriac in antiquity, see Laurence Totelin, "Mithradates' Antidote: A Pharmacological Ghost," *Early Science and Medicine* 9, no.1 (2004): 1–19.

6. Prospero Alpini, *De medicina Aegyptiorum* (Venice: Franciscum de Franciscis Senensem, 1591), 133v.

7. Jean-Pierre Bénézet and Jean Flahaut, *Pharmacie et Médicament en Méditerranée Occidentale (XIIIe-XVIe Siècles)* (H. Champion; Slatkine, 1999), 680.

8. On the importance of theriac in Renaissance era botanical research, see Richard Palmer, "Pharmacy in the Republic of Venice in the Sixteenth Century," in *The Medical Renaissance of the Sixteenth century*, ed. Andrew Wear, R. K. French, and Iain M. Lonie (Cambridge University Press, 1985); Paula Findlen, *Possessing Nature: Museums, Collecting, and Scientific Culture in Early Modern Italy* (University of California Press, 1994), 272–77. See also, Brian W. Ogilvie, *The Science of Describing: Natural History in Renaissance Europe* (University of Chicago Press, 2006).

9. See Sebastian Kroupa's chapter in this volume; Linda A. Newson, *Making Medicines in Early Colonial Lima, Peru: Apothecaries, Science and Society* (Brill, 2017), 138–86; Paula Ronderos, *El Dilema de los Rótulos: Lectura del Inventario de una Botica Santafereña de Comienzos del Siglo XVII* (Editorial Pontifica Universidad Javeriana, 2007); Benjamin Breen, "The Flip Side of the Pharmacopeia: Poisons in the Atlantic World," in *Drugs on the Page: Pharmacopoeias and Healing Knowledge in the Early Modern Atlantic World*, ed. Matthew Crawford and Joseph Gabriel (University of Pittsburgh Press, 2019).

10. On substitutes in Galenic pharmacy, see Alain Touwaide, "*Quid pro Quo*: Revisiting the Practice of Substitution in Ancient Pharmacy," in *Herbs and Healers from the Ancient Mediterranean Through the Medieval West*, ed. Anne van Arsdall and Timothy Graham (Ashgate, 2012); Samir Boumediene and Valentina Pugliano, "La route des succédanés. Les remèdes exotiques, l'innovation médicale et le marché des substituts au XVIe siècle," *Revue d'histoire moderne contemporaine* 66, no. 3 (2019): 24–54.

11. Crawford and Gabriel, eds., *Drugs on the Page*.

12. For a thorough overview of theriac exports, see Di Gennaro, *State Drug*, 204–13 and relative bibliography.

13. On the genesis of the *Antidotario romano*, see Elisa Andretta, *Roma medica: anatomie d'un système médical au XVIe siècle* (École française de Rome, 2011), 165–76; Alexandra Kolega, "Speziali, spagirici, droghieri e ciarlatani. L'offerta terapeutica a Roma tra Seicento e Settecento," *Roma Moderna e Contemporanea*, 3 (1998): 322–23. On the work's editions, see E. Cingolani and Leonardo Colapinto, *Dagli antidotari alle moderne farmacopee* (Di Renzo, 2000), 53–58.

14. Kolega, "Speziali, spagirici, droghieri e ciarlatani," 323; Cingolani and Colapinto, *Dagli antidotari*, 53–54.

15. *Antidotario romano. Tradotto da latino in volgare da Ippolito Ceccarelli romano spetiale all'insegna della vecchia. Con l'aggiunta dell'elettione di semplici, prattica delle compositioni, et vn trattato dell'apparato della teriaca, et ragione de suoi ingredienti* (Rome: Bartolomeo Zanetti, 1612). The book was republished in 1619, 1624, 1635 (Milan), 1637 (Messina), 1639, 1651, 1664 (Venice), 1668, 1675, and 1678 (Venice).

16. In Kolega, "Speziali, spagirici, droghieri e ciarlatani," 331, from Archivio di Stato di Roma, Università b. 9.

17. *Antidotario romano* (1612 edition).

18. *Antidotario romano* (1619 edition), 37.

19. Alix Cooper, *Inventing the Indigenous: Local Knowledge and Natural History in Early Modern Europe* (Cambridge University Press, 2007).

20. Biblioteca Universitaria Bologna (hereafter BUB), ms Aldrovandi 70, fol. 58r. On Aldrovandi's interest for American flora and fauna, see Mario Cermenati, *Ulisse Aldrovandi e l'America con frammenti e note esplicative* (Enrico Voghera, 1906).

21. BUB, ms Aldrovandi 6, vol. ii, fol. 73r.

22. *Antidotario romano* (1619 edition), 38.

23. *Antidotario romano commentato dal dottor Pietro Castelli* (Messina: appresso la vedova di Gio. Francesco Bianco stampatore camerale, 1637), *dedicatoria*.

24. *Antidotario romano* (1619 edition), 183.

25. Alpini, *De medicina Aegyptiorum*, 133v.

26. *Antidotario romano* (1619 edition), 295–302.

27. On the new scholarly methods to study materia medica, see Paula Findlen, *Possessing Nature Museums, Collecting, and Scientific Culture in Early Modern Italy* (University of California Press, 1996); and Ogilvie, *Science of Describing*. On apothecaries involved in materia medica studies, see Pamela Smith, *The Body of the Artisan: Art and Experience in the Scientific Revolution* (University of Chicago Press, 2004); Deborah E. Harkness, *The Jewel House: Elizabethan London and the Scientific Revolution* (Yale University Press, 2007); Florike Egmond, "Apothecaries as Experts and Brokers in the Sixteenth-Century Network of the Naturalist Carolus Clusius," *History of Universities* 23 (2008): 59–91; and Sabine Anagnostou, Florike Egmond, and Christoph Friedrich, eds., *A Passion for Plants: Materia Medica and Botany in Scientific Networks from the 16th to the 18th Centuries* (Wissenschaftliche Verlagsgesellschaft, 2011). On Venetian apothecaries, see Richard Palmer, "Pharmacy in the Republic of Venice"; and Valentina Pugliano, "Botanical Artisans: Apothecaries and the Study of Nature in Venice and London, 1550–1610," PhD diss., University of Oxford, 2012.

28. The full title reads: *Antidotario romano latino, e volgare. Tradotto da Ippolito Ceccarelli. Li ragionamenti, e le aggiunte dell'elettione de' semplici, e pratica delle compositioni. Con le annotationi del sig. Pietro Castelli romano. E trattati della teriaca romana, e della teriaca egittia. E nuoua aggiunta di molte ricette ultimamente publicate dal Collegio de'medici di Roma. Dedicato all'illustrissimo, e reuerendissimo monsignor Fausto Poli, arciuescovo d'Amasia, e maggiordomo di N. S. Urbano Ottavo* (Rome: Pietro Antonio Facciotti, 1651), 58–65.

29. See for example, Bartolomeo Maranta, *Della theriaca et del mithridato* (Venice: Marcantonio Olmo, 1572); Evangelista Quattrami, *Tractatus perutilis atque necessarius ad Theriacam, Mitridaticumque Antidotum componendam* (Ferrara: Victorius Baldinus, 1597). For medical literature about theriac, see Barbara Di Gennaro Splendore, "The Triumph of Theriac. Print, Apothecary Publications, and the Commodification of Ancient Antidotes (1497–1800)," in "Printing Medical Knowledge: Vernacular Genres, Reception, and Dissemination," ed. Sabrina Minuzzi, special issue, *Nuncius* 36, no. 2 (2021): 431–70; Valentina Pugliano, "Pharmacy, Testing, and the Language of Truth in Renaissance Italy," *Bulletin of the History of Medicine* 91, no. 2 (2017): 233–73.

30. On the history of balsam, see Shimshon Ben-Yehoshua, Carole Borowitz, and Lumír Ondřej Hanuš, "Frankincense, Myrrh, and Balm of Gilead: Ancient Spices of Southern Arabia and Judea," *Horticultural Reviews*, ed. Jules Janick (John Wiley & Sons, 2011); Marcus Milwright, "Balsam in the Mediaeval Mediterranean: A Case Study of Information and Commodity Exchange," *Journal of Mediterranean Archaeology* 14, no. 1 (2001): 3–23; Marcus Milwright, "The Balsam of Maṭariyya: An Exploration of a Medieval Panacea," *Bulletin of the School of Oriental and African Studies* 66, no. 2 (2003): 193–209; Elly R. Truitt, "The Virtues of Balm in Late Medieval Literature," *Early Science and Medicine* 14, no. 6 (2009): 711–36; and Barbara Di Gennaro Splendore, "Mediterranean Botany: Making Cross-Cultural Knowledge About Materia Medica in the Sixteenth Century," in *Plants in 16th and 17th Century: Botany Between Medicine and Science*, ed. Fabrizio Baldassarri (De Gruyter, 2023).

31. Francesco Perla, *De orientali opobalsamo nuper in theriace confectione adhibito, et inter romanos medicos controverso* (Rome: Ludovico Grignani, 1641), 15.

32. Perla, *De orientali opobalsamo*, 16.

33. J. P. Griffin, "Venetian Treacle and the Foundation of Medicines Regulations," *British Journal of Clinical Pharmacology* 58, no. 3 (2004): 317–25.

34. Archivio Nobile Collegio Chimico Farmaceutico, San Lorenzo in Miranda, Roma, Busta 99 III 7.21 Mazzo D Armario B Parte II; Busta 103 III 7.33 Mazzo F Armario B Parte II.

35. Perla, *De orientali opobalsamo*, 16–17.

36. Elisa Andretta and Federica Favino, "Scientific and Medical Knowledge in Early Modern Rome," in *A Companion to Early Modern Rome, 1492–1692*, ed. Pamela M. Jones, Barbara Wisch, and Simon Ditchfield (Brill, 2019).

37. Giaquinto Trivulzio, *Ragguaglio primo venuto di Parnaso l'anno 1640*. (Trento: Santo Zanetti, 1640), 1v.

38. Paolo Zacchia, *Dell'opobalsamo orientale. Lettera all'Ill. ed Ecc. Collicola*, in Pietro Castelli, *Opobalsamum Examinatum, defensum, iudicatum, absolutum et laudatum* (Venice: Apud Petrum Tomasinum, 1640), 126–27.

39. On the War of Castro, see William Nassau Weech, *Urban VIII: Being the Lothian Prize Essay for 1903* (Archibald Constable, 1905), 86–88; Rosario Russo, "Urbano VIII," *Dizionario Biografico degli Italian*, https://www.treccani.it/enciclopedia/papa-urbano-viii_(Dizionario-Biografico)/, and Omero Masnovo, "Ducato di Castro" in *Enciclopedia Italiana Treccani*, online at https://www.treccani.it/enciclopedia/ducato-di-castro_%28Enciclopedia-Italiana%29/; Georg Lutz, in "Urbano VIII," in *Enciclopedia dei Papi online*, http://www.treccani.it/enciclopedia/urbano-viii_%28Enciclopedia-dei-Papi%29/, all accessed June 11, 2020.

40. "Le industrie, il commercio, le imposte sotto i pontefici Pio VI e Pio VII sino al 1815," *La Civiltà Cattolica* 57, 4 (1906): 442–43.

41. For example, see Stefano de Gasparis, *Liquoris artificialis pro opobalsamo Orientali in conficienda theriaca Romae adhibiti physica oppugnatio* (Rome: Ex tipographia Antonii Landini, 1640); Baldo Baldi, *Opobalsami orientalis in conficienda theriaca Romæ adhibiti. Medicae propugnationes* (Rome: Ex typographia Reuerendæ Cameræ Apostolicæ, 1640); Giuseppe Donzelli, *Additio apologetica ad suam de opobalsamo orientali synopsim* (Naples: Ottavio Beltrani, 1640).

42. See, for example, *Lettera piacevole di mastro Granchio Lalli aiutante di Cucina a mastro Marforio in Roma* . . . (Florence: Accorto Sferzaimperiti, 1640); *Distruttione del falso Ragguaglio primo venuto di Parnaso sopra il falso Balsamo d'Arabia, con il quale li Spetiali di Roma, A. Manfredi & V. Panutio hanno composto la loro Triaca, l'anno 1639* (Constanz: per gl'heredi di Verità Frustamat, 1640). Indications of place and publisher are most likely fake.

43. Laurie Nussdorfer, "Print and Pageantry in Baroque Rome," *Sixteenth Century Journal* 29, no. 2 (1998): 451.

44. Filippo de Vivo, *Information and Communication in Venice: Rethinking Early Modern Politics* (Oxford University Press, 2007).

45. Nussdorfer, "Print and Pageantry in Baroque Rome."

46. Biblioteca Corsiniana (Accademia dei Lincei), Rome, Ms. VI (4), fol. 227r, October 5, 1643.

47. Alberto Merola, "Barberini, Francesco," in *Dizionario Biografico degli*

Italiani [orig. 1964], accessed June 11, 2020, http://www.treccani.it/enciclopedia/francesco-barberini_%28Dizionario-Biografico%29/.

48. Di Gennaro Splendore, "Triumph of Theriac."

49. See, for example, Hannah Murphy, *A New Order of Medicine: The Rise of Physicians in Reformation Nuremberg* (University of Pittsburgh Press, 2019).

50. Viviane Machado Caminha São Bento, "Emplastros: os medicamentos das boticas jesuítas no auxílio do cotidiano na América Portuguesa," *Revista História e Cultura* 3, no. 2 (2014): 303–4; Serafim Leite, *Artes e Ofícios dos Jesuítas no Brasil, 1549–1760* (Edições Brotéria, 1953), 2, 583; 1, 87–88, 298. On *Triaga Brasilica*, see also N. A. Pereira, R. J. Jaccoud, and W. B. Mors, "Triaga Brasilica: Renewed Interest in a Seventeenth-Century Panacea," *Toxicon: Official Journal of the International Society on Toxinology* 34, no. 5 (1996): 311–16; Di Gennaro, *The State Drug*, 205–6.

51. C. G. Uragoda and K. D. Paranavitana, "Dutch Pharmacopoeia of 1757: Probably the Earliest Such Document from Sri Lanka," *Journal of the Royal Asiatic Society of Sri Lanka*, New Series, Vol. 51 (2005): 1.

52. Harold John Cook, *Matters of Exchange: Commerce, Medicine, and Science in the Dutch Golden Age* (Yale University Press, 2007), 305, 308–9.

53. Hermanus Nicolaas Grimm, *Insulae Ceyloniae thesaurus medicus laboratorium* (Amsterdam: Henricum & Theodorum, 1679).

54. Uragoda and Paranavitana, "Dutch Pharmacopoeia of 1757," 13–14, 17.

55. Uragoda and Paranavitana, "Dutch Pharmacopoeia of 1757," 13–14.

56. Pereira, Jaccoud, and Mors, "Triaga Brasilica," 312. Similar to *Triaga brasilica*, in Europe new products derived from theriac, like orvietan, were often secret and protected by a privilege. By contrast, the recipe of *Theriaca magna* was never secret, and what added prestige to it was precisely the public certification of theriac's production.

57. On this image, see Ernest Wickersheimer, "La thériaque céleste dite de 'Strasbourg,'" *Revue d'Histoire de la Pharmacie* 8, no. 25 (1920): 157–58.

58. Wickersheimer, "La thériaque céleste dite de 'Strasbourg.'"

59. Joseph Du Chesne, *La Reformation des Theriaqves et Antidotes opiatiqves* (Paris: chez Claude Morel, 1608). There are two different editions of this book with the same frontispiece; I used the online copy from the Universitad Complutense de Madrid, https://babel.hathitrust.org/cgi/pt?id=ucm.532772782x&view=1up&seq=4, accessed July 8, 2020.

60. Du Chesne, *Reformation des Theriaqves*, 133.

61. Du Chesne, *Rreformation des Theriaqves*, 135.

62. Du Chesne, *Reformation des Theriaqves*, 143.

63. Greiff celebrated his productions by publishing customary pamphlets in Tübingen; see Greiff, *Decas Nobilissimorum Medicamentorum, Galeno-Chymico modo compositorum* (1641) and *Kurtze Beschreibung Deß Chymischen oder Himmelischen Theriacs* (1652). Wickersheimer, "La thériaque céleste dite de 'Strasbourg,'" 154.

64. Wickersheimer, "La thériaque céleste dite de 'Strasbourg,'" 155.

65. Gary Leiser and Michael Dols, "Evliyā Chelebi's Description of Medicine in Seventeenth-Century Egypt: Part I: Introduction," *Sudhoffs Archiv* 71, no. 2 (1987): 197–216; Gary Leiser and Michael Dols, "Evliyā Chelebi's Description of Medicine in Seventeenth-Century Egypt: Part II: Text," *Sudhoffs Archiv* 72, no. 1 (1988): 67–68.

66. Jean-Baptiste Tavernier, *Nouvelle relation de l'intérieur du serrail du Grand Seigneur* (Paris: Varennes, 1675), 189.

67. Jean-Baptiste Tavernier, *Parte prima de' viaggi nella Turchia, Persia et India* (Rome: Giuseppe Corvo, 1632), 165–66.

68. Harun K. Küçük, *Science without Leisure: Practical Naturalism in Istanbul, 1660–1732*, (University of Pittsburgh Press, 2019), 161–66.

69. John-Paul A. Ghobrial, *The Whispers of Cities: Information Flows in Istanbul, London, and Paris in the Age of William Trumbull* (Oxford University Press, 2013), 69.

70. Mary Lucille Shay, *The Ottoman Empire from 1720 to 1734 as Revealed in Despatches of the Venetian Baili* (University of Illinois Press, 1944), 45, 48, 51, 52, 54.

71. Francesco Foscari and Filippo Maria Paladini, eds., *Dispacci da Costantinopoli 1757–1762* (La Malcontenta, 2007), 27–28, 48.

72. E. C. Spary, *Eating the Enlightenment: Food and the Sciences in Paris, 1670–1760* (University of Chicago Press, 2012), 55–57.

73. Carlo Antonio Marin, *Storia civile e politica del commercio de' Veneziani* (Venice: Coleti, 1808), 343, 347.

74. Kiaya was a lieutenant, while the *Reis Effendi*, or *Reis ül-Küttab*, was the head of the scribes. This role acquired more power over time, but at the end of the eighteenth century it was still a rather junior position.

75. I would like to thank Prof. Harun Küçük for his thoughts on this question.

4. COMPETING MEDICAL SUBSTANCES DURING AN EPIDEMIC

1. The literature on the history of cinchona bark is abundant, but especially important are Saul Jarcho, *Quinine's Predecessor: Francesco Torti and the Early History of Cinchona* (Johns Hopkins University Press, 1993); Andreas-Holger Maehle, *Drugs on Trial: Experimental Pharmacology and Therapeutic Innovation in the Eighteenth Century* (Rodopi, 1999), chapter 4; Samir Boumediene, *La colonisation du savoir. Une histoire de plantes médicinales du "Nouveau Monde" (1492–1750)* (Éditions des Mondes à Faire, 2016); and Matthew James Crawford, *The Andean Wonder Drug: Cinchona Bark and Imperial Science in the Spanish Atlantic, 1630–1800* (University of Pittsburgh Press, 2016). For rhubarb, see Clifford M. Foust, *Rhubarb: The Wondrous Drug* (Princeton University Press, 1992); Erika Monahan, "Locating Rhubarb: Early Modernity's Relevant Obscurity," in *Early Modern Things: Objects and Their Histories, 1500–1800*, ed. Paula Findlen (Routledge, 2013); Stefan Heßbrüggen-Walter, "Problems with Rhubarb: Accommodating Experience in Aristotelian Theories of Science," *Early Science and Medicine* 19, no. 4 (2014): 317–40; and Matthew P. Romaniello, "True Rhubarb? Trading Eurasian Botanical and Medical Knowledge in the Eighteenth Century," *Journal of Global History* 11, no. 1 (2016): 3–23. For opium, see Maehle, *Drugs on Trial*, chapter 3; Hans Derks, *History of the Opium Problem: The Assault on the East, ca. 1500–1950* (Brill, 2012); and Saskia Klerk, "The Trouble with Opium. Taste, Reason and Experience in Late Galenic Pharmacology with Special Regard to the University of Leiden (1575–1625)," *Early Science and Medicine* 19, no. 4 (2014): 287–316.

2. I assembled these types of advertisements in two databases, which was the basis for most of my PhD research; see Wouter Klein, "New Drugs for the Dutch Republic: The Commodification of Fever Remedies in the Netherlands (c. 1650–1800)"

(PhD diss., Utrecht University, 2018). The contents and creation process of both databases are discussed extensively in the dissertation.

3. Louise Hill Curth, "Medical Advertising in the Popular Press: Almanacs and the Growth of Proprietary Medicines," in *From Physick to Pharmacology: Five Hundred Years of British Drug Retailing*, ed. Louise Hill Curth (Ashgate, 2006).

4. I discuss some remedies for which dozens of advertisements can be found in Dutch newspapers in my dissertation: see Klein, "New Drugs for the Dutch Republic," esp. chapter 3.

5. Key publications in this respect are, for France, Colin Jones, "The Great Chain of Buying: Medical Advertisement, the Bourgeois Public Sphere, and the Origins of the French Revolution," *American Historical Review* 101, no. 1 (1996): 13–40; for England, Lisa Forman Cody, "'No Cure, No Money,' or the Invisible Hand of Quackery: The Language of Commerce, Credit, and Cash in Eighteenth-Century British Medical Advertisements," *Studies in Eighteenth Century Culture* 28 (1999): 103–30; and Elizabeth Lane Furdell, *Publishing and Medicine in Early Modern England* (University of Rochester Press, 2002), esp. chapter 7; and, for the Dutch Republic, Frank Huisman, "Gezondheid te Koop: Zelfmedicatie en Medische Advertenties in de Groninger en Ommelander Courant," *Focaal* 21 (1993): 90-130.

6. See Roberta Mullini, *Healing Words: The Printed Handbills of Early Modern London Quacks* (Peter Lang, 2015), 84 and 138–39n59. Apart from the difference in publishing frequency, the main difference between handbills and advertisements in this period seems to be the absence of images in the latter (see Mullini, 141).

7. Klein, "New Drugs for the Dutch Republic," 193–44 and 274.

8. Klein, "New Drugs for the Dutch Republic," introduction.

9. The best-known studies of the Spanish Atlantic trade in cinchona bark consistently refer to cinchona bark as *cascarilla*, in line with their sources: Antonio García-Baquero González, *Cádiz y el Atlántico (1717–1778): El Comercio Colonial Español bajo el Monopolio Gaditano*, 2 vols. (Escuela de Estudios Hispano-America, 1976); and John Fisher, *Commercial Relations Between Spain and Spanish America in the Era of Free Trade, 1778–1796* (University of Liverpool, Centre for Latin American Studies, 1985).

10. Boumediene, *Colonisation du savoir*, 192–93.

11. Pierre Pomet, *Histoire Generale des Drogues* (Paris: Jean-Baptiste Loyson and Augustin Pillon, and Palais: Estienne Ducastin, 1694), appendix. Pierre Bonnet Bourdelot should not be confused with his better-known uncle Pierre Bourdelot (1610–85), personal physician to the prince de Condé, who is best remembered as the organizer of the Académie Bourdelot in the 1640s. I thank Emma Spary for supplying this information.

12. Nicolas Lémery, *Dictionaire ou Traité universel des Drogues Simples. Troisième Edition* (Aux Dépens de la Compagnie, 1716), 203.

13. Willem van Ranouw, "Vierde Verhandeling van de byzondere Natuurlyke Historischryvers, en in dezelve de Natuurlyke Historie van de Kina-Kina," *Kabinet der natuurlyke Historien, Wetenschappen, Konsten en Handwerken* 6 (1722). Van Raat's letter and description are at 136–45. Van Ranouw describes Van Raat in Dutch as "distinguished merchant of Rotterdam in drugs, and excellent lover and connoisseur of

natural histories" (*voornaam Koopman tot Rotterdam in de Drogeryen, en uitmuntend Liefhebber en kender der natuurlyke Historien*), 136.

14. Van Ranouw, "Vierde Verhandeling," 146–47.

15. Van Ranouw, "Vierde Verhandeling," 137, 126.

16. Bernard Le Bovier de Fontenelle, "Sur le Chacril," *Histoire de l'Academie Royale des Sciences. Année M.DCCXIX* (1721), 54.

17. Van Ranouw, "Vierde Verhandeling," 138.

18. Mark Catesby, *The Natural History of Carolina, Florida and the Bahama Islands* (London: W. Innys, R. Manby, Hauksbee, and the author, 1731–43), 2:46: "An Ricinoides Aeleagni folio?"

19. Philippus Adolphus Boehmer, *De Cortice Cascarillae eiusque Insignibus in Medicina Viribus* (Halle: Officina Grunertiana, 1738). Apparently, then, Catesby's image was already available before the official publication of his second volume in 1743.

20. Charles Marie de La Condamine, "Sur l'Arbre du Quinquina," *Histoire de l'Académie Royale des Sciences, Année 1738* (1740), after 244.

21. Boulduc, "Sur le Chacril," 55. See also Jarcho, *Quinine's Predecessor*, 63–64. It was a great asset for a remedy to be approved by the king's personal physician. The *poudre fébrifuge* of the Chevalier de Guiller was likewise approved by Fagon in 1713; see Justin Rivest, "Testing Drugs and Attesting Cures: Pharmaceutical Monopolies and Military Contracts in Eighteenth-Century France," *Bulletin of the History of Medicine* 91, no. 2 (2017): 362–90.

22. Étienne François Geoffroy, *Tractatus de Materia Medica, sive De Medicamentorum Simplicium Historiâ, virtute, delectu & usu* (Paris: Joannes Desaint & Caroli Saillant, 1741), 2:202–8. The work was also translated into English: Ralph Thicknesse, *A Treatise on Foreign Vegetables* (London: J. Clarke, J. Whiston, and S. Baker, 1749), 99–105.

23. Stadsarchief, Amsterdam, 5069, inv. nos. 1–190. These are the records of public auctions of consumer products by official brokers (*makelaars*) in Amsterdam (1721–1808). The records are incomplete before 1736. Seba's name occurs eleven times in the records for three auctions (dated August 15, 1726; September 12, 1726; and October 20, 1727).

24. Stadsarchief, Amsterdam, 5069, inv. no. 5 (October 20, 1727). Seroons, *ceroenen* in Dutch, was one of the most common trade units for cinchona bark. The word derives from the Spanish *zurrónes*, which referred to animal skins, in which the bark was wrapped, and then sewn together on both ends. It is unclear what the exact weight of three seroons would have been. See also Boumediene, *Colonisation du savoir*, 194.

25. Stadsarchief, Amsterdam, 5069, inv. no. 4 (September 12, 1726).

26. To substantiate this claim, a query was formulated in Delpher (see note 2) for instances of fever and spelling variants in Dutch (*koorts / koortsen / koortzen / koors*), in newspaper articles between July 1727 and June 1728, which yielded fifty results.

27. *Europische Mercurius* (Amsterdam: Hendrik van Damme Jansz, 1728), 54.

28. *Europische Mercurius*, 54. The mortality figures for Amsterdam were derived from *Statistiek der Bevolking van Amsterdam tot 1921 / Statistique Démographique de la Ville d'Amsterdam jusqu'à l'Année 1921* (Johannes Müller, 1923), 179.

NOTES TO PAGE 115

29. J. Z. Kannegieter, "Hevige Sterfte te Amsterdam in 1727," *Maandblad Amstelodamum* 59, no. 3 (1972): 49–56.

30. Considering Holland's history as a marshy area, the outbreak of fever might have been an instance of endemic malaria. This is debatable, but Henk Brouwer regards the epidemic as such, and he includes the cities of Harlingen and Leeuwarden in Friesland in his analysis. See Brouwer, "Malaria in Nederland in de Achttiende en Negentiende eeuw," *Tijdschrift voor Sociale Geschiedenis* 9, no. 2 (1983): 144. Cf. notes 40 and 44.

31. J. Buisman, *Duizend Jaar Weer, Wind en Water in de Lage Landen, Deel 5: 1675–1750* (Van Wijnen, 2006), 542–44. Yearly mortality figures for Hoorn are unknown. Luuc Kooijmans discusses mortality in Hoorn on the basis of burials in the Grote Kerk (circa one third of all burials in the city). See Kooijmans, *Onder Regenten: De Elite in een Hollandse Stad, Hoorn 1700–1780* (Bataafsche Leeuw, 1985), 17–18 and 235n17. Kooijmans gives data with ten-year intervals, but there is a peak during 1720–29 (222).

32. Renatus Theodorus Henricus Willemsen, *Enkhuizen tijdens de Republiek: een Economisch-Historisch Onderzoek naar de Stad en Samenleving van de 16e tot de 19e Eeuw* (Verloren, 1988), 105–7 and 113 (fig. 4.6). The epidemic was most severe in the months of October–November 1727 and April 1728, with well over one hundred burials during each month. Willemsen identifies the mortality peak as epidemic dysentery (112–16).

33. The yearly death rate in Leiden was extracted from archival material (Regionaal Archief, Leiden, 0501A, I.B.1.5.3, inv. nos. 1332–35). The general trend is discussed by C. A. Davids, "De Migratiebeweging in Leiden in de Achttiende Eeuw," in *Armoede en Sociale Spanning: Sociaal-Historische Studies over Leiden in de Achttiende Eeuw*, ed. H. A. Diederiks, D. J. Noordam, and H. D. Tjalsma (Verloren, 1985), 143 (fig. 8.3); and Dirk Jaap Noordam, "Demografische Ontwikkelingen," in *Leiden: De Geschiedenis van een Hollandse Stad, Deel 2, 1574–1795*, ed. Simon Groenveld (Stichting Geschiedschrijving Leiden, 2003), 47–50. Neither of them has analyzed the mortality peak of 1727–28 in detail. For the "Noorderkwartier," see A. M. van der Woude, *Het Noorderkwartier: Een Regionaal Historisch Onderzoek in de Demografische en Economische Geschiedenis van Westelijk Nederland van de Late Middeleeuwen tot het Begin van de Negentiende Eeuw* (H. Veenman & Zonen, 1972), 1:197–208, where he also discusses the relationship between high mortality and epidemic diseases, but not specifically with reference to 1727–28.

34. The data for Delft in figure 4.5 was derived from the work of Thera Wijsenbeek-Olthuis, which only includes burials in the Oude Kerk and Nieuwe Kerk. See Wijsenbeek-Olthuis, *Achter de Gevels van Delft: Bezit en Bestaan van Rijk en Arm in een Periode van Achteruitgang (1700–1800)* (Verloren, 1987), 410 (Bijlage 6). For Rotterdam, see G. J. Mentink and A. M. van der Woude, *De Demografische Ontwikkeling te Rotterdam en Cool in de 17e en 18e Eeuw: Een Methodisch en Analyserend Onderzoek van de Retroacta van de Burgerlijke Stand van Rotterdam en Cool* (Gemeentearchief Rotterdam, 1965), 126–27 (the figures come from the column "Totaal III"). Willibrord Rutten relates most eighteenth-century mortality peaks to smallpox, except for the fever epidemic of 1727–28, which he identifies as caused by malaria. However, he again treats the peak in Leiden in 1727 as correlating with a smallpox outbreak. See Rutten, *"De Vreselijkste Aller Harpijen": Pokkenepidemieën en Pokkenbestrijding in Nederland in*

de Achttiende en Negentiende Eeuw: Een Sociaal-Historische en Historisch-Demografische Studie (Landbouwuniversiteit Wageningen; HES, 1997), 75–76.

35. Justin Rivest offers another discussion of the meanings of secrecy in early modern pharmacy. See Rivest, "Beyond the Pharmacopoeia? Secret Remedies, Exclusive Privileges, and Trademarks in Early Modern France," in *Drugs on the Page: Pharmacopoeias and Healing Knowledge in the Early Modern Atlantic World*, ed. Matthew James Crawford and Joseph M. Gabriel (University of Pittsburgh Press, 2019). For a broader view of secrecy in early modern science, see Elaine Leong and Alisha Rankin, eds., *Secrets and Knowledge in Medicine and Science, 1500–1800* (Ashgate, 2011).

36. The database in which these advertisements were assembled is discussed in my dissertation; see Klein, "New Drugs for the Dutch Republic," introduction.

37. *Leydse Courant*, November 24, 1727.

38. *Amsterdamse Courant*, October 23, 1727.

39. Kannegieter, "Hevige Sterfte te Amsterdam in 1727."

40. Jan de Vries and Ad van der Woude, *The First Modern Economy: Success, Failure, and Perseverance of the Dutch Economy, 1500–1815* (Cambridge University Press, 1997), 49. It is unclear what their identification of malaria is based on. Cf. notes 30 and 44.

41. Boerhaave to Bassand, November 7, 1727, in *Boerhaave's Correspondence*, ed. G. A. Lindeboom (Leiden: Brill, 1962–79), 2:248–51. This and the following quotes from Boerhaave's correspondence are Lindeboom's translations. Boerhaave's account of the epidemic is also discussed in Luuc Kooijmans, *Het Orakel: De Man die de Geneeskunde Opnieuw Uitvond: Herman Boerhaave 1669–1738* (Balans, 2011), 230–31.

42. Boerhaave to Bassand, January 23, 1728, in *Boerhaave's Correspondence*, 2:252–57.

43. Boerhaave to Bassand, January 23, 1728.

44. Boerhaave to Bassand, January 23, 1728. G. A. Lindeboom argues that Boerhaave's disease (and, by implication, the epidemic) was an instance of typhoid or paratyphoid fever. See Lindeboom, "Boerhaaves Krankheiten," *Sudhoffs Archiv für Geschichte der Medizin und der Naturwissenschaften* 39, no. 2 (1955): 170. Cf. notes 30 and 40.

45. Kooijmans, *Het Orakel*, 286. It is unclear from which source this anecdote derives.

46. Boerhaave to Bassand, December 8, 1733, in *Boerhaave's Correspondence*, 2:324–29.

47. Boerhaave to Bassand, November 2, 1714, in *Boerhaave's Correspondence*, 2:112–17. The remedy contained, apart from cinchona bark, rind of caper root, pomegranate rind, tamarisk bark, oak leaves, and pomegranate blossom.

48. Boerhaave to Roëll, June 3, 1733, in *Boerhaave's Correspondence*, 3:46–51. This remedy contained the same ingredients as the one in note 47, except for the oak leaves. Extra ingredients in the later recipe were cinnamon bark, red sandalwood, and "oxylapath" (i.e., sorrel) root.

49. Boerhaave to Bassand, April 26, 1717, in *Boerhaave's Correspondence*, 2:144–57. The same batch also included other exotics: ipecacuanha root, rhubarb, and opium.

50. For a discussion of Boerhaave's views on fever, see Leonard G. Wilson,

"Fevers," in *Companion Encyclopedia of the History of Medicine*, ed. W. F. Bynum and Roy Porter (Routledge, 1997), 1:397–98.

51. Most of the sources discussed in this section and more are to be found in Henricus Franciscus Thyssen, *Over de Herfstkoortsen te Amsterdam, bijzonder over die van het jaar 1826* (P. Meyer Warnars, 1827), 26–28, esp. 28n19.

52. Gerard van Swieten, *Constitutiones Epidemicae et Morbi Potissimum Lugduni-Batavorum observati ex eiusdem adversariis*, ed. Maximilian Stoll (Vienna and Leipzig: Rudolphum Graefferum, 1782), 1:38–39.

53. Salomon de Monchy, "Verhandeling van de Oorzaaken, Genezing en Voorbehoeding der gewoone Ziekten van ons Scheepsvolk, 't geen naar de West-Indien vaart," *Verhandelingen uitgegeven door de Hollandsche Maatschappye der Weetenschappen te Haarlem* 6, no. 1 (1761): 111.

54. Théodore Tronchin, "Verhandeling over het Kolyk van Poitou," *Uitgezogte Verhandelingen uit de Nieuwste Werken van de Societeiten der Wetenschappen in Europa en van andere Geleerde Mannen* 2 (1757): 598–99 (my translation). The fever is described as a "bilious fever" (*galkoorts*).

55. A modern history of the brokers of Amsterdam has yet to be written, but see Hendrik van Malsen, *Geschiedenis van het Makelaarsgild te Amsterdam 1578–1933* (W. ten Have, 1933); and Klein, "New Drugs for the Dutch Republic," chapter 2.

56. The real contents (in terms of weight) of any of these trade units requires further research, so an analysis of the number of trade units is currently the closest we can get to the trade "volume" of cinchona bark, that was auctioned in Amsterdam.

57. One example can be found in the *Leydse Courant* (February 1, 1734), concerning the arrival of 220,000 pounds of cinchona bark in Spain. The news item mentions that letters from Panama had already complained about the bad condition (*slecht vertier*) of the shipment.

58. John Gray and William Arrot, "An Account of the Peruvian or Jesuits Bark," *Philosophical Transactions* 40, no. 446 (1738): 82. Cf. Maehle, *Drugs on Trial*, 276.

59. Crawford, *Andean Wonder Drug*, chapter 3.

60. Overall, however, it is reasonable to assume that a lot of bark may have been of inferior quality. According to Jarcho (*Quinine's Predecessor*, 203–4), there are numerous reports about spoilage of bark in eighteenth-century Spanish records.

61. The first appeared in the *Amsterdamse Courant* on December 16, 1727.

62. Other examples of alternatives for cinchona bark, with clearly different characteristics, are discussed in Rivest, "Testing Drugs"; and Pratik Chakrabarti, "Empire and Alternatives: Swietenia febrifuga and the Cinchona Substitutes," *Medical History* 54, no. 1 (2010): 75–94.

63. Stadsarchief, Amsterdam, 5069, inv. no. 9 (August 7, 1737).

64. *Pharmacopoea Almeriana galeno-chymica* (Alkmaar: Joannes van Beyeren, 1723), 8.

65. *Leydse Courant*, July 15, 1737.

5. THE MEDICAL RECEPTION OF SASSAFRAS IN EARLY MODERN ENGLISH PRINT

1. Daniela Bleichmar, "Books, Bodies and Fields: Sixteenth-Century Transatlantic Encounters with the New World," in *Colonial Botany: Science, Commerce and Politics*

in Early Modern World, ed. Londa Schiebinger and Claudia Swan (University of Pennsylvania Press, 2007); Antonio Barrera, "Empire and Knowledge: Reporting from the New World," *Colonial Latin American Review* 15, no. 1 (2006): 39–54; Matthew James Crawford, *The Andean Wonder Drug: Cinchona Bark and Imperial Science in the Spanish Atlantic, 1630–1800* (University of Pittsburgh Press, 2016); Samir Boumediene, *La Colonisation du savoir. Une histoire des plantes médicinales du "Nouveau Monde" (1492–1750)* (Les Éditions des Mondes à Faire, 2016).

2. For a larger study of 182 texts considering the reception of sassafras in literary, medical, religious, travel, geographical, and political texts, see Katrina Maydom, "New World Drugs in England's Early Empire" (PhD diss., University of Cambridge, 2019), Chapter 3. For a large-scale study of remedies in general in early modern English print, see Andrew Wear, *Knowledge and Practice in English Medicine, 1550–1680* (Cambridge University Press, 2000).

3. There was a long-standing tradition of adopting new simples into the Galenic framework from other parts of the world before European encounters with the Americas. See the chapter by Paula de Vos in this volume for a discussion of these processes.

4. On the development of European botanical knowledge about sassafras, see Clare Griffin, "Disentangling Commodity Histories: *Pauame* and Sassafras in the Early Modern Global World," *Journal of Global History* 15, no. 1 (2020): 1–18.

5. *Oxford English Dictionary* (Oxford: Oxford University Press, 2020), under "sassafras." https://www.oed.com/dictionary/sassafras_n.

6. Griffin, "Disentangling Commodity Histories," 3.

7. Russell M. Magnaghi, "Sassafras and its Role in Early America, 1562–1662," *Terrae Incognitae* 29, no. 1 (1997): 10–21.

8. For an account of the Spanish and French encounters with sassafras, see Griffin, "Disentangling Commodity Histories."

9. There were many debates over the extent to which plants from different regions could be substituted for each other in the early modern period, for example Canadian and Tartary ginseng and American and Chinese sarsaparilla. See Christopher M. Parsons, "The Natural History of Colonial Science: Joseph-François Lafitau's Discovery of Ginseng and Its Afterlives," *William & Mary Quarterly* 73, no. 1 (2016): 37–72; and Anna E. Winterbottom, "Of the China Root: A Case Study of the Early Modern Circulation of Materia Medica," *Social History of Medicine* 28, no. 1 (2014): 22–44. European merchants and government officials became wary of drugs such as "wild" or "bastard" cinchona and cinnamon. See Boumediene, *La colonisation du savoir*. While some early modern English texts warned about fake and adulterated sassafras, there were no concerns recorded about sassafras from different regions being less effective. On problems of drug authentication in the sixteenth century, see Valentina Pugliano, "Pharmacy, Testing, and the Language of Truth in Renaissance Italy," *Bulletin of the History of Medicine* 91, no. 2 (2017): 233–73.

10. Nicolás Monardes, *Joyfull Nevves Out of the Newe Founde Worlde*, trans. John Frampton (London: Willyam Norton, 1577).

11. See, for example, Bleichmar, "Books, Bodies, and Fields"; and Donald Beecher, "The Legacy of John Frampton: Elizabethan Trader and Translator," *Renaissance Studies* 20, no. 3 (2006): 320–39.

12. Wouter Klein and Toine Pieters, "The Hidden History of a Famous Drug: Tracing the Medical and Public Acculturation of Peruvian Bark in Early Modern Western Europe (c. 1650–1720)," *Journal of the History of Medicine and Allied Sciences* 71, no. 4 (2016): 400–421.

13. Harold J. Cook, and Timothy D. Walker, "Circulation of Medicine in the Early Modern Atlantic World," *Social History of Medicine* 26, no. 3 (2013): 337–51; Winterbottom, "Of the China Root"; Wouter Klein, Kalliopi Zervanou, Marijn Koolen, Peter van den Hooff, Frans Wiering, Wouter Alink, and Toine Pieters, "Creating Time Capsules for Historical Research in the Early Modern Period: Reconstructing Trajectories of Plant Medicines," *HistoInformatics* 2017, no. 1 (2017).

14. Paul Freedman, *Out of the East: Spices and the Medieval Imagination* (Yale University Press, 2008), chapter 2; Heidi Hausse, "European Theories and Local Therapies: Mordexi and Galenism in the East Indies, 1500–1700," *Journal of Early Modern History* 18, nos. 1–2 (2014): 121–40.

15. Anthony Grafton, *New Worlds, Ancient Texts: The Power of Tradition and the Shock of Discovery* (Harvard University Press, 1995).

16. For an analysis of the different forms of medical writing in early modern England, see Irma Taavitsainen, Peter Murray Jones, Päivi Pahta, Turo Hiltunen, Ville Marttila, Maura Ratia, Carla Suhr, and Jukka Tyrkkö, "Medical Texts in 1500–1700 and the Corpus of Early Modern English Medical Texts," in *Medical Writing in Early Modern English*, ed. Irma Taavitsainen and Päivi Pahta (Cambridge University Press, 2011).

17. The most recent update of EEBO at the time of the search was completed in November 2017, and more texts may have been transcribed since my study. At this time, EEBO included transcribed copies of 44 percent of extant texts published in England and its territories in all languages until 1700. EEBO has digitized 96 percent of the titles in the ESTC and has transcribed 46 percent of these printed works.

18. Discussions of sassafras were found by searching for several different spellings and abbreviations: "sassafras," "sassafra," "sassafr," "sassaf," "sassa," and "sasa," with the variant option enabled to locate orthographical alternatives, such as "sasafras," "sassaphras," and "sarsafras." I also searched all variants in which the letter "f" was exchanged for the "s" and vice versa. The most common spellings in both English and Latin texts were "sassafras" and "sassaphras," and the most frequent abbreviation was "sassafr." We should be aware, however, that there may be instances in which "sassafras" was spelled in such a manner that it was not found by the search. It is difficult for us to know the extent to which this may have occurred, but no substantially different spellings or names were used in the 114 texts consulted. We can therefore be confident that any missing references to sassafras do not significantly alter the results reported in this chapter.

19. Figures taken from Maydom, "New World Drugs," 100.

20. In this analysis, I account for *Lignum vitae* and other synonyms within the category of "guaiacum."

21. Maydom, "New World Drugs," 48.

22. Patrick Wallis, "Exotic Drugs and English Medicine: England's Drug Trade, c. 1550–c. 1800," *Social History of Medicine* 25, no. 1 (2012): 20–46; Maydom, "New World Drugs," chapter 1.

23. By "neuritick," Harvey referred to the ability to treat "sinewy diseases," ones caused by problems with the joints, such as palsies and gouts (Gideon Harvey, *The Disease of London: or, A New Discovery of the Scorvey* (London: T. James for W. Thackery, 1675).

24. Daniel Sennert, *Two Treatises*, trans. Nicholas Culpeper and Abdiah Cole (London: John Streater and William Whitwood, 1660), 43.

25. Tobias Venner, *A Briefe and Accvrate Treatise, Concerning, The Taking of the Fume of Tobacco* (London: W. I. for Richard Moore, 1621), unpaginated.

26. Jean de Renou, *A Medicinal Dispensatory, Containing the Whole Body of Physick: Discovering the Natures, Properties, and Vertues of Vegetables, Minerals, & Animals*, trans. Richard Tomlinson (London: Jo. Streater and Ja. Cottrel, 1657), 287.

27. For other examples of the attribution of humoral qualities to plants incorporated into early modern European medicine, see de Vos, this volume; Gianamar Giovannetti-Singh, "Galenizing the New World: Joseph-François Lafitau's 'Galenization' of Canadian Ginseng, ca. 1716–1724," *Notes and Records of the Royal Society* 75, no. 1 (2021): 59–72; and Matthew P. Romaniello, "Who Should Smoke? Tobacco and the Humoral Body in Early Modern England," *Social History of Alcohol and Drugs* 27, no. 2 (2013): 156–73.

28. Daniel Sennert, *Nine Books of Physick and Chirurgery* (London: J. M. for Lodowick Lloyd, 1658), 314.

29. Lazare Rivière, *The Universal Body of Physick in Five Books* (London: H. Eversden, 1657), 360.

30. William Coles, *Adam in Eden* (London: J. Streeter for Nathaniel Brooke, 1657), 307.

31. Rivière, *Universal Body of Physick*, 360.

32. On the debate between Galenists and Chymists in seventeenth-century English medicine, see Piyo M. Rattansi, "The Helmontian-Galenist Controversy in Restoration England," *Ambix* 12, no. 1 (1964): 1–23; Harold J. Cook, "The Society of Chemical Physicians, the New Philosophy, and the Restoration Court," *Bulletin of the History of Medicine* 61, no. 1 (1987): 61–77; Antonio Clericuzio, "From van Helmont to Boyle: A Study of the Transmission of Helmontian Chemical and Medical Theories in Seventeenth-century England," *British Journal for the History of Science* 26, no. 3 (1993): 303–34; and Wear, *Knowledge and Practice*, chapters 8 and 9.

33. Jan Baptista van Helmont, *Ortus Medicinae. Id est, initia physicæ inavdita. Progressus medicinæ novus, in Morburum ultionem, ad Vitam longam* (Amsterdam: Ludovicum Elzevirium, 1648). This was a collection of works published posthumously by van Helmont's son, Franciscus Mercurius van Helmont, from his manuscripts.

34. Jean Baptiste van Helmont, *Deliramenta Catarrhi: Or, The Incongruities, Impossibilities, and Absurdities Couched Under the Vulgar Opinion of Defluxions*, trans. Walter Charleton (London: E. G. for William Lee, 1650).

35. George Thomson, *Galeno-pale, or, A Chymical Trial of the Galenists* (London, 1665), 47.

36. Thomson, *Galeno-pale*, 47.

37. Paracelsus mentions this in the first sentence of the prologue to the *Herbarius*: "Because I see that the medicines of the German nation come from far off lands at

great cost and with much care, effort, and travail, I have been moved to ask whether Germany might not itself be in command of medicines, and whether, without the foreign sort, these may exist also in its own." Paracelsus, "The 'Herbarius' of Paracelsus," trans. Bruce T. Moran, *Pharmacy in History* 35, no. 3 (1993): 99–127. For an analysis of how European medical writers, including Paracelsus, began to take a greater interest in "local" flora and fauna in response to increasing exotic imports, see Alix Cooper, *Inventing the Indigenous: Local Knowledge and Natural History in Early Modern Europe* (Cambridge University Press, 2007).

38. Wear, *Knowledge and Practice*, chap. 8.

39. Everard Maynwaring, *Medicus Absolutus. Adespotos. The Compleat Physitian* (London: for the Booksellers, 1668).

40. Jonathan Barry, "The 'Compleat Physician' and Experimentation in Medicines: Everard Maynwaring (c. 1629–1713) and the Restoration Debate on Medical Practice in London," *Medical History* 62, no. 2 (2018): 155–76.

41. On the failure of the chymists to reform medical theory and practice, see Charles Webster, *The Great Instauration: Science, Medicine and Reform, 1626–1660* (Duckworth, 1975); Harold J. Cook, *The Decline of the Old Medical Regime in Stuart London* (Cornell University Press, 1986); and Andrew Wear, *Knowledge and Practice*.

42. Monardes, *Joyfull Nevves*.

43. Richard Hakluyt, *The Principal Navigations, Voyages, Traffiqves and Discoveries of the English Nation* (London: George Bishop, Ralph Newberie, and Robert Barker, 1599), 227.

44. John Gerard, *The Herball or Generall Historie of Plantes* (London, 1633), 1,525.

45. Oswald Croll, *Bazilica Chymica, & Praxis Chymiatricæ or Royal and Practical Chymistry in Three Treatises*, trans. John Hartman (London: for John Starkey and Thomas Passinger, 1670).

46. Croll, *Bazilica Chymica*.

47. André du Laurens, *A Discourse of the Preseruation of the Sight*, translated by Richard Surphlet (London: Felix Kingston for Ralph Iacson, 1599), 159.

48. Du Laurens, *Discourse of the Preservation of the Sight*, 159.

49. Monardes, *Joyfull Nevves*, 54.

50. Hakluyt, *Principal Navigations*, 259; Francis Bacon, *The Historie of Life and Death with Observations Naturall and Experimentall for the Prolonging of Life* (London: I. Okes for Humphrey Mosley, 1638), 253–54; Sennert, *Two Treatises*, 41. It is likely that Nicholas Culpeper translated the works of Sennert, and that these translations were later edited for publication by Abdiah Cole. See Mary Rhinelander McCarl, "Publishing the works of Nicholas Culpeper, Astrological Herbalist and Translator of Latin Medical Works In Seventeenth-Century London," *Canadian Bulletin of Medical History* 13, no. 2 (1996): 225–76.

51. Sennert, *Two Treatises*, 35.

52. Sennert, *Two Treatises*, 35–36.

53. William Trigg and Eugenius Philanthropos, *Dr. Trigg's Secrets, Arcana's & Panacea's* (London: R. D. for Dixy Page, 1665), 71; Joseph Blagrave, *Blagrave's Supplement or enlargement to Mr. Nich. Culpeppers English Physitian* (London: Obadiah

Blagrave, 1674), 90; Kenelm Digby, *Choice and Experimented Receipts in Physick and Chirurgery* (London: Andrew Clark for Henry Brome, 1675), 103.

54. W. M., *The Queens Closet Opened: Incomparable Secrets in Physick, Chyrurgery, Preserving, and Candying &c.* (London: Nath. Brooke, 1659), 281.

55. Bacon, *Historie of Life and Death*, 254.

56. For a discussion of Bacon's interest in prolonging longevity, see Guido Giglioni, "The Hidden Life of Matter: Techniques for Prolonging Life in the Writings of Francis Bacon," in *Francis Bacon and the Refiguring of Early Modern Thought: Essays to Commemorate The Advancement of Learning (1605–2005)*, ed. Julie Robin Solomon and Catherine Gimelli Martin (Routledge, 2005).

57. John Gascoigne, "The Royal Society, Natural History and the Peoples of the 'New World(s),' 1660–1800," *British Journal for the History of Science* 42 no. 4 (2009): 539–62; Raymond Phineas Stearns, *Science in the British Colonies of America* (University of Illinois Press, 1970), 101–6.

58. Gascoigne, "Royal Society," 545.

59. Peter Dear, "Totius in Verba: Rhetoric and Authority in the Early Royal Society," *Isis* 76, no. 2 (1985): 144–61; Steven Shapin, *A Social History of Truth: Civility and Science in Seventeenth-Century England* (University of Chicago Press, 1994).

60. Gascoigne, "Royal Society," 545.

61. Pierre Morel, *The Expert Doctors Dispensatory*, trans. anonymous (London: N. Brooke, 1657). Nicholas Culpeper is often credited as the translator of this work, but Poynter has presented evidence that this is a misattribution. F. N. L. Poynter, "Nicholas Culpeper and His Books," *Journal of the History of Medicine and Allied Sciences* 17, no. 1 (1962): 152–67.

62. Morel, *Expert Doctors Dispensatory*, 116.

63. Morel, *Expert Doctors Dispensatory*, 127.

64. For an example of the perceived importance of freshness for the potency of cinchona, see Crawford, *Andean Wonder Drug*, 146–47.

65. Nicholas Culpeper, *A Physicall Directory, or, A Translation of the London Dispensatory Made by the Colledge of Physicians in London* (London: Peter Cole, 1649), 319–20.

66. Moyse Charas, *The Royal Pharmacopoeea, Galenical and Chymical* (London: John Starkey and Moses Pitt, 1678), 34.

67. Robert Bayfield, *Tes Iatrikes Kartos, or, A Treatise de Morborum Capitis Essentiis & Pronosticis* (London: D. Maxwel and Richard Tomlins, 1663), 162.

68. Coles, *Adam in Eden*, 307.

69. Coles, *Adam in Eden*, 308.

70. For example, see John Tanner, *The Hidden Treasures of the Art of Physick* (London: George Sawbridge, 1659), 520.

71. Robert Boyle, *Some Considerations Touching the Usefulnesse of Experimental Naturall Philosophy* (Oxford: H. Hall, 1663); Nicaise Le Fèvre, *A Discourse upon Sr Walter Rawleigh's Great Cordial*, trans. Peter Belon (London: J. F., 1664).

72. For Raleigh's receipts, see Walter Raleigh, "Collection of Chemical Receipts," Sloane MS 359, British Library, London.

73. Boyle, *Some Considerations*, 317.

74. Boyle, *Some Considerations*, 319

75. Le Fèvre, *Discourse*, 44–45.

76. Harvey, *The Disease of London*.

77. For a discussion of diseases and symptoms in the early modern period, see Wear, *Knowledge and Practice*, chap. 3.

78. Diseases mentioned in four texts were: asthma, consumptions, cramps or stiches, hypochondriac disease, leprosy and wind-colic. In three texts: apoplexy, cancer, diabetes or pissing evil, fistula, looseness, lame or cripples, the megrim, phthisis, preserving sight, tumors, weakness of the sinews. In two texts: nephritic, swoonings, and the mother. In one text: bruises, buboes and parotids, chin-cough, congealed blood by a fall, difficult travels of women, enterocele or hernia, over-flowing of milk in women's breasts, gangrene, gutta serena or amaurosis, herpes miliar, hysterical, involuntary pissing, measles, pleurisy, poisonous wound from arrows, stings or venomous bites, prevention of miscarriages, scabies, small pox, sore eyes, spots and pustules, suffusion, the worms, the burning of the urine, toothache, ulcer of the reins and bladder, urinary constipation, venereal impotency, and vulva fallen.

79. On early modern disease classification and the development of nosology, see Margaret DeLacy, "Nosology, Mortality, and Disease Theory in the Eighteenth Century," *Journal of the History of Medicine and Allied Sciences* 54, no. 2 (1999): 261–84; and Volker Hess and J. Andrew Mendelsohn, "Sauvages' Paperwork: How Disease Classification Arose from Scholarly Note-Taking," *Early Science and Medicine* 19, no. 5 (2014): 471–503.

80. See also Charles Manning and Merrill Moore, "Sassafras and Syphilis," *New England Quarterly* 9, no. 3 (1936): 473–75; Magnaghi, "Sassafras and Its Role in Early America."

81. Giovanni Benedetto Sinibaldi, *Rare Verities. The Cabinet of Venus Unlocked, and Her Secrets Laid Open*, trans. anonymous (London: P. Briggs, 1658), 43.

82. Monardes, *Joyfull Nevves*.

83. Some of the texts were focused on one disease, so their exclusion of other diseases did not reflect a rejection but indicated that they were not part of the subject of that particular treatise.

84. Alexander Read, *The Chirurgicall Lectures of Tumors and Vlcers* (London: I. H. for Francis Constable and E. B., 1635).

85. Maydom, "New World Drugs," chapter 2.

6. ACCUMULATION OF THE EXOTIC IN THE *PALESTRA PHARMACEUTICA* (MADRID, 1706) OF FÉLIX PALACIOS

1. See, for example, Harold Cook, *Matters of Exchange: Commerce, Medicine, and Science in the Dutch Golden Age* (Yale University Press, 2007); Judith Carney and Richard Nicholas Rosomoff, *In the Shadow of Slavery: Africa's Botanical Legacy in the Atlantic World* (University of California Press, 2011); Timothy D. Walker, "The Medicines Trade in the Portuguese Atlantic World: Acquisition and Dissemination of Healing Knowledge from Brazil (c. 1580–1800)," *Social History of Medicine* 26, no. 3 (2013): 403–31; Timothy D. Walker, "Acquisition and Circulation of Medical Knowledge within the Early Modern Portuguese Colonial Empire," in *Science in the Spanish and*

Portuguese Empires, 1500–1800, ed. Daniela Bleichmar et al. (Stanford University Press, 2009); Benjamin Breen, "Portugal, Early Modern Globalization and the Origins of the Global Drug Trade," *Perspectives on Europe* 42, no.1 (2012): 84–88; Pratik Chakrabarti, "Empire and Alternatives: *Swietenia febrifuga* and the Cinchona Substitutes," *Medical History* 54, no.1 (2010): 75–94; Anna Winterbottom, "Of the China Root: A Case Study of the Early Modern Circulation of Materia Medica," *Social History of Medicine* 28, no. 1 (2015): 22–44; Anna Winterbottom, *Hybrid Knowledge in the Early East India World* (Palgrave Macmillan, 2016); Londa Schiebinger and Claudia Swan, eds., *Colonial Botany: Science, Commerce, and Politics in the Early Modern World* (University of Pennsylvania, 2005); and Steven J. Harris, "Long-Distance Corporations, Big Sciences, and the Geography of Knowledge," *Configurations* 6, no. 2 (1998): 269–304.

2. For the Atlantic world and American products specifically, see J. Worth Estes, "The European Reception of the First Drugs from the New World," *Pharmacy in History* 37, no. 1 (1995): 3–23; Teresa Huguet-Termes, "New World Materia Medica in Spanish Renaissance Medicine: From Scholarly Reception to Practical Impact," *Medical History* 45, no. 3 (2001): 359–76; Matthew Crawford, *The Andean Wonder Drug* (University of Pittsburgh Press, 2016); Londa Schiebinger, *Plants and Empire: Colonial Bioprospecting in the Atlantic World* (Cambridge: Harvard University Press, 2007); Daniela Bleichmar, *Visible Empire: Botanical Expeditions and Visual Culture in the Hispanic Enlightenment* (University of Chicago Press, 2012); James Delbourgo and Nicholas Dew, eds., *Science and Empire in the Atlantic World* (Routledge, 2008); Marcy Norton, *Sacred Gifts, Profane Pleasures: A History of Tobacco and Chocolate in the Atlantic World* (Cornell University Press, 2010); Sabine Anagnostou, "Jesuits in Spanish America: Contributions to the Exploration of the American Materia Medica," *Pharmacy in History* 41, no. 1 (2005): 3–17.

3. Patrick Wallis, "Exotic Drugs and English Medicine: England's Drug Trade, c.1550-c.1800," *Social History of Medicine* 25, no. 1 (2012): 20–46; and Patrick Wallis, "Consumption, Retailing, and Medicine in Early-Modern London," *Economic History Review* 61, no. 1 (2008): 26–53. See also Mark Jenner and Patrick Wallis, eds., *Medicine and the Market in England and its Colonies, c.1450–c.1850* (Palgrave Macmillan, 2007).

4. Brian W. Ogilvie, "The Many Books of Nature: Renaissance Naturalists and Information Overload," *Journal of the History of Ideas* 64, no. 1 (2003): 29–40; Brian W. Ogilvie, *The Science of Describing: Natural History in Renaissance Europe* (University of Chicago Press, 2006).

5. World historians have noted that the master narrative, as typically taught in the history of Western civilization courses, recognizes societies in Mesopotamia, Egypt, and Greece as fundamental founders of Western civilization, but only in the ancient period. The focus moves progressively farther west in subsequent periods, and largely ignores the crucial role of Arabic contributions to the formation of Western society. Thus Asia Minor, North Africa, the Near East, and the Middle East disappear as the narrative moves through what these historians have called the European "tunnel of time." For some examples of this argumentation, see essays by Marshall Hogdson and André Gunder Frank in *The New World History: A Teacher's Companion*, ed. Ross E. Dunn (Bedford/St. Martin's, 2000). Here I am making the argument that ancient Greek and medieval Islamic societies formed the basis for Galenic pharmacy, which

was then adopted and appropriated by medieval Europeans—but that we should, on the one hand, recognize Islamic contributions, not just Greek ones, as fundamental to European medicine, and on the other hand, recognize that Europe, especially northern Europe, developed separately from Indo-Mediterranean society in many ways, though Spain and Italy served as important conduits between the two. For further discussion, see Paula DeVos, *Compound Remedies: Galenic Pharmacy from the Ancient Mediterranean to New Spain* (University of Pittsburgh Press, 2020).

6. Real Academia Española, *Diccionario de Autoridades*, vol. 3 (1732), under "exotico–extrangero, advenedizo, peregrino."

7. Helen Anne Curry, "Naturalising the Exotic and Exoticising the Naturalised: Horticulture, Natural History and the Rosy Periwinkle," *Environment and History* 18, no. 33 (2012): 343–65. Curry uses the phrase to refer to the transplantation of plants, but I use it here to denote the transplantation of the Galenic pharmaceutical tradition—its ideas and practices as well as the materials it employed.

8. Félix Palacios, *Palestra pharmaceutica* (1792), "Discurso Preliminar," 3. The translation was meant to correct what Palacios saw as a faulty and plagiarized version in Assín y Ongoz's *Florilegio Theorico-Practico: Nuevo Curso Químico, En Que Se Contiene Quatro Reflexiones Generales, La Primera Sobre La Fisico-Mecanica ... Y Las Otras Tres Sobre Los Reynos Mineral, Vejetal Y Animal* (Madrid: por Antonio Gonçalez de Reyes, 1712). The first edition was published in 1706 in Madrid: Félix Palacios, *Palestra Pharmaceutica en la qual se trata de la eleccion de los simples, sus preparaciones chymicas, y galenicas, y de las mas selectas composiciones antiguas, y modernas* (Madrid: J. Garcia Infançon, 1706).

9. I have discussed Palacios and his work in my dissertation, "The Art of Pharmacy in Seventeenth- and Eighteenth-Century Mexico" (PhD diss., University of California, Berkeley, 2001), in *Compound Remedies*, and in "Pharmacopoeias and the Textual Tradition in Galenic Pharmacy," in *Drugs on the Page: Pharmacopoeias and Healing Knowledge in the Early Modern Atlantic World*, ed. Matthew J. Crawford and Joseph M. Gabriel (University of Pittsburgh Press, 2019).

10. Allen G. Debus has written extensively about medical chemistry and the "chemico-Galenic compromise." See, among others, *The Chemical Philosophy: Paracelsian Science and Medicine in the Sixteenth and Seventeenth Centuries*, rev. ed. (Dover Publications, 2002; orig. 1977); *The English Paracelsians* (F. Watts, 1966); and *The French Paracelsians: The Chemical Challenge to Medical and Scientific Tradition in Early Modern France* (Cambridge University Press, 1977). More recently, see Jennifer Rampling and Peter Jones, eds., *Alchemy and Medicine from Antiquity to the Enlightenment* (Routledge, 2021).

11. For Palacios's biography, see Guillermo Folch Jou, "Félix Palacios y Bayá," *Boletín de la Sociedad Española de Historia de la Farmacia* 7 (1956): 97–114.

12. Galen and Galenic medicine have been the subject of scores of books and articles. Especially helpful for understanding widespread and long-term impact are Owsei Temkin, *Galenism: Rise and Decline of a Medical Philosophy* (Cornell University Press, 1973); Owsei Temkin, *"On Second Thought" and Other Essays in the History of Medicine and Science* (Johns Hopkins University Press, 2002); Owsei Temkin, *The Double Face of Janus and Other Essays in the History of Medicine* (Johns Hopkins University

Press, 2006); Luis Garcia Ballester, *Galen and Galenism: Theory and Medical Practice from Antiquity to the European Renaissance*, ed. Jon Arrizbalaga et al. (Ashgate, 2002); John Scarborough, "The Galenic Question," *Sudhoff's Archiv* 65, no. 1 (1981): 1–31. On the various topics of Galenic medicine and natural philosophy, see Christopher Gill et al., eds., *Galen and the World of Knowledge* (Cambridge University Press, 2012); R. J. Hankinson, ed., *The Cambridge Companion to Galen* (Cambridge University Press, 2008).

13. For more on Mesue and his impact, see Paula De Vos, "'The Prince of Medicine': Yūḥannā Ibn Māsawaih and the Foundations of the Western Pharmaceutical Tradition," *Isis* 104, no. 4, (2013): 667–712; also De Vos, *Compound Remedies*.

14. Palacios, *Palestra Pharmaceutica* (1706 edition): " Al Senior Doctor D. Diego Matheo Zapata" (preliminary pages/dedications). For this essay, I used several different editions of the work, including Palacios, *Palestra pharmaceutica* (Madrid: por los Herederos de la Viuda de Juan Garcia Infançon, 1753); Palacios, *Palestra pharmaceutica* (Madrid: por Joachin Ibarra, 1763); Palacios; *Palestra pharmaceutica* (Madrid: Viuda de Don Joaquin Ibarra, 1792).

15. Palacios, *Palestra pharmaceutica* (1706), 1.

16. See Paula De Vos, "European Materia Medica in Historical Texts: Longevity of a Tradition and Implications for Future Use," *Journal of Ethnopharmacology* 132, no. 1 (2010): 28–47; as well as De Vos, *Compound Remedies*.

17. See Martin Jones et al., "Food Globalization in Prehistory," *World Archaeology* 43, no. 4 (2011): 665–75; and Nicole Boivin et al., "Old World Globalization and the Columbian Exchange: Comparison and Contrast," *World Archaeology* 44, no. 3 (2012): 452–69. See also Carl O. Sauer, *Agricultural Origins and Dispersals* (American Geographical Society, 1952).

18. John Riddle, *Dioscorides on Pharmacy and Medicine* (University of Texas Press, 1985), xviii–xix; Vivian Nutton, "Ancient Mediterranean Pharmacology and Cultural Transfer," *European Review* 16, no. 2 (2008): 212; and Charles Singer, "The Herbal in Antiquity and Its Transmission to Later Ages," *Journal of Hellenic Studies* 47, no. 1 (1927): 1–52. Thus the number of medicinal substances in use had expanded since the time of Hippocrates, whose texts discussed about two hundred fifty different substances—see Laurence Totelin, *Hippocratic Recipes: Oral and Written Transmission of Pharmacological Knowledge in Fifth- and Fourth-Century Greece* (Brill, 2009) and John Riddle, "Folk Tradition and Folk Medicine: Recognition of Drugs in Classical Antiquity," in *Folklore and Folk Medicines*, ed. John Scarborough (American Institute for History of Pharmacy, 1987).

19. See De Vos, "European Materia Medica."

20. For the widespread and sophisticated nature of exchange within the Islamic empires, see Edmund Burke III, "Islam at the Center: Technological Complexes and the Roots of Modernity," *Journal of World History* 20, no. 2 (2009): 165–86; John Obert Voll, "Islam as a Special World-System," *Journal of World History* 5, no. 2 (1994): 213–26; Marshall G. S. Hodgson, "The Role of Islam in World History," *International Journal of Middle East Studies* 1, no. 2 (1970): 99–123; Marshall G. S. Hodgson, "Hemispheric Interregional History as an Approach to World History," in Dunn, *New World History*; Anna Akasoy et al., eds., *Islamic Crosspolinations: Interactions in the Medieval*

Middle East (E. J. W. Gibb Memorial Trust, 2007); Leigh Chipman, "Islamic Pharmacy and the Mamlūk and Mongol Realms: Theory and Practice," *Asian Medicine* 3 (2007): 265–78.

21. Andrew Watson also discusses the major transformation of Afro-Eurasian agriculture under the Islamic empires, where irrigation, land use, and plant exchange underwent significant intensification—and also would have led to greater knowledge, cultivation, and use of previously unknown or exotic spices. See Watson, *Agricultural Innovation in the Early Islamic World: The Diffusion of Crops and Farming Techniques, 700–1100* (Cambridge University Press, 2008).

22. In this case, "identifiable" means two things: first, whether I was able to identify what the plant was, and second, if I could determine its native habitat. In some cases, there were so many species listed for certain substances that it was impossible to determine native origins.

23. On the ways in which foreign or "exotic" remedies were domesticated in Europe, see Barbara di Gennaro Splendore's chapter in this volume.

24. See Jonathan Pereira, *Elements of Materia Medica and Therapeutics* (Philadelphia: Lea and Blanchard, 1843), 3:629; William Dymock et al., *Pharmacographia Indica* (London: Kegan Paul, Trench & Trubner, 1893), 496; and *The Seven Books of Paulus Aegineta*, trans. Francis Adams (London: Printed for the Sydenham Society, 1847), vol. 3, book 7, 114–16.

25. For further discussion, see DeVos, *Compound Remedies*.

26. *Seven Books of Paulus Aegineta*, vol. 3, book 7, 439–40.

27. James Innes Miller, *The Spice Trade of the Roman Empire* (Clarendon Press, 1969); Martin Levey, "Babylonian Chemistry: A Study of Arabic and Second Millenium B. C. Perfumes," *Osiris* 12 (1956): 377–78. For culinary and medical use of spices, see Felipe Fernández Armesto and Benjamin Sacks, "The Global Exchange of Food and Drugs," in *The Oxford Handbook of the History of Consumption*, ed. Frank Tenenbaum (Oxford University Press, 2012); Alain Touwaide and E. Appetiti, "Food and Medicines in the Mediterranean Tradition. A Systematic Analysis of the Earliest Extant Body of Textual Evidence," *Journal of Ethnopharmacology* 167 (2015): 11–29.

28. Miller, *Spice Trade of the Roman Empire*, 42–43; and P. N. Ravindran et al., eds., *Cinnamon and Cassia: The Genus Cinnamomum* (CRC Press, 2003). Both spices were taken from the bark of the tree, which was stripped, dried, and rolled into tubes; both were widely known to Roman authors (Pliny, etc.), and had a long history of use and value to the ancient world. Cassia, for example, was included in the earliest Chinese herbals—as early as 2700 BCE.

29. Miller, *Spice Trade of the Roman Empire*, 80–81. There were several varieties of pepper, including long pepper—the pungent tiny seeds of its long, spike-like fruit—and its less pungent relative, black pepper, the dried unripe seeds or berries of the pepper tree. The latter also produced white pepper from the dried ripe seed.

30. Miller, *Spice Trade of the Roman Empire*, 83, 91.

31. Miller, *Spice Trade of the Roman Empire*, 63.

32. Antonio Pigafetta, *The Voyage of Magellan*, trans. Paula Spurlin Paige (Prentice-Hall, 1969), 121. See also Robin A. Donkin, *Between East and West: The Moluccas and the Traffic in Spices Up to the Arrival of Europeans* (American Philosophical

Society, 2003); Ian Burnet, *The Spice Islands* (Rosenberg Publishing, 2011); and Miller, *Spice Trade of the Roman Empire*, 58–59.

33. It is important to note that Dioscorides named nutmeg, but not mace. The introduction of both to Western materia medica is attributed to exchanges in the medieval Islamic world.

34. *Seven Books of Paulus Aegineta*, vol. 3, book 7, 89, 249–50. See also, for poppy, Roy Porter and Mikuláš Teich, eds., *Drugs and Narcotics in History* (Cambridge University Press, 1997); and for tormentil, Graeme Tobyn et al., *The Western Herbal Tradition: 2000 Years of Medicinal Plant Knowledge* (Elsevier, 2011).

35. See John M. Riddle, *Contraception and Abortion from the Ancient World to the Renaissance* (Harvard University Press, 1994); John M. Riddle, *Eve's Herbs: A History of Contraception and Abortion in the West* (Harvard University Press, 1997); John M. Riddle, "Oral Contraceptives and Early-Term Abortifacients During Classical Antiquity and the Middle Ages," *Past and Present* 132 (1991): 3–32; John M. Riddle, "Pseudo-Dioscorides' 'Ex Herbis Femininis' and Early Medieval Medical Botany," *Journal of the History of Biology* 14, no. 1 (1981): 43–81; Monica H. Green, *Women's Healthcare in the Medieval West: Texts and Contexts* (Ashgate, 2000).

36. John Hill, *A History of the Materia Medica* (London: T. Longman, C. Hitch, and L. Hawes, 1751), 373, 603–604; Christopher Vasey, *Natural Remedies for Inflammation* (Healing Arts Press, 2014).

37. *Seven Books of Paulus Aegineta*, vol. 3, book 7, 369–70; Andreas Lardos et al., "Resins and Gums in Historical *Iatrosophia* Texts from Cyprus—A Botanical and Medico-Pharmacological Approach," *Frontiers in Pharmacology* 2, no. 32 (2011), doi:10.3389/fphar.2011.00032.

38. See Miller, *Spice Trade of the Roman Empire*; Lardos et al., "Resins and Gums in Historical *Iatrosophia* Texts from Cyprus"; *Seven Books of Paulus Aegineta*, vol. 3, book 7, 362, 341, 208–9, 294, and 374–75. There is some confusion over the exact identity of storax—where it comes from, the identity of its solid versus its liquid form, and its relationship to benzoin and liquidambar, or sweet gum. See Pereira, *Elements of Materia Medica*, 2:567.

39. Roy Genders, *Perfume Through the Ages* (G. P. Putnam's Sons, 1972).

40. Dioscorides, *De materia medica*, ed. and trans. Lily Y. Beck (Olms-Wiedmann, 2017).

41. Hill, *History of the Materia Medica*, 372, 419, 440, 588–9.

42. Marina Heilmeyer, *Ancient Herbs* (Getty Publications, 2007); Andrew Dalby, *Food in the Ancient World from A to Z* (Routledge, 2013), 132; Christopher A. Faraone and Dirk Obbink, eds., *Magika Hiera: Ancient Greek Magic and Religion* (Oxford University Press, 1991); Dioscorides, *De materia medica*.

43. See Miller, *Spice Trade of the Roman Empire*, 98–99; Ben-Yehoshua et al., "Frankincense, Myrrh, and Balm of Gilead: Ancient Spices of Southern Arabia and Judea," *Horticultural Reviews* 39 (2011): 1–76.

44. Gus W. Van Beek, "Frankincense and Myrrh in Ancient South Arabia," *Journal of the American Oriental Society* 78, no. 3 (1958): 144.

45. Van Beek, "Frankincense and Myrrh," 141. See also Ben-Yehoshua et al., "Frankincense, Myrrh, and Balm of Gilead."

46. Riddle, "Oral Contraceptives." See also Ben-Yehoshua et al., "Frankincense, Myrrh, and Balm of Gilead."

47. On asafoetida, see Riddle, "Oral Contraceptives"; Dioscorides, *De materia medica*. On gum tragacanth, see Miller, *Spice Trade of the Roman Empire*, 100. On gum ammoniac, see *Seven Books of Paulus Aegineta*, vol. 3, book 7, 184. See also Hill, *History of the Materia Medica*, 741–42; Edward Balfour, ed., *Cyclopædia of India and of Eastern and Southern Asia, Commercial, Industrial and Scientific: Products of the Mineral, Vegetable and Animal Kingdoms, Useful Arts and Manufactures*, vol. 1, 2nd ed. (Madras: Scottish and Adelphi Presses, 1871). Also known as acacia gum, gum arabic is still used as a thickener, binder, and emulsifying agent for a variety of commercial goods.

48. *Seven Books of Paulus Aegineta*, vol. 3, book 7, 427–29; Miller, *Spice Trade of the Roman Empire*, 40–41, 53. See also R. A. Donkin, *Dragon's Brain Perfume: An Historical Geography of Camphor* (Brill, 1999).

49. *Seven Books of Paulus Aegineta*, vol. 3, book 7, 433–34.

50. *Seven Books of Paulus Aegineta*, vol. 3, book 7, 448–50.

51. *Seven Books of Paulus Aegineta*, vol. 3, book 7, 455–56.

52. Paul Freedman, *Out of the East: Spices and the Medieval Imagination* (Yale University Press, 2008).

53. *Seven Books of Paulus Aegineta*, vol. 3, book 7, 434–46.

54. Miller, *Spice Trade of the Roman Empire*, 69; *Seven Books of Paulus Aegineta*, vol. 3, book 7, 70.

55. Efraim Lev and Zohar Amar, *Practical Materia Medica of the Medieval Eastern Mediterranean According to the Cairo Genizah* (Brill, 2007), 429, 537; José Léon de Mendoza, "El Mangle," *Anales del Instituto Médico Nacional* (Mexico: Instituto Médico Nacional, 1900), vol. 4, 323–32; *Seven Books of Paulus Aegineta*, vol. 3, book 7, 358.

56. Nicolás Monardes, *La historia medicinal de las cosas que se traen de nuestras Indias Occidentales (1565–1574)* (Ministerio de Sanidad y Consumo, 1989): fos. 41r–49r.

57. Francisco del Paso y Troncoso, *Papeles de Nueva España: Geografía y estadística* (Establicimiento tip. "Sucesores de Rivadeneyra," 1905), vol. 4, 130.

58. Jan Elferlink, "Ethnobotany of the Aztecs," in *Encyclopaedia of the History of Science, Technology, and Medicine in Non-Western Cultures*, vol. 1, A–K, ed. Helaine Selin Francisco del Paso y Troncoso, *Papeles de Nueva España: Geografía y estadística* (Establicimiento tip. "Sucesores de Rivadeneyra," 1905), vol. 6, 280.

59. Monardes, *Historia medicinal*, fols. 18r–22v.

60. Monardes, *Historia medicinal*, fol. 25r.

61. F. Martinez Cortes, *Pegamentos, gomas, y resinas en el México prehispanico* (Resistol, 1970); F. N. Howes, *Vegetable gums and resins* (Chronica Botanica Company, 1949); J. Lagenheim, *Plant Resins* (Timber Press, 2003), F. F. Berdan and P. R. Anawalt, *The Essential Codex Mendoza* (University of California Press, 1997); R. J. Case, et al., "Chemistry and Ethnobotany of Commercial Incense Copals, Copal Blanco, Copal Oro, and Copal Negro, of North America," *Economic Botany* 57 (2003): 189–202.

62. Antonio de Alcedo and George Alexander Thompson, *The Geographical and Historical Dictionary of America and the West Indies* (London, Oxford and Cambridge:

J. Carpenter et al., 1812–15), 5:96. Tacamahaca has also been identified as a type of resin from the *Bursera* genus, used to treat tumors and muscle spasm (Macdiel Acevedo et al., "Cytotoxic and Anti-inflammatory Activities of *Bursera* species from Mexico," *Journal of Clinical Toxicology* 5, no. 1 (2015): 1). See also William Gates, who says tacamahaca was also called *copal-ihyac*, or "smelling like copal." Gates, *An Aztec Herbal: The Classic Codex of 1552* (Courier Corporation, 2000).

63. Monardes, *Historia medicinal*, fols. 3r–5r.

64. For further discussion of these techniques and their limited use in Galenic pharmacy from the medieval period, see De Vos, "Rosewater and Philosophers' Oil: Thermo-chemical Processing in Medieval and Early Modern Spanish Pharmacy," *Centaurus* 60, no. 3 (2018): 159–72. It should be noted that there is an important medieval work that discusses these kinds operations and procedures in detail for use in medicine: a tenth-century Arabic text that formed the twenty-eighth chapter of a treatise titled the *Kitab al-Tasrif*, written by the Cordoban physician Abulcasis/al-Zahrāwī (936–1013). This chapter, called the *Liber servitoris* in its Latin translation, provides clear and detailed instructions for operations and procedures to be carried out in recipes for formulating medicines from plant, mineral, and animal materials. The overall treatise fits clearly within the Galenic tradition, but the *Liber servitoris* is somewhat unusual in its major focus on thermochemical procedures. See al-Zahrāwī/Abulcasis, *Servidor de Abulcasis*, trans. Alonso Rodrigues de Tudela (Valladolid, 1515).

65. For further discussion of solvent extraction, see Frederick Holmes, "Analysis by Fire and Solvent Extractions: The Metamorphosis of a Tradition," *Isis* 62, no. 2 (1971): 129–48.

66. See Lawrence Principe, *The Secrets of Alchemy* (University of Chicago Press, 2013) for an overview of the history of alchemy.

67. Ancient authors also included action-based purgatives, antidotes, and emetics (medicines classified by their effect on the body) that I do not discuss here, but these kinds of remedies are discussed in Totelin, *Hippocratic Recipes*.

68. For scholarly translations with introductions to Ibn Sahl's formularies, see Oliver Kahl, *Sābūr ibn Sahl, The Small Dispensatory* (Brill, 2003); and Oliver Kahl, *Sābūr ibn Sahl's Dispensatory in the Recension of the 'Aḍudī Hospital* (Brill, 2009).

69. Leigh Chipman, *The World of Pharmacy and Pharmacists in Mamluk Cairo* (Brill, 2010), 105–7; Laura Mason, *Sugar-Plums and Sherbets: The Prehistory of Sweets* (Prospect Books, 1998).

70. See De Vos, "'Prince of Medicine.'"

7. A SCIENCE OF THINGS

This paper was presented at various occasions: first in Cambridge during a workshop organized by Emma Spary and Justin Rivest in April 2017, later in September 2017 at the European Rural History Congress in Louvain. The majority of the arguments can be found in Chapters 3–7 of my book: Samir Boumediene, *La Colonisation du savoir. Une histoire des plantes médicinales du "Nouveau Monde" (1492–1750)* (Les éditions des mondes à faire, 2016). I also addressed the topic in a paper published in Linda Newson, ed., *Cultural Worlds of the Jesuits in Colonial Latin America* (University of London

Press, 2020). I thank Emma Spary and Justin Rivest for their comments and suggestions. This chapter owes them a lot and is a part of the research I had the pleasure to conduct with them and Laia Portet on the Leverhulme Trust Research Project Grant 2014–289, "Selling the Exotic," at the University of Cambridge.

1. This project was sent several times between 1686 and 1690. See Archivum Romanum Societatis Iesu (henceforth ARSI), N.R.-Q. 15, doc. 26, fol. 134r; Archivo de la Antigua Provincia Jesuítica de Quito (henceforth APQ), VI/540a.

2. APQ, VI/520; APQ, VI/523; ARSI, N.R.-Q. 15, doc. 31.

3. ARSI, N.R.-Q. 15, doc. 31, fol. 230v.

4. The Spanish word *cascarilla*, meaning "little bark," was the basis of several febrifuges. Usually it referred to a plant from the genus *Croton*, but it could also designate Peruvian bark, which belongs to the genus *Cinchona*. For a detailed discussion, see Samir Boumediene, *La Colonisation du savoir. Une histoire des plantes médicinales du "Nouveau Monde" (1492–1750)* (Les éditions des mondes à faire, 2016), chapter 4. My usage throughout this essay will therefore differ from that of Wouter Klein for the Dutch case.

5. José Luis Valverde, *Presencia de la Compañía de Jesús en el desarrollo de la farmacia* (Universidad de Granada, 1978); María Elena del Río Huas and Manuel Revuelta González, "Enfermerías y boticas en las casas de la Compañía en Madrid siglos XVI-XIX," *Archivum Historicum Societatis Iesu* 64 (1995): 46–48; Sabine Anagnostou, "Jesuit Missionaries in Spanish America and the Transfer of Medical-Pharmaceutical Knowledge," *Archives internationales d'histoire des sciences* 52, no. 148 (2002): 176–97; Luis Martín, *The Intellectual Conquest of Peru: The Jesuit College of San Pablo, 1568–1767* (Fordham University Press, 1968); Mordechai Feingold, ed., *Jesuit Science and the Republic of Letters* (Harvard University Press, 2003); Mordechai Feingold, ed., *The New Science and Jesuit Science: Seventeenth Century Perspectives* (Kluwer Academic, 2003); John W. O'Malley (S.J) et al., eds., *The Jesuits II. Cultures, Sciences and the Arts, 1540–1773* (University of Toronto Press, 2006); Florence C. Hsia, *Sojourners in a Strange Land: Jesuits and Their Scientific Missions in Late Imperial China* (University of Chicago Press, 2011); Miguel de Asúa, *Science in the Vanished Arcadia: Knowledge of Nature in the Jesuit Missions of Paraguay and Río de La Plata* (Brill, 2014).

6. Juan de Esteyneffer, *Florilegio Medicinal...* (Madrid: por Alonso Balvas, 1729); Jos V. M. Welie, "Ignatius of Loyola on Medical Education. Or: Should Today's Jesuits Continue to Run Health Sciences Schools?," *Early Science and Medicine* 8, no. 1 (2003): 25–43.

7. Real Academia de la Historia (henceforth RAH), 9/3702, fols. 258r-v.

8. Josep M. Barnadas and Manuel Plaza, eds., *Mojos, seis relaciones jesuíticas: geografía, etnografía, evangelización, 1670–1763* (Historia Boliviana, 2005), 64.

9. See below, for instance, the example of the pharmacy of the Jesuit College of San Pablo in Lima.

10. Juan Riera Palermo and Albi Romero Guadalupe, "Productos medicinales en La Flota a Indias de 1509," *Llull: Revista de la Sociedad Española de Historia de las Ciencias y de las Técnicas* 19, no. 37 (1996): 551–59; Saul Jarcho, "Medicine in Sixteenth Century New Spain as Illustrated by the Writings of Bravo, Barfan, and Vargas Machuca," *Bulletin of the History of Medicine* 31, no. 5 (1957): 425–41.

11. Nicolás Monardes, *Dos libros: el vno trata de todas las cosas q[ue] trae[n] de*

n[uest]ras Indias Occide[n]tales . . . (Sevilla: Sebastian Trugillo, 1565); Bernardo de Vargas Machuca, *Milicia y descripción de las Indias* (Madrid: Victoriano Suárez, 1892).

12. Fernão Cardim, *Tratados da Terra e Gente do Brasil*, ed. Batista Caetano et al. (J. Leite & Cia, 1925), 73–74. On the later history of ipecacuanha root, see E. C. Spary, "Publishing Virtue: Medical Entrepreneurship and Reputation in the Republic of Letters," *Centaurus* 62, no. 3 (2020): 498–521.

13. See, for instance, the *Confessionario para los curas de Indios*, published by the Provincial Council of Lima in 1583, and translated to Quechua. Questions about poisons and abortifacients appear in the questionnaire referring to the Fifth Commandment. They are repeated in other handbooks, such as Juan de Valdivia's *Confessionario breve en la lengua del Reyno de Chile* (1606), in Spanish and Mapudungun. The use of abortifacients is a central theme of Londa Schiebinger, *Plants and Empire: Colonial Bioprospecting in the Atlantic World* (Harvard University Press, 2004).

14. Archivo Arzobispal de Lima (henceforth AAL), Idolatrías y Hechicerías, I-17.

15. Andrés I. Prieto, *Missionary Scientists: Jesuit Science in Spanish South America, 1570–1810* (Vanderbilt University Press, 2011), 58.

16. Luis Millones-Figueroa and Domingo Ledezma, eds., *El saber de los Jesuitas. Historias naturales y el Nuevo Mundo* (Vervuert-Iberoamericana, 2005); Christophe Giudicelli, "Le cabinet en campagne, chroniques jésuites de 'pacification,' Tucumán et Nouvelle Biscaye XVIIe siècle," *e-Spania. Revue interdisciplinaire d'études hispaniques médiévales et modernes*, no. 26 (February 2017).

17. Georges Baudot, *Utopie et histoire au Mexique. Les premiers chroniqueurs de la civilisation mexicaine (1520–1569)* (Privat, 1976).

18. François de Dainville, *L'Éducation des Jésuites (XVIe-XVIII siècles)* (Minuit, 1978); Cinthia Gannett and John Brereton, *Traditions of Eloquence: The Jesuits and Modern Rhetorical Studies* (Fordham University Press, 2016).

19. On Jesuits and new natural knowledge, see the *Journal of Jesuit Studies*, vol. 7, no. 2 (2020); also Ofer Gal and Raz Chen-Morris, *Baroque Science* (University of Chicago Press, 2014); and José Ramón Marcaida López, *Arte y ciencia en el Barroco español: historia natural, coleccionismo y cultura visual* (Marcial Pons, 2014).

20. See, for instance, Ana Carolina de Carvalho Viotti and Jean Marcel Carvalho França, *Coleção de várias receitas e segredos particulares das principais boticas da nossa Companhia de Portugal, da Índia, de Macau, e do Brasil* (Edições Loyola Jesuitas, 2019).

21. Valverde, *Presencia*; Sabine Anagnostou, "The International Transfer of Medicinal Drugs by the Society of Jesus (Sixteenth to Eighteenth Centuries) and Connections with the Work of Carolus Clusius," in *Carolus Clusius: Towards a Cultural History of a Renaissance Naturalist*, ed. Florike Egmond et al. (Royal Netherlands Academy of Arts and Sciences, 2007), 293–312; Timothy Walker, "The Medicines Trade in the Portuguese Atlantic World: Acquisition and Dissemination of Healing Knowledge from Brazil (c. 1580–1800)," *Social History of Medicine* 26, no. 3 (2013): 403–31; Ana Carolina de Carvaiho Viotti, "A Study on the Jesuits' Apothecary Shops and Medical Recipes in the Portuguese Empire (Seventeenth and Eighteenth Centuries)," *Historia unisinos*, 23, no. 3 (2019): 464–74.

22. RAH, 9/3631.

23. RAH, 9/3671, no. 65.

24. RAH, 9/3823. For another example, see RAH, 9/3426, no. 2.

25. Steven J. Harris, "Long-Distance Corporations, Big Sciences, and the Geography of Knowledge," *Configurations* 6, no. 2 (1998): 269–304.

26. "To Gaspar Berze, by commission, Rome, February 24, 1554," Ignatius Loyola, *Letters and Instructions*, ed. John W. Padberg and John L. McCarthy, trans. Martin E. Palmer (Institute of Jesuit Sources, 1996), 472–74.

27. Edmond Lamalle, "La Documentation d'histoire missionnaire dans le fondo Gesuitico aux archives romaines de la Comp. de Jesus," *Euntes Docete* 21 (1968): 160–62; Aliocha Maldavsky, "Société urbaine et désir de mission. Les ressorts de la mobilité missionnaire jésuite à Milan au début du XVIIe siècle," *Revue d'histoire moderne et contemporaine* 56, no. 3 (2009): 7–32.

28. Caroline Cunill and Francisco Quijano, "Los procuradores de las Indias en el Imperio hispánico: reflexiones en torno a procesos de mediación, negociación y representación," *Nuevo Mundo Mundos Nuevos*, February 24, 2020, https://journals.openedition.org/nuevomundo/79934.

29. Felix Zubillaga, "El procurador de las Indias Occidentales," *Archivum historicum Societatis Iesu*, no. 22 (1953): 367–417; Agustín Galán García, *El "Oficio de Indias" de Sevilla y la organización económica y misional de la compañía de Jesús: 1566–1767* (Fundación Fondo de cultura de Sevilla, 1995); J. Gabriel Martínez-Serna, "Procurators and the Making of Jesuits' Atlantic Network," in *Soundings in Atlantic History: Latent Structures and Intellectual Currents, 1500–1830*, ed. Bernard Baylin and Patricia L. Denault (Harvard University Press, 2009).

30. Sébastien Malaprade, "The College of San Hermenegildo in Seville. The Centerpiece of a Global Financial Network," in *Trade and Finance in Global Missions (16th-18th Centuries)*, ed. Hélène Vu Thanh and Ines G. Županov (Brill, 2020).

31. Archivo general de la Nación, Lima (henceforth AGNL), Jesuitas, PR 1/119, doc. 2019, fols. 107r–108v.

32. Luisa Elena Alcalá, "'De compras por Europa.' Procuradores jesuitas y cultura material en Nueva España," *Goya: Revista de arte*, no. 318 (2007): 141–58.

33. AGNL, Jesuitas, PR 1/8, doc. 508, fols. 22r–32v.

34. RAH, 9/3686, fol. 224r.

35. The request was repeated several times by Juan de Lugo, on January 13, 1635, February 19, 1635, and May 26, 1636: RAH, 9/3686, fols. 225r-v, 263r-v; RAH, 9/3684, fols. 353r-v.

36. Archivo histórico nacional de España (henceforth AHNE), Jesuitas, leg. 121, no. 16. Another example can be found in a 1636 letter from Superior General Muzio Vitelleschi to the Superior Provincial of Toledo: "I cannot determine who has to send the history of Father Juan Eusebio [Nieremberg] here; it's easy to avoid the double inconvenience of cost and insecurity by handing the work over to the procurator, if no-one else from the Company can bring it sooner." See RAH, 9/7259.

37. RAH, 9/3692, fol. 670r.

38. Francisco Javier Alegre, *Historia de la Compañía de Jesus en Nueva España*, ed. Carlos Maria de Bustamante, Vol. 1 (México: impr. de J. M. de Lara, 1841), 125; AGNL, Jesuitas, PR 1/14, doc. 710, fol. 70r.

39. See for instance RAH, 9/3687, fol. 158r-v; RAH, 9/3788, fol. 381r.

40. RAH, 9/7259.

41. Father Juan Chacón to Father Rafael Pereira, July 1635, RAH, 9/3684, fols. 331r, 332r-v.

42. See the letters by Diego de Carrión, Juan del Marmol, Martín de Fonseca and Joan de Pina in RAH, 9/7333.

43. Juan del Marmol to Martín de Fonseca, Seville, 18 February 1641: RAH, 9/7260.

44. RAH, 9/3687, fol. 38v.

45. See AHNE, Inquisición, leg. 5345 exp. 2, doc. 1; AGNL, Jesuitas, PR 1/14, doc. 710, fols. 41R, 70r; PR 1/16, doc. 734, fol. 6; PR 1/8, doc. 492; AGNL, PR 1/8, doc. 508.

46. AGNL, Jesuitas, PR 1/119, doc. 2019, fol. 108r.

47. On cacao and tobacco, see also Marcy Norton, *Sacred Gifts, Profane Pleasures: A History of Tobacco and Chocolate in the Atlantic World* (Cornell University Press, 2008).

48. Giuseppe Piras, *Martin de Funes S.I. (1560–1611) e gli inizi delle riduzioni dei gesuiti nel Paraguay* (Edizioni di Storia e letteratura, 1998), 41–102; Maldavsky, "Société urbaine et désir de mission," 28.

49. Biblioteca Estense di Modena, Ms. gamma.h.1.21 = cam.0338, fol. 73; Ms. gamma.h.1.21 = cam.0338, fols. 5, 47, 72; Ms. gamma.h.1.22 = cam.0339.

50. RAH, 9/3788, fol. 160r.

51. In Rome, for instance, the Jesuits obtained the right to practice medicine from Pope Gregory XIII in 1576.

52. See for instance ARSI, F.G. 1143, "Conti e Ricevute Della Spetieria, Casa Professa."

53. Justin Rivest, "Beyond the Pharmacopoeia? Secret Remedies, Exclusive Privileges, and Trademarks in Early Modern France," in *Drugs on the Page: Pharmacopoeias and Healing Knowledge in the Early Modern Atlantic World*, ed. Joseph M. Gabriel and Matthew Crawford (University of Pittsburgh Press, 2019), 81–100.

54. See, for a later example, an "Account of the roots of a fertilizing liana" written on April 13, 1751, by the Jesuit Nicolás de la Torre (RAH, 9/3687, fol. 38r).

55. RAH, 9/7290.

56. Enrique Otte, "Cartas privadas de puebla del siglo XVI," *Jahrbuch für Geschichte Lateinamerikas*, no. 3 (1966): 42–43.

57. "Methodo de usar los polvos de la corteza del Arbol llamado Choch," in RAH, 9/3426, no. 2. See also RAH, 9/3671, no. 65; RAH, 9/3693, fols. 527r–528r.

58. Sebastiano Bado, *Anastasis Corticis Peruuiae, seu Chinae Chinae defensio* (Genova: Pietro Giovanni Calenzani, 1663).

59. Arthur Robert Steele, *Flores para el rey. La expedición de Ruiz y Pavón y la Flora del Perú, 1777–1788* (Ediciones del Serbal, 1982); Alba Moya, *El Árbol de la Vida. Esplendor y muerte en los Andes Ecuatorianos. El auge de la cascarilla en el siglo XVIII* (FLACSO, 1990); Saul Jarcho, *Quinine's Predecessor: Francesco Torti and the Early History of Cinchona* (Johns Hopkins University Press, 1993); Matthew James Crawford,

The Andean Wonder Drug: Cinchona Bark and Imperial Science in the Spanish Atlantic, 1630–1800 (University of Pittsburgh Press, 2016).

60. Charles-Marie de La Condamine, "Sur l'arbre du Quinquina," in *Histoire de l'Academie Royale des Sciences. Année M.DCCXXXVIII* (Paris: Imprimerie Royale, 1640), 238.

61. Fernando I. Ortiz Crespo, *La corteza del Árbol sin nombre. Hacia una historia congruente del descubrimiento y difusión de la quina* (Fundación Fernando Ortiz Crespo, 2002); Francisco Guerra, "El descubrimiento de la quina," *Medicina e historia*, no. 69 (1977): 7–25.

62. Juan de Vega's involvement has been contested by several historians, on the grounds that he was still in Lima in the 1650s, teaching medicine. While this is true, Vega did come to Seville in 1642, and requested to return to Peru in 1643 (Archivo General de Indias, Seville, Spain, Contratación 5425, no. 35).

63. Gaspar Caldera de Heredia, *De pulvere febrifugo Occidentalis Indiae, 1663 de Gaspar Caldera de Heredia y la introduccíon de la quina en Europa*, ed. José María López Piñero and Francisco Calero (CSIC, 1992).

64. Charles Gibson, ed., *The Black Legend: Anti-Spanish Attitudes in the Old World and the New* (Alfred A. Knopf, 1971); William S. Maltby, *The Black Legend in England: The Development of Anti-Spanish Sentiment, 1558–1660* (Duke University Press, 1971); Margaret Rich Greer et al., eds., *Rereading the Black Legend: The Discourses of Religious and Racial Difference in the Renaissance Empires* (University of Chicago Press, 2007); Jocelyn Nigel Hillgarth, *The Mirror of Spain, 1500–1700: The Formation of a Myth* (University of Michigan Press, 2000).

65. María del Carmen Martínez Martín, "Búsqueda y Hallazgo de las ruinas de Logroño en la región de los Jíbaros (siglos XVI-XIX)," in *Estudios sobre América: siglos XVI-XX*, ed. Antonio Gutiérrez Escudero and María Luisa Laviana Cuetos (Asociación Española de Americanistas, 2005), 89–106.

66. Archivo nacional de Historia, Quito (henceforth ANHQ), F. Especial III/129, fols. 10r–18v; AGNL, Jesuitas, PR 1/11, doc. 586; Archivo nacional de Chile (henceforth ANC), Jesuitas 438, fol. 324v. Pantoja was also credited with introducing *quina quina* seeds (probably from *Myroxylon*) at Madrid's Colegio Imperial. See Tomás Murillo y Velarde, *Aprobacion de ingenios y curacion de hipochondricos* (Zaragoza: Diego de Ormer, 1672), fol. 133v.

67. APQ, VI/526; ARSI, N.R.-Q. 15, doc. 31, fol. 231v.

68. Stanis Pérez, "Louis XIV et le Quinquina," *Vesalius* 9, no. 2 (2003): 25–30.

69. Anonymous, *Les admirables qualitez du Kinkina* (Paris: Martin Jouvenel, 1689), 39; Samir Boumediene, "L'Acclimatation portuaire des savoirs sur le lointain. Les drogues exotiques à Séville, Cadix et Livourne (XVIe-XVIIe siècles)," in *Les Savoirs-mondes. Mobilité et circulation des savoirs depuis le Moyen Âge*, ed. Pilar Bernaldo González and Liliane Hilaire-Pérez (Presses universitaires de Rennes, 2015).

70. ARSI, N.R.-Q. 15, doc. 31, fol. 231v.

71. See, for instance, ANHQ, Gobierno VII/17, fols. 28, 71, 171–93.

72. Caldera de Heredia, *De pulvere febrifugo*, 35; Linda A. Newson, *Making Medicines in Early Colonial Lima, Peru* (Brill, 2017), 168.

73. On Salubrino, see ARSI, Peru 4, fols. 99v, 151v, 199r, 270r, 330v, 358v; ARSI,

Peru 15, fols. 188v–189r, 196v–198v; ARSI, Vitae 24, fol. 267v; AGNL, Jesuitas, PR 1/119, doc. 2019, 24r, 55r, 136r–38; AGNL, Jesuitas, PR 1/16, doc. 738. On Claude Chicaut, see ARSI, Peru 4, fols. 332r, 360r, 453r; ARSI, Peru 5, fol. 11r; ANC, Jesuitas 438, fol. 228r-v; AHNE, Jesuitas, leg. 121, doc. 21.

74. Caldera de Heredia, *De pulvere febrifugo*, 36–38.

75. Archivo de la Diputación de Sevilla (henceforth ADPS), leg. 117, and Amor de Dios, leg. 73, leg. 111. For more precise references, see Boumediene, *La Colonisation*, 458n53. The suppliers at these institutions were apothecaries like Blas Sánchez or Diego Gómez Duarte, the latter of whom, according to Caldera de Heredia, was an associate of Juan de Vega. See Diego Salado Garcés, *Apologético Discurso* (Sevilla: Thomas Lopez de Haro, 1678), 1.

76. ADPS, Real Jardín Botánico de Madrid, Mutis IV, leg. 11, 51, fol. 2v. Tafur was close to Cardinal de Lugo, writing a preface for Lugo's *Notas sobre los privilegios* (1645). On Tafur, see Enrique Torres Saldamando, *Los antiguos Jesuitas del Perú. Biografías y apuntes para su historia* (Lima: Impr. liberal, 1882), 294.

77. ANC, Jesuitas 438, fol. 244r; AGNL, Jesuitas, PR 1/1, doc. 69; PR 1/6, doc. 440 and doc. 442, fol. 2r, 22v; PR 1/8, doc. 508; PR 1/11, doc. 586, fol. 22v. See also ANC, Jesuitas 438, 324v-325v.

78. In around 1690, a box of bark sent to Europe without being entrusted to the procurators was seized by English corsairs. See Antonio Bastidas to Pedro Bermudo, General Procurator of the Indies in Madrid, Popayán, November 16, 1690, in RAH, 9/7263.

79. Bado, *Anastasis Corticis Peruuiae*, 240–41.

80. RAH, 9/3702, fol. 989r.

81. RAH, 994r.

82. RAH, 996r.

83. From 1647 onward, Pucciarini sent samples to a number of hospitals in Rome, Genoa, Florence and other cities; see Bado, *Anastasis Corticis Peruuiae*, 240–41.

84. Pietro De Angelis, *La spezieria dell'Arcispedale di Santo Spirito in Saxia e la lotta contro la malaria* (Coluzza, 1954), 101–3.

85. Michael R. McVaugh, "Determining a Drug's Properties: Medieval Experimental Protocols," *Bulletin of the History of Medicine* 91, no. 2 (2017): 190, https://doi.org/10.1353/bhm.2017.0024.

86. Bado, *Anastasis Corticis Peruuiae*, 240–241; Bibliothèque nationale de France, TE151–1220, "Modo Di Adoprare La Corteccia Chiamata Della Febre."

87. RAH, 9/7263, sf.

88. Bibliothèque d'étude et du patrimoine de Toulouse, Ms 763, "Recueil de Recettes Pharmaceutiques et Culinaires," 71–72; Jean-Jacques Chifflet, *Pulvis Febrifvgvs orbis Americani* ([Louvain], 1653), 4–5; Rolandus Sturmius, *Febrifugi Peruviani Vindiciarum . . . exhibens* (Delft: Petrus Oosterhout, 1659); Thomas Bartholin, *Historiarum Anatomicarum & Medicarum Rariorum Centuria V. & VI* (Copenhague: Henricus Gödianus, 1661).

89. Jarcho, *Quinine's Predecessor*; Andreas Holger Maehle, "Peruvian Bark: From Specific Febrifuge to Universal Remedy," *Clio Medica* 74, no. 4 (2000): 223–309.

90. Guy Patin, *Lettres de Gui Patin*, ed. Joseph-Henri Reveilé-Parisse, 3 vols. (Paris: J-B Baillière, 1846), 1:197; 2:103, 112, 121, 262; 3:19, 666.

91. For a Jansenist critique of Jesuit involvement in pharmacy, see also Antoine Arnauld, *La morale pratique des jésuites* (Cologne: Gervinus Quentel, 1669), 61.

92. See, e.g., Antoine de Jussieu, "Recherches d'un Spécifique contre la Dysenterie, Indiqué par les anciens Auteurs sous le nom de MACER, auquel l'Ecorce d'un Arbre de Cayenne, appellé *Simarouba*, peut être comparé & substitué," in *Histoire de l'Academie Royale des Sciences. Année M. DCCXXIX* (Paris: Imprimerie Royale, 1731), "Mémoires", 32–40; British Library (henceforth BL), Sloane 4050, fol. 117r-v; BL, Sloane 4053, fol. 347r; BL, Sloane 4068, fol. 257.

93. On April 30, 1724, John Burnet, a physician to the South Sea Company, wrote a letter to Hans Sloane in which he affirmed, "I should be well satisfy'd if the Royall Society & S.º Sea Comp.ʸ would send me a Missionary (as the Jesuits do) from this to Portobello, Panama, Lima, Potosi & so home by way of Buenos Ayres." See BL, Sloane 4047, fol. 330v.

8. MASTERS OF THE EXOTIC?

The author and the research team gratefully acknowledge funding support from the Leverhulme Trust (RPG-2014, "Selling the Exotic in Paris and Versailles, 1670–1730") and the Faculty of History, University of Cambridge.

1. On probate inventories as an historical source, see Giorgio Riello, "'Things Seen and Unseen': The Material Culture of Early Modern Inventories and Their Representation of Domestic Interiors," in *Early Modern Things: Objects and Their Histories, 1500–1800*, ed. Paula Findlen (Routledge, 2013).

2. Jon Stobart, *Sugar and Spice: Grocers and Groceries in Provincial England, 1650–1830* (Oxford University Press, 2013); Anne E. McCants, "Poor Consumers as Global Consumers: The Diffusion of Tea and Coffee Drinking in the Eighteenth Century," *Economic History Review* 61, no. S1 (2013): 172–200.

3. See Annick Pardailhé-Galabrun, *La Naissance de l'intime: 3000 foyers parisiens XVIIe—XVIIIe siècles* (Presses Universitaires de France, 1988); Daniel Roche, *A History of Everyday Things: The Birth of Consumption in France, 1600–1800* (Cambridge University Press, 2000).

4. For a summary of inheritance law, see Ralph E. Giesey, "Rules of Inheritance and Strategies of Mobility in Pre-Revolutionary France," *American Historical Review* 82, no. 2 (1977): 271–89.

5. Emile Coornaert, *Les Corporations en France avant 1789*, 2nd ed. (Les Éditions Ouvrières, 1968); René de Lespinasse, *Histoire générale de Paris. Les métiers et corporations de la ville de Paris*, I: *XIVe—XVIIIe siècle: Ordonnances générales, métiers de l'alimentation* (Imprimerie Nationale, 1886). For a succinct overview, see Gail Bossenga, "Protecting Merchants: Guilds and Commercial Capitalism in Eighteenth-Century France," *French Historical Studies* 15, no. 4 (1988): 693–703.

6. The following search tool was used: https://www.siv.archives-nationales.culture.gouv.fr/siv/cms/content/display.action?uuid=Accueil1RootUuid&onglet=1. A list of inventories is at the end of chapter 8.

7. McCants, "Poor Consumers as Global Consumers"; Anne E. McCants, "Exotic Goods, Popular Consumption, and the Standard of Living: Thinking About

Globalization in the Early Modern World," *Journal of World History* 18, no. 4 (2007): 433–62. For the English case, see Patrick Wallis, "Exotic Drugs and English Medicine: England's Drug Trade, c. 1550–c. 1800," *Social History of Medicine* 25, no. 1 (2012): 20–46, the first study to make substantive use of port records.

    8. Woodruff D. Smith, *Consumption and the Making of Respectability, 1600–1800* (Routledge, 2002); Markman Ellis et al., *Empire of Tea: The Asian Leaf that Conquered the World* (Reaktion, 2015); Julia Landweber, "'This Marvelous Bean': Adopting Coffee into Old Regime French Culture and Diet," *French Historical Studies* 38, no. 2 (2015): 193–223.

    9. On cinchona, see Samir Boumediene, *La Colonisation du savoir. Une histoire des plantes médicinales du 'Nouveau Monde' (1492–1750)* (Les Éditions des Mondes à Faire, 2016); Saul Jarcho, *Quinine's Predecessor: Francesco Torti and the Early History of Cinchona* (Johns Hopkins University Press, 1993); Matthew J. Crawford, *The Andean Wonder Drug: Cinchona Bark and Imperial Science in the Spanish Atlantic, 1630–1800* (University of Pittsburgh Press, 2016). On ginseng, see Christopher M. Parsons, *A Not-So-New World: Empire and Environment in French Colonial North America* (University of Pennsylvania Press, 2018), chapter 6; on ipecacuanha, see, e.g., E.C. Spary, "Publishing Virtue: Medical Entrepreneurship and Reputation in the Republic of Letters," *Centaurus* 62, no. 3 (2020): 498–521, https://doi.org/10.1111/1600-0498.12291; Teunis Willem van Heiningen, "La Dynastie des Helvétius," *Histoire des sciences médicales* 48, no. 4 (2014): 447–56. On dyeing, see especially Agustí Nieto-Galán, *Colouring Textiles: A History of Natural Dyestuffs in Industrial Europe* (Kluwer Academic, 2001); Jutta Wimmler, *The Sun King's Atlantic: Drugs, Demons and Dyestuffs in the Atlantic World, 1640–1730* (Brill, 2017); Liza Oliver, *Art, Trade, and Imperialism in Early Modern French India* (Amsterdam University Press, 2019).

    10. See especially Phillip P. Boucher, *France and the American Tropics to 1700: Tropics of Discontent?* (Johns Hopkins University Press, 2008).

    11. BIUS-P, Registre 3, pièce 88: sentence de police, April 29, 1701.

    12. BIUS-P, Registre 11, pièces 53–54, April 29 and May 9, 1701: "Memoire ou catalogue des remedes de Chimie que les apoticaires de paris doiuent preparer vendre et debiter a l'exclusion des marchands epiciers;" pièce 61; Boîte AT 72, February 28, 1710.

    13. Nicolas Delamare, *Traité de la Police*, 2nd ed., 4 vols. (Paris: Michel Brunet and Jean-François Herissant, 1719–1738), 1:618: "Le Commerce des Epiciers n'est pas moins délicat ni moins important à la santé, que celui des Apotiquaires: si ceux-ci composent les remedes, ce sont ceux-là qui fournissent la plus grande partie des drogues & des ingrediens qui entrent dans ces compositions. Ce sont eux qui tirent des païs les plus éloignez, & qui en font le debit: il y a peu d'Apotiquaires qui fassent, & même qui puissent faire ce commerce éloigné, & ces voyages de long cours."

    14. Most Parisian guild archives were destroyed in 1871, but those of the apothecaries had been rehoused during their legal separation from the grocers. They are now at the BIUS-P in Paris. Bénédicte Dehillerin and Jean-Pierre Goubert, "À la conquête du monopole pharmaceutique: Le Collège de Pharmacie de Paris (1777–1796)," *Historical Reflections/Réflexions Historiques* 9, nos. 1–2 (1982): 233–48.

15. E. C. Spary, *Feeding France: New Sciences of Food, 1760–1815* (Cambridge University Press, 2014), chapter 4. On domestic healing, see Susan Broomhall, *Women's Medical Work in Early Modern France* (Manchester University Press, 2004); Leigh Whaley, *Women and the Practice of Medical Care in Early Modern Europe, 1400–1800* (Palgrave Macmillan, 2011), chapter 8.

16. Six out of twenty-seven inventories coupled the titles of grocer and apothecary.

17. For full details, see the end of chapter 8. Earlier lists of 1655 and 1671 unfortunately lack addresses. See Christian Warolin, "Les Apôthicaires et la maîtrise d'épicerie à Paris. I. Deux listes de reception en 1655 et en 1671," *Revue d'histoire de la pharmacie* 78, no. 286 (1990): 295–302.

18. Daniel Roche, *The People of Paris: An Essay in Popular Culture in the 18th Century* (University of California Press, 1987), 20. On shopkeeping in Paris, see especially Natacha Coquery, *Tenir boutique à Paris au XVIIIe siècle. Luxe et demi-luxe* (Éditions du CTHS, 2011). Archival records show that shops could change location during a merchant's lifetime.

19. Natacha Coquery, *L'Hôtel aristocratique. Le marché du luxe à Paris au XVIIIe siècle* (Publications de la Sorbonne, 1998), 49–67, 206–8.

20. Jean Nagle, *Luxe et charité. Le Faubourg Saint-Germain et l'argent* (Perrin, 1994).

21. See E. C. Spary, *Taking Drugs in the Sun King's Reign* (forthcoming).

22. Pierre Pomet, *Histoire generale des Drogues, traitant des Plantes, des Animaux, & des Minéraux* (Paris: Jean-Baptiste Loyson and Augustin Pillon, 1694), is the most comprehensive contemporary work on drugs. For translations of names into English, we used the following sources: Randle Cotgrave and James Howell, *A French and English Dictionary* (London: Anthony Dolle, 1673); James Sutherland, *Hortus Medicus Edinburgensis* (Edinburgh: Heir of Andrew Anderson, 1683); Joseph Pitton de Tournefort, *Materia Medica; Or, a Description of Simple Medicines generally us'd in Physick...*, 2nd ed. (London: W. H. for Andrew Bell, 1716); Philip Miller, *The Gardeners Dictionary*, 4th ed., 3 vols. (London: John and James Rivington, 1754–57); Philipp Andreas Nemnich, *An Universal European Dictionary of Merchandise* (London: J. Remnant et al., 1799); N.-J.-B.-G. Guibourt, *Histoire abrégée des drogues simples* (Paris: L. Colas et al., 1820); Joseph Wilson, *A French and English Dictionary* (London: Joseph Ogle Robinson, 1833). Plant names were variably applied, and we have not attempted to "translate" contemporary or common names into Linnaean ones.

23. In identifying places of origin, we again relied on Pomet, *Histoire generale*, part 1, which described contemporary practice in France, especially useful in cases where drugs were available from various locations. For example, ginger, originating botanically in Asia, was supplied to France from the Antilles by this juncture (Pomet, *Histoire generale*, part 1, 61). We only deviated from this in a few cases where Pomet's own information was lacking—for example, rhubarb.

24. *Statuts et Ordonnances pour les Marchands Apothicaires-Espiciers, et les Marchands Espiciers de la Ville, Fauxbourgs, & Banlieuë de Paris* (Paris: Jean-Baptiste Coignard, 1716), ¶17: "sont bien souvent les Marchands contraints de faire de longs & perilleux voyages és Pays & Royaumes étrangers, où ils hazardent leurs vies & leurs biens..."

25. On the Robins, see Gérard Cusset, "Sur les jardins botaniques parisiens au XVIe siècle," *Journal d'agriculture traditionelle et de botanique appliquée* 13, nos. 8–9 (1966): 385–404; Ernest-Théodore Hamy, "Vespasien Robin, Arboriste du Roy, premier sous-démonstrateur de botanique du Jardin royal des plantes (1635–1662)," *Nouvelles Archives du Muséum d'Histoire Naturelle*, 3rd series, 8 (1896): 1–24. The Robins are often characterized as "botanists"; but in contemporary legal documents they self-identified as "herboristes" (Minutier Central, Archives nationales de France, minutes de Claude Le Vasseur, étude XXXV, passim).

26. Pomet, *Histoire generale*, 1694, part 1, 161: "nous ne vendons pas à Paris de ces sortes de Plantes, à cause des Herboristes." See, similarly, Nicolas de Blegny, *Le Livre commode des adresses de Paris pour 1692*, 2 vols. (Paris: Daffis, 1878), 1:165. Charles de Saint-Germain described the herbalists as specializing in roots, while grocers sold flowers, seeds, and other simples. See Saint-Germain, *Medecin Royal ov le parfait Medecin Charitable* (Paris: Cardin Besongne & Augustin Besongne, 1668), 533.

27. Alice Stroup, *A Company of Scientists: Botany, Patronage and Community at the Seventeenth-Century Parisian Royal Academy of Sciences* (University of California Press, 1990).

28. BnF, Nouvelles Acquisitions Françaises 5147: "Académie Royale des Sciences. XV. Registre des dépenses faites pour le laboratoire (1667–1699)."

29. Women's involvement in herbal medicine was of long standing (see Whaley, *Women and the Practice of Medical Care*, chapter 2). Among those describing themselves as "herbalists" in legal documents of the earlier seventeenth century are several women, such as Claude Lalandre, who rented half a shop near the cemetery of Saints-Innocents in 1632 "afin qu'elle y débite ses herbes médicales" (so that she could sell her medicinal herbs there; Minutier Central, Archives nationales de France, étude XXXV, 210, December 19, 1632). The dependency of leading metropolitan *savants* and grocers upon herbalists—rather than vice versa—implies that Jacques Bruslons de Savary's 1723 definition of them as "pauvres femmes établies la plupart dans des échopes au coin des rues, particulièrement près des boutiques des Apoticaires les plus achalandez" (poor women mostly installed in corner shops, especially close to the best-frequented apothecary shops) may have reflected market and gender polemics more than reality. Bruslons de Savary is quoted in Émilie-Anne Pépy, "Les Femmes et les plantes: Accès négocié à la botanique savante et résistance des savoirs vernaculaires (France, XVIIIe siècle)," *Genre & Histoire* 22 (2018), 13, online at: http://journals.openedition.org/genrehistoire/3654.

30. BnF, Bibliothèque de l'Arsénal, Fonds Bastille, Ms. 10348, fol. 400: "Elle... vit dans Lad Maison quil y fust fait Un grand fourneau au milieu d'une chambre dans Lequel Il y eust du feu Jour et nuit, pendant Un temps Considerable pour faire des distillations Et que C'estoit Un grand homme noir picotté de verolle qu'elle Enttendit appeller lachaboissiere qui trauailloit ausd distillations Jour & nuit & lequel disoit que C'estoient des Eaüx que lon distilloit pour Le Roy, Quil venoit dans lad maison un homme qui se disoit medecin Et qu'on appelloit Monsieur delorme... Que Led Lachaboissiere pendant led temps enuoyá plusieurs fois Ledit Jardinier Chercher des herbes pour distiller, Et Enuoyé elle Colignon aussy Une fois en Chercher dans Le bois de Vinciennes & autour des Minimes & vers le menilmontant... Et outre celá á veu aussy

porter aud La Chaboissiére dans ledit temps de plusieurs autres sortes d'herbes qu'elle ne Connoist pas, Et que cestoient des femmes qui les apportoient de la halle, Touttes lesquelles herbes Led Lachaboissiere distilloit."

31. Giselle Ollivier, "Généalogie des familles Poisson du XVIème au XVIIIème siècle," online at https://fr.readkong.com/page/genealogie-des-familles-poisson-du-xvieme-au-xviiieme-siecle-7740633. Delorme's correspondence is in numerous collections, including BnF, Ms. Français 2392, Nouvelles Acquisitions Françaises 6343, Bibliothèque de l'Arsenal, Ms. 5132 and elsewhere.

32. See also BnF, Ms. Français 21738, fols. 241–4: *Tarif povr la Taxe des Drogves*.

33. Saint-Germain, *Medecin Royal*, 533.

34. Allen G. Debus, *The French Paracelsians: The Chemical Challenge to Medical and Scientific Tradition in Early Modern France* (Cambridge University Press, 1991).

35. See, e.g., BnF, Ms. Français 21738, fols. 257–60: *Declaration du Roy, portant réünion au Corps des Marchands Epiciers, des Offices de Maistres & Gardes de leur Communauté; avec desunion du Corps de l'Epicerie de celuy de l'Apotiquairerie* (Paris: Estienne Michallet, 1691).

36. Christopher M. Parsons, "Consuming Canada: *Capillaire du Canada* in the French Atlantic World," in *Drugs on the Page. Pharmacopoeias and Healing Knowledge in the Early Modern Atlantic World*, ed. Matthew James Crawford and Joseph M. Gabriel (University of Pittsburgh Press, 2019).

37. Paul Dorveaux, "Les Délibérations de la Compagnie des marchands épiciers-apothicaires de Paris (suite)," *Bulletin de la Société d'Histoire de la Pharmacie* 13, no. 48 (1925): 169–76.

38. The Crown made sporadic attempts to fix prices of compound remedies, but not simples (*Tarif povr la Taxe des Drogves des Apoticaires Epiciers, avec l'Ordonnance de Monsieur le Lieutenant General de Police, pour la vente d'icelles*, September 25, 1678, in BnF, Ms. Français 21737, fols. 241–44).

39. The work of Sandra Cavallo and Tessa Storey (*Healthy Living in Late Renaissance Italy* [Oxford University Press, 2013]) highlights the advantages, for the history of early modern medicine, of broad analytical categories in investigating probate inventories.

40. On guaiacum, see especially Kevin Siena, "The 'Foul Disease' and Privacy: The Effects of Venereal Disease and Patient Demand on the Medical Marketplace in Early Modern London," *Bulletin of the History of Medicine* 75, no. 2 (2001): 199–224. Its status as a by-product of the timber trade is mentioned in Pomet, *Histoire generale*, part 1, 114–15.

41. Kristof Glamann, *Dutch-Asiatic Trade, 1620–1740* (Danish Science Press, 1958), 108–9.

42. David Bulbeck, *Southeast Asian Exports Since the 14th Century: Cloves, Pepper, Coffee, and Sugar* (KITLV Press, 1998), 20.

43. For a full list of inventories used, see the end of chapter 8.

44. Wilson, *French and English Dictionary*, 288: "Garble." The term was medieval in origin; see Roy Shipperbottom, "Paradise Lost: The Adulteration of Spices," in *Spicing up the Palate: Studies of Flavourings—Ancient and Modern. Proceedings of the Oxford Symposium on Food and Cookery 1992* (Prospect Books, 1992), 247; Paul Freedman,

Out of the East: Spices and the Medieval Imagination (Yale University Press, 2008), 124. Pomet, *Histoire generale*, part 1, 147, claimed that senna garble "n'est le plus souvent que de la terre, ou des feuïlles d'une plante que les Colporteurs appellent Ourdon, qui se trouve par hazard, ou qui a été mis exprés dans les coufles ou balles de Sené" (is commonly nothing other than dirt, or the leaves of a plant that hawkers call *ourdon*, which has entered bales of senna either by chance or on purpose). On "ourdon," see anon., *Dictionnaire portatif*, vol. 7 (Avignon: Louis-Chambeau, 1761), 196.

    45. *Dictionnaire de l'Académie française*, 4th ed., vol. 1 (Paris: Brunet, 1762): "Épluchure": "Les ordures que l'on ôte de quelque chose qu'on épluche"; *Dictionnaire de l'Académie française*, "Éplucher": "Nettoyer en séparant avec la main les ordures & ce qu'il y a de mauvais, de gâté."

    46. Valentina Pugliano, "The Pharmacy Shop," in *Worlds of Natural History*, ed. H. A. Curry et al. (Cambridge University Press, 2018). See also Daniel Roche, "Négoce et culture dans la France du XVIIIe siècle," *Revue d'histoire moderne et contemporaine* 25 (1978): 375–95.

    47. E. C. Spary, "Pierre Pomet's Parisian Cabinet: Revisiting the Visible and the Invisible in Early Modern Collections," in *From Private to Public: Natural Collections and Museums*, ed. M. Beretta (Science History Publications, 2005).

    48. Pomet, *Histoire generale*, part 1, 32–33, 55, 207, 273, 303.

    49. Pomet, *Histoire generale*, part 1, 125, 162, 218, 246, 256, and unpaginated "Appendix."

    50. See especially Teresa Huguet-Termes, "New World Materia Medica in Spanish Renaissance Medicine: From Scholarly Reception to Practical Impact," *Medical History* 45, no. 3 (2001). Studies of individual drugs are too numerous to list here.

    51. Pomet, *Histoire generale*, part 1, 11; Spary, "Publishing Virtue."

    52. Pomet, *Histoire generale*, part 1, 86–88. On china root versus sarsaparilla, see Anna Winterbottom, "Of the China Root: A Case Study of the Early Modern Circulation of *Materia Medica*," *Social History of Medicine* 28, no. 1 (2015): 22–44.

    53. Bibliothèque Centrale du Muséum nationale d'Histoire naturelle, Paris, Ms. 253, fols. 53, 59–61; Pomet, *Histoire generale*, part 1, 68–9; Friedrich A. Flückiger and Daniel Hanbury, *Pharmacographia. A History of the Principal Drugs of Vegetable Origin Met with in Great Britain and British India*, 2nd ed. (London: Macmillan, 1879), 26.

    54. Adrian Helvetius, "Extrait d'une lettre . . . , touchant l'usage de la racine de *Parera-brava*," *Memoires pour l'Histoire des Sciences & des beaux Arts* (September 1704): 1425–34; Adrian Helvetius, "An Extract of a Letter from Dr. Helvetius at Paris," *Philosophical Transactions* 29 (1714–16): 365–67.

    55. [Bernard Le Bovier de Fontenelle,] "Sur le Pareira Brava," *Histoire de l'Academie Royale des Sciences. Année M.DCCIX* (Paris: Jean Boudot, 1712), "Histoire," 56.

    56. Anon. [attrib. François Martialot], *Nouvelle Instruction pour les Confitures, les Liqueurs et les Fruits* (Paris: Charles Sercy, 1692), 270: "Le Thé n'est pas si commun que le Caffé à cause de son prix qui est beaucoup plus cher."

    57. Tim Ecott, *Vanilla: Travels in Search of America's Most Popular Flavor* (Grove Press, 2004), chapter 2; Marcy Norton, *Sacred Gifts, Profane Pleasures: A History of Tobacco and Chocolate in the Atlantic World* (Cornell University Press, 2008), chapter 7.

    58. *Arrest du Conseil d'Etat du Roy, concernant la vente du Caffé, du Thé, du Sorbec*

& du Chocolat (Paris: Estienne Michallet, 1692): "tous les Caffez tant en féves qu'en poudres, le Thé, les Sorbecs, & les Chocolats, tant en pain, roullots, tablettes, pastilles, que de toute maniere qu'il soit mis; ensemble les drogues dont il est composé, comme le Cacao & la Vanille ... défenses à toutes personnes de s'immiscer en la composition, vente & debit ... desdites drogues & marchandises." This document was incorporated in the papers of the police lieutenant Nicolas Delamare, in BnF, Ms. Français 21663, fols. 286r–7v. Later evidence comes from port records for the Bordeaux area in the 1710s, where vanilla is listed several times as both an import and an export good (Archives départementales de la Gironde, Mss. C 4268–4269).

59. BnF, Ms. Français 21663, fols. 289r–92v. Philip Withington has drawn attention to the curious absence of actual coffee until after 1700 even in England, where coffeehouses supposedly abounded after 1650. See Withington, "Where Was the Coffee in Early Modern England?," *Journal of Modern History* 92 (2020): 65–66.

60. For details of why this profitable monopoly collapsed, see E. C. Spary, *Eating the Enlightenment: Food and the Sciences in Paris* (University of Chicago Press, 2012), 105.

61. On Challiou, see Jean Costentin and Pierre Delaveau, *Café, thé, chocolat: Les bienfaits pour le cerveau et pour le corps* (Odile Jacob, 2010), 102; on French chocolate, see Suzanne Perkins, "Is It a Chocolate Pot? Chocolate and Its Accoutrements in France from Cookbook to Collectible," in *Chocolate: History, Culture, and Heritage*, eds. H.-Y. Shapiro and L. E. Grivetti (John Wiley & Sons, 2009); Sophie D. Coe and Michael Coe, *The True History of Chocolate* (Thames and Hudson, 1996). Ina Baghdiantz-McCabe identifies sales by non-guild members protected by élite patrons as a possible reason for the failure of Damanne's monopoly. See Baghdiantz-McCabe, *Orientalism in Early Modern France: Eurasian Trade, Exoticism, and the Ancien Régime* (Berg, 2008), 193. On Basque chocolatiers, see André Constantin, "A propos du chocolat de Bayonne," *Bulletin du Musée Basque* 5 (1933): 404–14.

62. BnF, Bibliothèque de l'Arsenal, Ms. 6535: "États des recettes et dépenses de la maison d'Élisabeth d'Orléans, duchesse de Guise, 1672–1696," fols. 146r, 149v, 223r.

63. Antoine Furetière, *Dictionnaire universel, contenant generalement tous les Mots François tant vieux que modernes*, 3 vols. (La Haye and Rotterdam: Arnout and Reinier Leers, 1690–91), I: "Banille"; see, similarly, anon., *Nouvelle Instruction pour les Confitures*, 273.

64. Nicolas Lémery, *Traité Universel des Drogues simples* (Paris: Laurent d'Houry, 1698), 55: "Ce fruit est fort rare en Europe, & quand on en a on le garde pour la curiosité."

9. CONSUMING THE EXOTIC IN EIGHTEENTH-CENTURY SWITZERLAND

The author gratefully acknowledges funding from the Research Grants Council of the Hong Kong Special Administrative Region (Project Nos. HKU 17401614 and 37000418), a residential fellowship at the Institute for Advanced Study, Nantes, France (2018–19), archival material provided by Josep Camarasa and Neus Ibáñez, Institut Botanic Barcelona, Coline Gosciniak, Bibliothèque Carnegie, Reims, and research by Felix Yeung Shing Hay. In Switzerland, I was assisted by Nicolas Fumeaux, Conservatoire et Jardin Botaniques, Geneva; staff of the Archives de l'État, Neuchâtel;

Olivier Girardbille, Archives de la Ville, Neuchâtel; Lorraine Filippozzi, Archives Communales, Vevey; Charline Dekens, Bibliothèques et Archives, Lausanne; Catherine Guanzini, Archives de la Ville, Yverdon-les-Bains; Gilles Jeanmonod, Archives Cantonales Vaudoises; and Pierre de Blonay, Fondation de Château de Blonay.

1. Bezoars are calculi of undigested organic and inorganic material containing minerals, formed around a foreign body in the stomachs of goats, monkeys, and porcupines. The stones are obtained by slaughtering the animal. The term "bezoar" derives from Farsi, *pá zahar*, meaning to expel poison. Garcia da Orta additionally links the Farsi word for goat, *zahar*, to "bezoar" and "bazaar," among others (*Colóquios dos simples he cousas medicinais da Índia* [Goa: n.p., 1563]), colloquy 45; see also W. D. Ian Rolfe, "Materia Medica in the Seventeenth-Century Paper Museum of Cassiano dal Pozzo," in *A History of Geology and Medicine*, ed. C. J. Duffin et al. (Geological Society, 2012); Peter Borschberg, "The Euro-Asian Trade in Bezoar Stones (Approx. 1500–1700)," in *Artistic and Cultural Exchanges Between Europe and Asia, 1400–1900: Rethinking Markets, Workshops and Collections*, ed. Michael North (Ashgate, 2010); and Maria Do Sameiro Barroso, "Bezoar Stones, Magic, Science and Art," *Geological Society of London* 375 (2013): 193–207.

2. The Andalusian physician Abū Marwān Abd al-Malik ibn Zuhr (ca. 1090–1162), known in Latin as Avenzoar, is credited with introducing bezoars to Europe. In modern-day India, bezoar goats are put to graze on *Acacia nilotica* (L.) Delile (the gum arabic tree), whose seeds, leaves, bark, and roots exhibit a variety of medicinal properties. See Misbahuddin Azhar Mustehasan et al., "Zahar Mohra (Bezoar) an Alexipharmic Unani Mineral Drug: Review Introduction," *Journal of Drug Delivery and Therapeutics* 10, no. 6 (2020): 237–38. These findings confirm the value of taking bezoars from animals fed on aromatic mountain herbs. Garcia da Orta, *Colóquios*, dialogue 45; Christopher J. Duffin, "Porcupine Stones," *Pharmaceutical Historian* 13, no. 1 (2012): 13–22; Barroso, "Bezoar Stones," 199. Pre-contact New World peoples revered bezoar stones harvested from Andean camelids, and the Spanish introduced New World bezoars, known as *Lapis occidentalis*, to Europe. See Luis Millones Figueroa, "The Bezoar Stone: A Natural Wonder in the New World," *Hispanófila* 171, no. 1 (2014): 139–56.

3. By wearing or displaying the coveted stones in their collections of curiosities, high-status individuals demonstrated that they were both important enough to be poisoned, and rich enough to ward off the threat by applying the "princely antidote."

4. Duffin, "Porcupine Stones," 18.

5. *Chandelor v. Lopus* (1603) 79 English Reports 3.

6. In contemporary usage, "Brahmin" referred to an "Indian" priest or philosopher. See "Bracmane, Bramine ou Bramin," *Dictionnaire de L'Académie française*, 4th ed., vol. 1 (Paris: Brunet, 1762).

7. Laurent Garcin, "Lettre . . . contenant des Réflexions sur les Remèdes en général, & en particulier sur les Vertus & les Usages des Pilules, qu'il a promis au Public," *Journal helvétique, ou, Annales littéraires et politiques de l'Europe et principalement de la Suisse* (October 1744): 342–43. The "Brahmin" Garcin refers to could have been a practitioner of Tamil Ayurvedic medicine, now officially designated by the Indian government as "Siddha medicine." See Brigitte Sébastia, "From Siddha Corpus to Siddha Medicine: Reflection on the Reduction of Siddha Knowledge Through Exploration

of Manuscripts," in *Local Health Traditions: Plurality and Marginality in South Asia*, ed. Arima Mishra (Orient BlackSwan, 2019). For a succinct summary of Siddha medicine's eighteenth-century European appropriation, see S. Jeyaseela Stephen, "The Circulation of Medical Knowledge Through Tamil Manuscripts in Early Modern Paris, Halle, Copenhagen, and London," in *Histories of Medicine and Healing in the Indian Ocean World: The Medieval and Early Modern Period*, ed. Anna Winterbottom and Facil Tesaye, vol. 1 (Palgrave MacMillan, 2016). Isaac Augustijn Rumpf (1673–1723), the governor of Ceylon, should not be confused with Georg Eberhard Rumpf or Rumphius (1627–1702), a German employee of the Dutch East India Company (VOC) and the author of works on the natural history of Ambon, e.g., *Herbarium Amboinense*, ed. Johannes Burman (Amsterdam: F. Changuion, J. Catuffe, and H. Uytwerf, 1741).

8. Garcin, "Lettre ... contenant des Réflexions," 343n. (Emphasis added.)

9. See Benjamin Breen, "Drugs and Transcultural Exchanges in the Portuguese Colonial World," *LSE Trading Medicines Working Paper* (2013), 13–15. Online at: https://www.academia.edu/19588825/Drugs_and_Transcultural_Exchanges_in_the_Portuguese_Colonial_World_Working_Paper_.

10. Anne Goldgar, *Impolite Learning: Conduct and Community in the Republic of Letters, 1680–1750* (Yale University Press, 1995).

11. On co-construction of knowledge, see Kapil Raj, "Go-Betweens, Travelers, and Cultural Translators," in *A Companion to the History of Science*, ed. Bernard Lightman (Wiley Blackwell, 2016); and Phillip B. Wagoner, "Precolonial Intellectuals and Production of Colonial Knowledge," *Comparative Studies in Society and History* 45, no. 4 (2003): 783–86.

12. Lissa Roberts, "Situating Science in Global History: Local Exchanges and Networks of Circulation," *Itinerario* 33, no. 1 (2009): 9–30.

13. Londa Schiebinger, *Secret Cures of Slaves: People, Plants, and Medicine in the Eighteenth-Century Atlantic World* (Stanford University Press, 2017); Daniel Rood, "Toward a Global Labor History of Science," in *Global Scientific Practice in an Age of Revolutions, 1750–1850*, ed. Patrick Manning and Daniel Rood (University of Pittsburgh Press, 2016).

14. Sébastia, "From Siddha Corpus to Siddha Medicine," 139.

15. Geoffrey Gunn, *History Without Borders: The Making of an Asian World Region (1000–1800)* (Hong Kong University Press, 2011).

16. Patrick Manning, "Introduction: Building Global Perspectives in History of Science: The Era from 1750 to 1850," in Manning and Rood, *Global Scientific Practice*, 17.

17. Brigitte Sébastia, "Preserving Identity or Promoting Safety? The Issue of Mercury in Siddha Medicine: A Brake on the Crossing of Frontiers," *Études asiatiques/Asiatische Studien* 69, no. 4 (2015): 939.

18. On heavy metals in modern Siddha medicine, see Sébastia, "Preserving Identity or Promoting Safety?," passim. On mercury as a prized European pharmaceutical, see Andrew Cunningham, "Mercury: 'One of the Most Valuable Drugs We Have' (1937)," in *It All Depends on the Dose: Poisons and Medicines in European History*, edited by Ole P. Grell et al. (Routledge, 2018).

19. B. V. Subbarayappa, "Siddha Medicine," in *Medicine and Life Sciences in India*, edited by B. V. Subbarayappa (Project of History of Indian Science, Philosophy and Culture, 2001), vol. 4, part 2, 429–51.

20. Joseph S. Alter, "Āyurvedic Acupuncture—Transnational Nationalism: Ambivalence About the Origin and Authenticity of Medical Knowledge," in *Asian Medicine and Globalization*, edited by Joseph S. Alter (University of Pennsylvania Press, 2005).

21. Jean Filliozat, "La Médecine indienne et l'expansion bouddhique en Extrême Orient," *Journal asiatique* 224 (1934): 301–7.

22. Alter, "Āyurvedic Acupuncture"; Stefanie Gänger, "World Trade in Medicinal Plants from Spanish America, 1717–1815," *Medical History* 59, no. 1 (2015): 44–62.

23. The term comes from Kenneth Pomeranz, *The Great Divergence: China, Europe, and the Making of the Modern World Economy* (Princeton University Press, 2001).

24. Unfortunately, Bridel's account contains glaring inaccuracies, including Garcin's death year, which was 1751, not 1752, as claimed not only by Bridel, but also by most biographical sources. See *Registre des morts*, vol. 3, 11 janvier 1740 à 29 décembre 1818. AVN C 101.04.03.03, fol. 83, Archives de la Ville, Neuchâtel. Similarly, Garcin's birth year is questionable: even though traditionally given as 1683, a birth year of 1680 or 1681 seems more plausible, since Garcin stated in a letter of 1749 that he was sixty-eight years old. SeeBritish Library, Sloane Add. Ms 28537, fols. 221–222v: letter, Laurent Garcin to Emanuel Mendes da Costa, July 30, 1749; also Frank Lequin, "Het personeel van de verenigde Oost-Indische compagnie in Azië in de achttiende eeuw, meer in het bijzonder in de vestiging Bengalen" (PhD diss., University of Leiden, 1982), 1:168, 314n97. His apprenticeship would therefore have commenced ca. 1695, pace Philippe-Sirice Bridel, "Biographie nationale," vol. 13 of *Le Conservateur suisse, ou recueil complet des étrennes helvétiennes* (Benjamin Corbaz: Lausanne, 1831), 99.

25. Garcin's name in the VOC archives was "Laurens Garcij," one of many variants of a common French surname, making it difficult to trace family connections. Nobody matching the known facts about Garcin surfaces in online Dutch church records (marriages, baptisms, funerals) and notarial archives.

26. German bombing of southern Holland during World War II obliterated archives that might have clarified Garcin's connections with Zeeland, where the Walloon community enjoyed a strong presence, resulting from the influx of Huguenot refugees in the wake of Louis XIV's 1685 revocation of the Edict of Nantes. See J. F. Bosher, "Huguenot Merchants and the Protestant International in the Seventeenth Century," *William and Mary Quarterly* 52, no. 1 (1995): 77–102.

27. Iris Bruijn, *Ship's Surgeons of the Dutch East India Company: Commerce and the Progress of Medicine in the Eighteenth Century* (Leiden University Press, 2009), 35. This enhanced surgical training might explain Bridel's claim that Garcin studied medicine. Although Garcin never matriculated for medical study in any Dutch university, he referred to himself, while still a surgeon, as a "Médecin-Observateur."

28. Laurent Garcin, "Lettre . . . à l'ocasion de quelques Remèdes nouveaux & expérimentés, qu'il a découverts dans ses Voïages des Indes," *Journal helvétique* (September 1744): 269.

29. After serving by his own account for sixteen years in a Dutch regiment, Garcin departed in May 1720 for the East Indies. This would mean he entered the army in ca. 1703–4, a supposition that cannot be confirmed due to incomplete enlistment records. See Bridel, "Biographie nationale," 99; Garcin, "Lettre... à l'ocasion de quelques Remèdes."

30. Laurent Garcin, "The Establishment of a New Genus of Plants, Called Salvadora, with Its Description," *Philosophical Transactions* 46 (1749–1750): 47–53.

31. Josep M. Camarasa and Neus Ibáñez, "Joan Salvador and James Petiver: The Last Years (1715–1718) of their Scientific Correspondence," *Archives of Natural History* 39, no. 2 (2012): 191–216.

32. Garcin, Manuscrits, 7:2, n.d., Ms E14, Special Collections, Kenneth Spencer Research Library, University of Kansas, Lawrence, Kansas.

33. The average age at death for ship's surgeons who departed for Asia before 1725 and settled there was just twenty-seven. If a surgeon survived to make a second voyage, his average age at death was thirty-eight, whereas it was fifty-eight if he did not make another voyage (Bruijn, *Ship's Surgeons*, 199–200).

34. Contrary to conventional belief, VOC ship's surgeons did not "spen[d] most of their time in the quest for new knowledge" (Pratik Chakrabarti, *Materials and Medicine: Trade, Conquest, and Therapeutics in the Eighteenth Century* [Manchester University Press, 2010], 7-8). The time-consuming work as ship's surgeon did not lend itself to botanizing, which was more often undertaken by land-based actors, like the physicians Jacobus Bontius (1592–1631) or Engelbert Kaempfer (1651–1716), the VOC commander Rheede tot Drakenstein, and the merchant Georg Everhard Rumphius (1627–1702).

35. Garcin, "Lettre... à l'ocasion de quelques Remèdes," 278.

36. Garcin, "Lettre... à l'ocasion de quelques Remèdes," 269 (author's emphasis).

37. Sea Muster Roll, Algemeen RijksArchief Inv. Nr. 12801, Nationaal Archief, The Hague.

38. Alexandra Cook, "Laurent Garcin, M.D. F.R.S.: A Forgotten Source for N. L. Burman's Flora Indica (1768)," *Harvard Papers in Botany* 21, no. 1 (July 2016): 31–53.

39. Louis Bourguet, "Lettre à Mr. Garcin...; Sur la Pétrification des Petits Crabes de Mer, de la Côte de Coromandel," *Journal helvétique* (September 19, 1740): 276–84. The elder Burman is credited with transforming Amsterdam's Hortus into a "scientific powerhouse." See Marie-Christine Skuncke, *Carl Peter Thunberg: Botanist and Physician* (Swedish Collegium for Advanced Study, 2014), 57.

40. These contributions were long obscured by Nicolaas Burman's use of multiple abbreviations to designate Garcin's specimens, and Burman junior's failure to attribute certain specimens, such as *Sida persica* (*Abutilon persicus* (Burm. f.) Merr.), to Garcin at all. Burman even published an illustration of this plant in his *Flora Indica: cui accedit series zoophytorum Indicorum, nec non prodromus florae Capensis* (Amsterdam: Apud Cornelium Haek, 1768), 148, tab. 47, fol. 1.

41. Personal communication, Martin Callmander and Nicolas Fumeaux, Conservatoire et jardin botaniques, Geneva, February 2019.

42. For Garcin's introductions, via Burman's *Flora Indica*, see Elmer D. Merrill, "A Review of the New Species of Plants Proposed by N. L. Burman in His Flora Indica,"

Philippine Journal of Science 19 (1921): 329–32. See also World Health Organization, "Herba Andrographidis," *WHO Monographs on Selected Medicinal Plants*, vol. 2 (Geneva, 2002), 12–24.

43. George W Staples and Fernand Jacquemoud, "Typification and Nomenclature of the Convolvulaceae in N. L. Burman's Flora Indica, with an Introduction to the Burman Collection at Geneva," *Candollea* 60, no. 2 (2005): 446; while correct, their assessment lacks quantitative support.

44. Garcin, "Lettre . . . contenant des Réflexions," 338. Patients did not necessarily agree, however: "A slow cure is not to the taste of many patients, who want to be cured from one day to the next." Garcin, "Lettre . . . contenant des Réflexions," 336–37. On the popularity of chemical remedies, see Marieke M. A. Hendriksen, "Boerhaave's Mineral Chemistry and its Influence on Eighteenth-Century Pharmacy in the Netherlands and England," *Ambix* 65, no. 4 (2018): 1–21.

45. Laurent Garcin, Manuscrits, fol. 7:13r., University of Kansas Libraries.

46. Garcin's successful examination for the medical degree *petit ordinaire* (versus the *grand ordinaire*, which required not only a thesis, but also completion of the full course of medical study) took place on July 18, 1731: "Catalogus . . . Academia Remensi, a natalibus Facultatis medicae (1550–1794)," Bibliothèque Carnegie (Bibliothèque municipale de Reims), Ms 1085, fol. 51. On January 14, 1732, Garcin paid 300 *livres faibles* for naturalization in the principality, conferred with the right to "purchase a bourgeoisie [*acheter une bourgeoisie*]" in the city of Neuchâtel. To attain the latter, he appeared three times before the city authorities, presented his bona fides as physician, Royal Society Fellow and corresponding member of the Académie royale des sciences, swore the required oath, and paid another 1,250 *livres faibles* ("Séance du 17ᵉ Mars 1732 en Conseil General et Ordinaire," in "Manvel de Conseil, N° 15, 1730–1732," fol. 131 (Archives de la ville, Neuchâtel, classmark AVN B 101.01.01.18). On this naturalization regime for religious refugees, see Alexandre de Chambrier, "Naturalisation des réfugiés français de Neuchâtel de la Révocation de l'Edit de Nantes à la Révolution française 1685–1794," *Musée neuchâtelois* 50 (September–October 1900): 232.

47. Boerhaave, for example, evinces a high regard for Garcin's scholarly merits, in a testimonial published in Pierre Massuet, "Lettre Vingt-Cinquième à Monsieur le docteur Pingré," vol. 4 of *Lettres Sérieuses et Badines sur les ouvrages des Savans et sur d'autres matières* (The Hague: Jean Van Duren, 1730), 447.

48. Laurent Garcin, "The Establishment of a New Genus of Plants, Called Salvadora;" Laurent Garcin, "The Settling of a new Genus of Plants, called after the Malayans, MANGOSTANS," *Philosophical Transactions* 38 (1733): 232–42; Laurent Garcin, "Letter from Dr. Laurence Garcin, of Neuchâtel, F.R.S. to Sir Hans Sloane Bart. late P.R.S. concerning the Cyprus of the Ancients," *Philosophical Transactions* 45 (1748): 564–78; Laurent Garcin, "Memoirs communicated by Mons. Garcin to Mons. St. Hyacinthe, F. R. S. containing a Description of a new Family of Plants called Oxyoïdes; some Remarks on the Family of Plants called Musa; and a Description of the Hirudinella Marina, or Sea Leäch," *Philosophical Transactions* 36 (1729–1730): 377–94. Garcin's meteorological reports included "Observations météorologiques et barométriques faites dans deux differentes latitudes, l'une Maritime & l'autre Alpine, comparées ensemble, & raportés à leur veritable cause," *Journal helvétique* (May 1740): 465–78.

49. Joëlle Kunz summarizes eighteenth-century Swiss emigration as a result of "too many men with too little work and too little space"; quoted in Alexandra Cook, *Jean-Jacques Rousseau and Botany: The Salutary Science* (Voltaire Foundation 2012), 63n32.

50. In 1729 Bosset purchased La Rochette, on what is now Avenue de la Gare, and on November 11, 1732, Garcin was reported as residing there (Archives de l'État, Neuchâtel, Neuchâtel Décès, 18 juillet 1706–2 mars 1737, EC 264, fol. 314).

51. Bourguet, "Lettre à Mr. Garcin"; Garcin, "Lettre . . . à l'ocasion de quelques Remèdes," 269.

52. British Library, Sloane Ms 4052, fols. 149, 201–206b: letter, Laurent Garcin to Sir Hans Sloane, July 19, 1732. While in Hulst, Garcin used the Maduran Pills to treat "fevers" ("Lettre . . . contenant des Réflexions," 346) and recorded meteorological observations: [Garcin], "Observations météorologiques et barométriques." However, the town's archives make no reference to his residing, much less exercising his profession there.

53. See Cook, "Laurent Garcin," 33.

54. Garcin, "Establishment of a New Genus," 47–53.

55. Jean-Antoine D'Ivernois, "Lettre à Monsieur Cartier ministre du St. Evangile & pasteur de l'église de la Chaux du Milieu," *Journal helvétique* (January 1742): 31. See also Louis Bourguet, "Lettre à Mr. D.C.M.A.N. sur la Nouvelle Edition du DICTIONNAIRE DE COMMERCE," *Journal helvétique, ou, annales littéraires et politiques de l'Europe et principalement de la Suisse* (February 28, 1742): 148–54.

56. Letter, Gagnebin to Albrecht von Haller, November 25, 1739, quoted in Marcel S. Jacquat, "Abraham Gagnebin: médecin (1707–1800)," in *Biographies neuchâteloises*, vol. 1, ed. Michel Schlup (G. Attinger, 1996), 101; my emphasis.

57. Laurent Garcin, "Lettre . . . à l'ocasion de quelques Remèdes," 270–73; Laurent Garcin, "Lettre . . . contenant des Réflexions"; Laurent Garcin, "Lettre . . . sur la nature et l'usage des Pilules Madurines, avec des Remarques intèressantes sur les efets du Thé," *Journal helvétique* (May 1747): 473–90.

58. Garcin, "Lettre . . . à l'ocasion de quelques Remèdes," 262; Subbarayappa, "Siddha Medicine," 436.

59. Garcin, "Lettre . . . contenant des Réflexions," 344. Nothing is known about the subjects of his drug trials.

60. Garcin, "Lettre . . . à l'ocasion de quelques Remèdes," 276.

61. Skuncke, *Carl Peter Thunberg*; Raj, "Go-Betweens, Travelers, and Cultural Translators;" also Rood, "Toward a Global Labor History."

62. No extant copy of this purported collection is known, and it may never have existed. See Kattungal Subramaniam Manilal, "The Implication of Hortus Malabaricus with the Botany and History of Peninsular India," in *Botany and History of Hortus Malabaricus*, edited by Kattungal Subramaniam Manilal (A. A. Balkema, 1980), 3. In a personal communication, Minakshi Menon (McGill University) suggests that "*Manhaninggatnam*" is most likely "a mis-transliteration of the name of one of the many medical lexicons in Sanskrit."

63. Bruijn, *Ship's Surgeons*, 212.

64. Garcin, "Lettre... à l'ocasion de quelques Remèdes," 269.

65. British Library, Sloane Add. Ms 28537, fols. 221–222v: letter, Laurent Garcin to Emanuel Mendes da Costa, July 30, 1749.

66. Garcin, "Lettre... contenant des Réflexions," 348; and anon., "Avis," *Journal helvétique* (April 1745): 380. The purveyors included: at Geneva, "Mme la veuve Maystre," probably Garcin's mother-in-law, Marie Maystre née Aigoin (born ca. 1673); at Morges, "Mrs la veuve Laval & Aigoin," probably Marie Laval and Marguerite Aigoin, relations of Garcin's wife, Marguerite Maystre (1702–1785); at Yverdon, Augustin de Miére; and at Basel, Jean Boukard. No further information is available concerning these last two individuals.

67. "Séance du 8ième Octobre 1694 en Conseil General et Ordinaire," in "Manvel de Conseil, N° 8, 1690–1697," fol. 253 (Archives de la ville, Neuchâtel, classmark AVN B 101.01.01.10).

68. [Laurent Garcin], "Avis sur les pilules madurines," *Journal helvétique* (April 1745), 380.

69. Olivier Lafont, "L'Évolution de la législation pharmaceutique des origines à la loi de Germinal an XI," *Revue d'histoire de la pharmacie* 339, no. 3 (2003): 370; François Ledermann, "La Pharmacie suisse: mille ans d'histoire et quelques particularités," *Revue d'histoire de la pharmacie* 363, no. 3 (2009): 295–302.

70. Garcin, "Lettre... contenant des Réflexions," 344–46.

71. Subbarayappa, "Siddha Medicine"; Sébastia, "Preserving Identity."

72. Garcin, "Lettre... contenant des Réflexions," 346, original emphasis.

73. Garcin, Lettre... sur la nature et l'usage des Pilules Madurines," 473–75.

74. Samir Boumediene, *La Colonisation du savoir. Une histoire des plantes médicinales du "Nouveau Monde" (1492–1750)* (Les Éditions des Mondes à Faire, 2016), 193.

75. Garcin, "Lettre... sur la nature et l'usage des Pilules Madurines," 475, author's emphasis.

76. Breen, "Drugs and Transcultural Exchanges," 15.

77. Garcin, "Lettre... à l'ocasion de quelques Remèdes," 276.

78. Garcin, "Lettre... sur la nature et l'usage des Pilules Madurines," 475. See also W. K. Soh, "Taxonomic Revision of *Cinnamomum* (Lauraceae) in Borneo," *Blumea* 56 (2011): 241–64.

79. *Cinnamomum verum, C. zeylanicum* and *C. tamala* are listed in "The Siddha Medicine Pharmacopeia of India," part I, vol. 1, 62, 66. Online at: https://pcimh.gov.in/show_content.php?lang=1&level=1&ls_id=57&lid=55, last updated September 22, 2021. See Garcin, "Lettre... sur la nature et l'usage des Pilules Madurines," 476.

80. Laurent Garcin, "Canelle ou Cannelle," in Jacques Savary des Bruslons, *Dictionnaire universelle de commerce*, edited and revised by Philemon-Louis Savary, vol. 1 (Geneva: Les Héritiers Cramer & Frères Philibert, 1742), 645–50; Boumediene, *La Colonisation du savoir*, 193.

81. See B. Sathiyathilaga et al., "Chemical Characterization of Siddha Herbo-Mineral Formulation Sangu Chunnam by Using Modern Techniques," *Indo American Journal of Pharmaceutical Research* 4, no. 9 (2014): 3662–68 (author's emphasis); and R. Sudha and S. Ayyasamy, "Chunnam: A Commended Dosage Form in Siddha

Medicine," *International Journal of Research in Ayurveda and Pharmacy* 4, no. 1 (2013): 3. Chunnam's alkaline pH is today vaunted for its beneficial health effects.

82. Garcin, "Lettre... sur la nature et l'usage des Pilules Madurines," 477.

83. Sébastia, "From Siddha Corpus to Siddha Medicine," 139.

84. Sudha and Ayyasamy, "Chunnam."

85. B. Rama Devi et al., "Herbomineral Formulation's Safety and Efficacy Employed in Siddha System of Medicine: A Review," *International Research Journal of Pharmacy* 10, no. 1 (2019): 16.

86. Sathiyathilaga et al., "Chemical Characterization of Siddha Herbo-Mineral Formulation."

87. Garcin, "Lettre... contenant des Réflexions," 336–7.

88. Garcin, "Lettre... sur la nature et l'usage des Pilules Madurines," 476; author's emphasis.

89. Sébastia, "Preserving Identity," 941.

90. Sudha and Ayyasamy, "Chunnam," 3; Sathiyathilaga et al., "Chemical Characterization of Siddha Herbo-Mineral Formulation." Garcin, "Lettre... sur la nature et l'usage des Pilules Madurines," 481.

91. Garcin, "Lettre... sur la nature et l'usage des Pilules Madurines," 476.

92. Boumediene, *La Colonisation du savoir*, 193, 240ff.

93. Sudha and Ayyasamy, "Chunnam," 2.

94. I have seen no indication that palm-leaf manuscripts brought to Halle were accompanied by drugs and/or knowledge of drug manufacturing techniques. See Stephen, "The Circulation of Medical Knowledge"; Josef N. Neumann, "Malabarischer Medicus—eine ethnomedizinischhistorische Quelle des frühen 18 Jahrhunderts," in *Mission und Forschung: Translokale Wissensforschung zwischen Indien und Europa im 18. und 19. Jahrhundert*, edited by Heike Leibau (Verlag der Franckeschen Stiftungen, 2010); Arno Lehmann, "Hallesche Mediziner und Medizinen am Anfang deutsch-indischer Beziehungen," *Wissenschaftliche Zeitschrift der Martin-Luther-Universität Halle. Mathematisch-Naturwissenschaftliche Reihe* 5, no. 2 (1955): 117–32.

95. Koen Vermeir and Daniel Margócsy, "States of Secrecy: An Introduction," *The British Journal for the History of Science* 45, no. 2 (2012): 161.

96. Garcin, "Lettre... à l'ocasion de quelques Remèdes," 269; author's emphasis.

97. Michael Hunter, "Robert Boyle and Secrecy," in *Secrets and Knowledge in Medicine and Science, 1500–1800*, edited by Elaine Leong and Alisha Rankin (Ashgate, 2011).

98. Garcin, "Lettre... sur la nature et l'usage des Pilules Madurines," 476–8.

99. See Wouter Klein's contribution to this volume.

100. A case in point is the Dutch-born remedy vendor Adrien (Adriaan) Helvétius (1662–1727): see E. C. Spary, "Publishing Virtue: Medical Entrepreneurship and Reputation in the Republic of Letters," *Centaurus* 62, no. 3 (2020): 498–521.

101. By 1734, Garcin was suffering from paralysis of the fingers, due to plant collecting in the cold. He consequently used his big toe and a "large woollen glove [gros gand de Laine]" to write. See Laurent Garcin, "Lettre à Mr. DANIEL BERNOULLI; Docteur en Médecine, Professeur en Anatomie & en Botanique, servant de Réponse aux dificultés de Mr. de MUSCHENBROEK sur les principales Causes des

Mouvements du Baromètre," *Journal helvétique*, June 1742, 37; British Library, Sloane Add. Ms 28537, fols. 221–222v: letter, Laurent Garcin to Emanuel Mendes da Costa, July 30, 1749.

102. Sébastia, "From Siddha Corpus to Siddha Medicine," 139.

103. Laurent Garcin, "Préface historique: Addition communiquée par Mr. le Docteur Garcin," in *Dictionnaire universelle de commerce*, vol. 1, XXXI.

104. See, e.g., Garcin, "Canelle."

105. Raj discusses this in "Go-Betweens, Travelers, and Cultural Translators," and "Thinking Without the Scientific Revolution: Global Interactions and the Construction of Knowledge," *Journal of Early Modern History* 21, no. 5 (2017): 445–58.

106. Rood, "Toward a Global Labor History of Science," 266.

10. MERCURIUS IN THE STOREHOUSE

1. Michael Bernhard Valentini (1657–1729), *I. Museum museorum, oder vollständige Schau-Bühne... aller Materialien und Specereyen. Zweyte Edition... + Ost-Indianische Send-Schreiben... Zweyte Edition. II. Museum museorum... Schaubuehne frembder Naturalien Zweyter Theil. + Anhang von verschiedenen Kunst- und Naturalien-Kammern. III. Ney-auffgerichtetes Ruest- und Zeughauss der Natur... Machinen und Instrumenten... Dritten Theils des Museum museorum* (Frankfurt am Main: J. D. Zunner & Zunner's Witwe, J. A. Jungen, 1714).

2. For an overview of how the term "Orientalism" has been applied in early modern studies, see Marcus Keller and Javier Irigoyen-Garcia, "Introduction," in *The Dialectics of Orientalism in Early Modern Europe*, ed. Marcus Keller and Javier Irigoyen-Garcia (Palgrave Macmillan, 2018).

3. In a discussion of renderings of the tropics, Denis Cosgrove defines three basic intellectual framings, or ontological tropics, as he puts it, used by Westerners to define and describe the "geographical spaces that lie astride the equator." These "spaces" are cosmographic, geographic, and environmental/ethnographic, and each of them constitutes an order of tropical reality. Cosgrove illustrates well how the concept of spatiality can be seen as an ontological category underlying representations of the world. Perceiving the world in this way seems so natural to us that it is hard to imagine other ways of doing it. See Cosgrove, "Tropic and Tropicality," in *Tropical Visions in an Age of Empire*, ed. Felix Driver and Luciana Martins (University of Chicago Press, 2005), esp. 198, 215.

4. Three influential studies that explore these issues in depth are Edward W. Said, *Orientalism* (Pantheon Books, 1978); Michael Adas, *Machines as the Measure of Men: Science, Technology and Ideologies of Western Dominance* (Cornell University Press, 1989); and Donald S. Lopez Jr., *Prisoners of Shangri-La: Tibetan Buddhism and the West* (University of Chicago Press, 1998). See also Saree Makdisi, *Making England Western: Occidentalism, Race, and Imperial Culture* (University of Chicago Press, 2014).

5. See the contributions by Sebestian Kroupa, Paula De Vos, Barbara Di Gennaro Splendore, and Samir Boumediene, this volume.

6. It should be noted that this argument differs in substantial ways from that presented in Lorraine Daston and Katharine Park, *Wonders and the Order of Nature, 1150–1750* (Zone Books, 1998). Daston and Park study elite culture, and make frequent use

of spatial metaphors such as "center" and "periphery." Although they acknowledge that imagination was shaped by fabulous objects, it seems to me that they assume that medieval and early modern Europeans utilized a form of mental/spatial maps to order their perceptions of the wondrous. See Daston and Park, 32, 67 (objects), 60, 173, 175 (spatial assumption).

7. Renata Ago, *Gusto for Things: A History of Objects in Seventeenth-Century Rome*, trans. B. Bouely and C. Tazzara (University of Chicago Press, 2013 [orig. 2006]), 3; Alix Cooper, *Inventing the Indigenous: Local Knowledge and Natural History in Early Modern Europe* (Cambridge University Press, 2007), 3–7. For a study of the medieval period, Paul Freedman, *Out of the East: Spices and the Medieval Imagination* (Yale University Press, 2008).

8. This contrast draws on Stephen Greenblatt, *Marvelous Possessions: The Wonder of the New World* (University of Chicago Press, 1991); and Kapil Raj, *Relocating Modern Science: Circulation and the Construction of Knowledge in South Asia and Europe, 1650–1900* (Palgrave Macmillan, 2007). See also Alfred W. Crosby, *Ecological Imperialism: The Biological Expansion of Europe, 900–1900* (Cambridge University Press, 1986).

9. William Shakespeare, *The Tempest*, act 2, scene 2, lines 25–34.

10. See also Markku Hokkanen and Kalle Kananoja, "Healers and Empires in Global History: Healing as Hybrid and Contested Knowledge" in *Healers and Empires in Global History: Healing as Hybrid and Contested Knowledge*, ed. M. Hokkanen and K. Kananoja (Palgrave Macmillan, 2019), 9.

11. Daston and Park, *Wonders and the Order of Nature*, 181–82, 321, 361.

12. On medical knowledge production in the Jesuit order, see also Kroupa's contribution to this volume.

13. Hjalmar Fors, "Making Sense of the World: The Creation and Transfer of Knowledge," in *The Routledge Companion to Cultural History in the Western World, 1250 to the Present*, ed. Alessandro Arcangeli et al. (Routledge, 2020); Christine R. Johnson, *The German Discovery of the World: Renaissance Encounters with the Strange and Marvelous* (University of Virginia Press, 2008), 14–15.

14. Valentini, *Museum Museorum*; Willem Piso and Georg Marggraf, *Historia Naturalis Brasiliae: In qua nontantum Plantae et Animalia, sed et Inddigenarum supra quingentas illustratur. Ioannes de Laet, in ordinem digessit & annotationes addidit* (Leiden: Franciscus Hackius, 1648). See also Benjamin Schmidt, "Inventing Exoticism: The Project of Dutch Geography and the Marketing of the World, circa 1700," in *Merchants and Marvels: Commerce, Science, and Art in Early Modern Europe*, ed. Pamela Smith and Paula Findlen (Routledge, 2002).

15. Paula Findlen, "Early Modern Things: Objects in Motion, 1500–1800," in *Early Modern Things: Objects and Their Histories, 1500–1800*, ed. Paula Findlen (Routledge, 2013), 8–11. Paula Findlen, "Inventing Nature: Commerce, Art, and Science in the Early Modern Cabinet of Curiosities," in *Merchants and Marvels*. Hjalmar Fors, "Medicine and the Making of a City: Spaces of Pharmacy and Scholarly Medicine in Seventeenth-Century Stockholm," *Isis* 107, no. 3 (2016): 475–82.

16. As an adjective, the word has been used in this sense since the late sixteenth century: *Oxford English Dictionary*, 3rd ed., https://www.oed.com/dictionary/china_adj?tab=factsheet#1311792620.

17. Valentini, *Museum Museorum*, 1:169–70.

18. Valentini, *Museum Museorum*, 1:255–56.

19. Georgius Franke, *Georgi Franci... Lexicon Vegetabilium usualium*... (Argentorati: Josiae Staedeli, 1672), 25–26. This is a very slight duodecimo volume.

20. Heinrich von Bergen, *Versuch einer Monographie der China* (Hamburg: Hartwig & Müller, 1826), 78–82. See also Johann Hübner, *Curieuses und reales Natur-Kunst-Berg-Gewerck und Handlungs-Lexikon* (Sachsen: Joh. Friedr. Gleditschen sel. Sohn, 1731).

21. Freedman, *Out of the East*, 95–103.

22. "Note XIX. Account of Cathay by a Turkish dervish, as related to Auger Gislen de Busbeck (Circa 1560)," in *Cathay and the Way Thither: Being a Collection of Medieval Notices of China*, trans. And ed. Henry Yule (Hakluyt Society, 1915), 1:296–98. Quoted from *Busbequii Epistolae* (Amsterdam, 1661), 326–30.

23. "Note XVIII. Hajji Mahomed's account of Cathay, as delivered to Messer Giov. Battista Ramusio (Circa 1550)," in *Cathay and the Way Thither*, 1:290–96.

24. On the importance of medical use of Chinese rhubarb root in Europe, see Clifford M. Foust, *Rhubarb: The Wondrous Drug* (Princeton University Press, 1992); Erika Monahan, "Locating Rhubarb: Early Modernity's Relevant Obscurity," in *Early Modern Things*.

25. "Note XVIII. Hajji Mahomed's account of Cathay," in *Cathay and the Way Thither*, 1:291. The passage concerns the land of Campion and the city of Succuir. According to Stephen G. Haw, Succuir is Suzhou, now Jiuquan, and Kampion is Ganzhou, now Zhangye. See Haw, *Marco Polo's China: A Venetian in the Realm of Khubilai Khan* (Routledge, 2006), 89–90.

26. "Note XVIII. Hajji Mahomed's account of Cathay," in *Cathay and the Way Thither*, 1:292. Mehmet's account is a clear exaggeration. Although never as popular in China as in Europe, rhubarb, or Da Huang was, and still is, used and appreciated in traditional Chinese medicine (Haw, *Marco Polo's China*, 124–26).

27. Matteo Ricci, *China in the Sixteenth Century: The Journals of Matthew Ricci, 1583–1610*, trans. Louis J. Gallagher, S. J. (Random House, 1953), 16.

28. See further Carla Nappi, "Surface Tension: Objectifying Ginseng in Chinese Early Modernity," in Findlen, *Early Modern Things*.

29. Of course, only truly sought-after objects were worth the immense effort involved in long-distance trade during the period.

30. I am grateful to Simon Schaffer for pointing out the usefulness of this concept for the study of global circulation of botanical objects. For a more typical (and classical) usage of the concept, see, e.g., Harry Collins, "Son of Seven Sexes: The Social Destruction of a Physical Phenomenon," *Social Studies of Science* 11, no. 1 (1981): 33–62.

31. See Daston and Park, who observe that the initial phase of European expansion, accompanied by increased enchantment, eventually gave way to commodification. *Wonders and the Order of Nature*, 166, 172, 219, 231, 309–10.

32. As the attentive reader may have noticed, the above analysis proceeds from that of Michel Foucault, *The Order of Things: An Archaeology of the Human Sciences* (Routledge, 2002).

33. Said, *Orientalism*. I used the Swedish translation: Edward W. Said, *Orientalism*, trans. Hans O. Sjöström (Stockholm: Ordfront, 1993).

34. See also Clare Griffin's contribution to this volume.

35. Keller and Irigoyen-Garcia, "Introduction," 2–4.

36. Said, *Orientalism* (Swedish ed.), 65, 70–71.

37. Schaffer observes that "the character of orientalist accounts was . . . closely connected with the properties of trade goods" ("The Poisoner's Dilemma" (unpublished manuscript), 6.) See also Samuel Thévoz, "The French for Shangri-La: Tibetan Landscape and French Explorers," *French Cultural Studies* 25, no. 2 (2014): 108.

38. Londa Schiebinger, *Plants and Empire: Colonial Bioprospecting in the Atlantic World* (Harvard University Press, 2004), 3, 5, 18.

39. Robert N. Proctor, "Agnotology: A Missing Term to Describe the Cultural Production of Ignorance (and its Study)," in *Agnotology: The Making and Unmaking of Ignorance*, ed. Robert N. Proctor and Londa Schiebinger (Stanford University Press, 2008), 27.

40. See, e.g., Sujit Sivasundaram, "Sciences and the Global: On Methods, Questions, and Theory," *Isis* 101, no. 1 (2010): 146–58. See also Qiong Zhang, "About God, Demons, and Miracles: The Jesuit Discourse on the Supernatural in Late Ming China," *Early Science and Medicine* 4, no. 1 (1999): 8–9, 29, 33–34.

SELECTED BIBLIOGRAPHY OF SECONDARY SOURCES

This bibliography presents a selection of significant studies on the history of drugs, with an emphasis on those published in recent years and key earlier studies, with a view to allowing the reader to explore this subject in more depth. A complete bibliography is available online on the book's webpage at www.upittpress.org. The sources below range over many aspects of the history of drugs, from consumption and procurement to trade, scientific and medical study, cultivation, nomenclature, and networks of exchange.

Ago, Renata. *Gusto for Things: A History of Objects in Seventeenth-Century Rome*. University of Chicago Press, 2013.

Amar, Zohar. *Arabian Drugs in Medieval Mediterranean Medicine*. University of Edinburgh Press, 2018.

Anagnostou, Sabine, Florike Egmond, and Christoph Friedrich, eds. *A Passion for Plants: Materia Medica and Botany in Scientific Networks from the 16th to the 18th Centuries*. Wissenschaftliche Verlagsgesellschaft, 2011.

Anagnostou, Sabine. "Jesuit Missionaries in Spanish America and the Transfer of Medical–Pharmaceutical Knowledge." *Archives internationales d'histoire des sciences* 52, no. 148 (2002): 176–97.

Anagnostou, Sabine. "Jesuits in Spanish America: Contributions to the Exploration of the American Materia Medica." *Pharmacy in History* 47, no. 1 (2005): 3–17.

Anderson, Stuart. *Pharmacopoeias, Drug Regulation, and Empires: Making Medicines Official in Britain's Imperial World, 1618–1968*. McGill-Queen's University Press, 2024.

SELECTED BIBLIOGRAPHY

Baldassarri, Fabrizio, ed. *Plants in 16th and 17th Century: Botany Between Medicine and Science*. De Gruyter, 2023.

Barrera Osorio, Antonio. *Experiencing Nature: The Spanish American Empire and the Early Scientific Revolution*. University of Texas Press, 2006.

Benedict, Carol. *Golden-Silk Smoke: A History of Tobacco in China, 1550–2010*. University of California Press, 2011.

Bénézet, Jean-Pierre. *Pharmacie et médicament en Méditerranée occidentale (XIIIe-XVIe siècles)*. H. Champion; Slatkine, 1999.

Bian, He. *Know Your Remedies: Pharmacy and Culture in Early Modern China*. Princeton University Press, 2020.

Bleichmar, Daniela. "Books, Bodies, and Fields: Sixteenth-Century Transatlantic Encounters with New World Materia Medica." In *Colonial Botany: Science, Commerce, and Politics in the Early Modern World*, edited by Londa Schiebinger and Claudia Swan. University of Pennsylvania Press, 2005.

Boumediene, Samir, and Valentina Pugliano. "La Route des succédanés. Les remèdes exotiques, l'innovation médicale et le marché des substituts au XVIe siècle." *Revue d'histoire moderne contemporaine* 66, no. 3 (2019): 24–54.

Boumediene, Samir. *La Colonisation du savoir. Une histoire des plantes médicinales du "Nouveau Monde" (1492–1750)*. Les Éditions des Mondes à Faire, 2016.

Bouras-Vallianatos, Petros, and Dionysios Ch. Stathakopoulos, eds. *Drugs in the Medieval Mediterranean: Transmission and Circulation of Pharmacological Knowledge*. Cambridge University Press, 2024.

Breen, Benjamin. *The Age of Intoxication: Origins of the Global Drug Trade*. University of Pennsylvania Press, 2019.

Breen, Benjamin. "The Failed Globalization of Psychedelic Drugs in the Early Modern World." *Historical Journal* 65, no. 1 (2022): 12–29.

Breen, Benjamin. "Portugal, Early Modern Globalization and the Origins of the Global Drug Trade." *Perspectives on Europe* 42, no. 1 (2012): 84–88.

Brockliss, Laurence, and Colin Jones. *The Medical World of Early Modern France*. Clarendon Press, 1997.

Brohard, Yvan. *Remèdes, onguents, poisons. Une histoire de la pharmacie*. Université Paris Descartes; Éditions de la Martinière, 2012.

Carney, Judith, and Richard Nicholas Rosomoff. *In the Shadow of Slavery: Africa's Botanical Legacy in the Atlantic World*. University of California Press, 2010.

Chakrabarti, Pratik. "Empire and Alternatives: *Swietenia febrifuga* and the Cinchona Substitutes." *Medical History* 54, no. 1 (2010): 75–94.

Chakrabarti, Pratik. *Materials and Medicine: Trade, Conquest, and Therapeutics in the Eighteenth Century*. Manchester University Press, 2010.

Charles, Loïc, and Paul Cheney. "The Colonial Machine Dismantled: Knowledge and Empire in the French Atlantic." *Past & Present* 219, no. 1 (2013): 127–63.

Chipman, Leigh. *The World of Pharmacy and Pharmacists in Mamluk Cairo*. Brill, 2010.

Cook, Alexandra. *Jean-Jacques Rousseau and Botany: The Salutary Science*. Voltaire Foundation, 2012.

Cook, Harold J. "Good Advice and Little Medicine: The Professional Authority of Early Modern English Physicians." *Journal of British Studies* 33, no. 1 (1994): 1–31.

Cook, Harold J. "Markets and Cultures: Medical Specifics and the Reconfiguration of the Body in Early Modern Europe." *Transactions of the Royal Historical Society* 21 (2011): 123–45.

Cook, Harold J. *Matters of Exchange: Commerce, Medicine, and Science in the Dutch Golden Age*. Yale University Press, 2007.

Cook, Harold J., ed. *Translation at Work: Chinese Medicine in the First Global Age*. Clio Medica: Studies in the History of Medicine and Health, vol. 100. Brill, 2020.

Cook, Harold J., and Timothy D. Walker. "Circulation of Medicine in the Early Modern Atlantic World." *Social History of Medicine* 26, no. 3 (2013): 337–51.

Cooper, Alix. *Inventing the Indigenous: Local Knowledge and Natural History in Early Modern Europe*. Cambridge University Press, 2007.

Courtwright, David T. *Forces of Habit: Drugs and the Making of the Modern World*. Harvard University Press, 2001.

Crawford, Matthew J. "An Empire's Extract: Chemical Manipulations of Cinchona Bark in the Eighteenth-century Spanish Atlantic World." *Osiris* 29, no. 1 (2014): 215–29.

Crawford, Matthew James. *The Andean Wonder Drug: Cinchona Bark and Imperial Science in the Spanish Atlantic, 1630–1800*. University of Pittsburgh Press, 2016.

Crawford, Matthew James, and Joseph M. Gabriel, eds. *Drugs on the Page: Pharmacopoeias and Healing Knowledge in the Early Modern Atlantic World*. University of Pittsburgh Press, 2019.

Crosby, Alfred W. *The Columbian Exchange: Biological and Cultural Consequences of 1492*. Greenwood, 1972.

Curry, Helen Anne, Nicholas Jardine, J. A. Secord, and E. C. Spary, eds. *Worlds of Natural History*. Cambridge University Press, 2018.

Curth, Louise Hill. "Medical Advertising in the Popular Press." *Pharmacy in History* 50, no. 1 (2008): 3–16.

Curth, Louise Hill, ed. *From Physick to Pharmacology: Five Hundred Years of British Drug Retailing*. Ashgate, 2006.

Daston, Lorraine, and Katharine Park. *Wonders and the Order of Nature, 1150–1750*. Zone Books, 1998.

SELECTED BIBLIOGRAPHY

Delbourgo, James, and Nicholas Dew, eds., *Science and Empire in the Atlantic World*. Routledge, 2008.

DeVos, Paula. *Compound Remedies: Galenic Pharmacy from the Ancient Mediterranean to New Spain*. Pittsburgh University Press, 2020.

DeVos, Paula. "The Past and Future of Early Modern Pharmacy History." *Pharmacy in History* 61, nos. 3–4 (2019): 154–59.

DeVos, Paula. "'The Prince of Medicine:' Yuhanna Ibn Masawaih and the Foundations of the Western Pharmaceutical Tradition." *Isis* 104, no. 4 (2013): 667–712.

Di Gennaro Splendore, Barbara. "The Triumph of Theriac." *Nuncius: Journal of the Material and Visual History of Science* 36, no. 2 (2021): 431–70.

DiMeo, Michelle, and Sara Pennell, eds. *Reading and Writing Recipe Books, 1550–1800*. Manchester University Press, 2013.

Dorner, Zachary. *Merchants of Medicines. The Commerce and Coercion of Health in Britain's Long Eighteenth Century*. University of Chicago Press, 2020.

Easterby-Smith, Sarah. "Recalcitrant Seeds: Material Culture and the Global History of Science." *Past & Present*, suppl. 14 (2019): 215–42.

Egmond, Florike. *The World of Carolus Clusius: Natural History in the Making, 1550–1610*. Routledge, 2015.

Ellis, Markman, Richard Coulton, and Matthew Mauger. *Empire of Tea: The Asian Leaf That Conquered the World*. Reaktion, 2015.

Estes, J. Worth. "The European Reception of the First Drugs From the New World." *Pharmacy in History* 37, no. 1 (1995): 3–23.

Fernández Armesto, Felipe, and Benjamin Sacks. "The Global Exchange of Food and Drugs." In *The Oxford Handbook of the History of Consumption*, edited by Frank Tenenbaum. Oxford University Press, 2012.

Findlen Paula, *Possessing Nature: Museums, Collecting, and Scientific Culture in Early Modern Italy*. University of California Press, 1994.

Findlen, Paula, ed., *Early Modern Things: Objects and Their Histories, 1500–1800*. 2nd ed. Routledge, 2021.

Flückiger, Friedrich A., and Daniel Hanbury. *Pharmacographia. A History of the Principal Drugs of Vegetable Origin, Met With in Great Britain and British India*. 2nd ed. Macmillan and Co., 1879.

Fors, Hjalmar. "Making Sense of the World: The Creation and Transfer of Knowledge." In *The Routledge Companion to Cultural History in the Western World, 1250 to the Present*, edited by Alessandro Arcangeli, Jörg Rogge, and Hannu Salmi. Routledge, 2020.

Fors, Hjalmar. "Medicine and the Making of a City: Spaces of Pharmacy and Scholarly Medicine in Seventeenth-Century Stockholm." *Isis* 107, no. 3 (2016): 473–94.

Foust, Clifford M. *Rhubarb: The Wondrous Drug*. Princeton University Press, 1992.

SELECTED BIBLIOGRAPHY

Freedman, Paul. *Out of the East: Spices and the Medieval Imagination.* Yale University Press, 2008.

Gänger, Stefanie. "World Trade in Medicinal Plants from Spanish America, 1717–1815." *Medical History* 59, no. 1 (2015): 44–62.

Gänger, Stefanie. *A Singular Remedy: Cinchona Across the Atlantic World, 1751–1820.* Cambridge University Press, 2020.

Gerritsen, Anne, and Giorgio Riello, eds. *The Global Lives of Things.* Routledge, 2016.

Giovannetti-Singh, Gianamar. "Galenizing the New World: Joseph-François Lafitau's 'Galenization' of Canadian Ginseng, ca. 1716–1724." *Notes and Records of the Royal Society* 75, no. 1 (2021): 59–72.

Gómez, Pablo F. *The Experiential Caribbean: Creating Knowledge and Healing in the Early Modern Atlantic.* University of North Carolina Press, 2017.

Goodman, Jordan, Paul E. Lovejoy, and Andrew Sherratt, eds. *Consuming Habits: Global and Historical Perspectives on How Cultures Define Drugs.* Routledge, 2007.

Grafton, Anthony, April Shelford, and Nancy G. Siraisi. *New Worlds, Ancient Texts: The Power of Tradition and the Shock of Discovery.* Belknap Press, 1992.

Grell, Ole Peter, Andrew Cunningham, and Jon Arrizabalaga, eds. *"It All Depends on the Dose": Poisons and Medicines in European History.* Routledge, 2018.

Griffin, Clare. "Disentangling Commodity Histories: *Pauame* and Sassafras in the Early Modern Global World." *Journal of Global History* 15, no. 1 (2020): 1–18.

Griffin, Clare. *Mixing Medicines: The Global Drug Trade and Early Modern Russia.* McGill-Queen's University Press, 2022.

Griffin, Clare. "Russia and the Medical Drug Trade in the Seventeenth Century." *Social History of Medicine* 31, no. 1 (2016): 2–23.

Gürbüzel, Aslıhan. "From New Spain to Damascus: Ottoman Religious Authorities and the Making of Medical Knowledge on Tobacco." *Early Science and Medicine* 26, nos. 5–6 (2021): 561–81.

Harris, Steven J. "Long-Distance Corporations, Big Sciences, and the Geography of Knowledge." *Configurations* 6, no. 2 (1998): 269–304.

Harris, Steven J. "Jesuit Scientific Activity in the Overseas Missions, 1540–1773." *Isis* 96, no. 1 (2005): 71–79.

Hausse, Heidi. "European Theories and Local Therapies: Mordexi and Galenism in the East Indies, 1500–1700." *Journal of Early Modern History* 18, nos. 1–2 (2014): 121–40.

Hendriksen, Marieke M. A. "Boerhaave's Mineral Chemistry and Its Influence on Eighteenth-Century Pharmacy in the Netherlands and England." *Ambix* 65, no. 4 (2018): 1–21.

Higby, Gregory, Elaine Condouris Stroud, and David L. Cowen, eds. *The History of Pharmacy: A Selected Annotated Bibliography.* Garland, 1995.

SELECTED BIBLIOGRAPHY

Huguet-Termes, Teresa. "New World Materia Medica in Spanish Renaissance Medicine: From Scholarly Reception to Practical Impact." *Medical History* 45, no. 3 (2001): 359–76.

Jarcho, Saul. *Quinine's Predecessor: Francesco Torti and the Early History of Cinchona.* Johns Hopkins University Press, 1993.

Jardine, N., J. A. Secord, and E. C. Spary, eds. *Cultures of Natural History.* Cambridge University Press, 1996.

Jenner, Mark, and Patrick Wallis, eds. *Medicine and the Market in England and its Colonies, c. 1450–c. 1850.* Palgrave, 2007.

Klein, Ursula, and E. C. Spary, eds. *Materials and Expertise in Early Modern Europe: Between Market and Laboratory.* University of Chicago Press, 2010.

Klein, Wouter, and Toine Pieters. "The Hidden History of a Famous Drug: Tracing the Medical and Public Acculturation of Peruvian Bark in Early Modern Western Europe (c. 1650–1720)." *Journal of the History of Medicine and Allied Sciences* 71, no. 4 (2016): 400–421.

Klein, Wouter. "New Drugs for the Dutch Republic: The Commodification of Fever Remedies in the Netherlands (c. 1650–1800)." PhD diss., Utrecht University, 2018.

Klerk, Saskia. "The Trouble with Opium. Taste, Reason and Experience in Late Galenic Pharmacology with Special Regard to the University of Leiden (1575–1625)." *Early Science and Medicine* 19, no. 4 (2014): 287–316.

Knight, Leah. *Of Books and Botany in Early Modern England: Sixteenth-Century Plants and Print Culture.* Routledge, 2016.

Kopytoff, Igor. "The Cultural Biography of Things: Commoditization as Process." In *The Social Life of Things: Commodities in Cultural Perspective*, edited by Arjun Appadurai. Cambridge University Press, 2014.

Koroloff, Rachel. "Travniki, Travniki, and Travniki: Herbals, Herbalists and Herbaria in Seventeenth- and Eighteenth-century Russia." *Vivliofika: E–Journal of Eighteenth-Century Russian Studies* 6 (2018): 58–76.

Kremers, Edward, and George Urdang. *Kremers and Urdang's History of Pharmacy*, edited by Glenn Sonnedecker. 4th ed. Lippincott, 1976.

Kroupa, Sebastian. "Georg Joseph Kamel (1661–1706): A Jesuit Pharmacist at the Frontiers of Colonial Empires." PhD diss., University of Cambridge, 2019.

Küçük, Harun K. *Science without Leisure: Practical Naturalism in Istanbul, 1660–1732.* University of Pittsburgh Press, 2019.

Landweber, Julia. "'This Marvelous Bean': Adopting Coffee into Old Regime French Culture and Diet." *French Historical Studies* 38, no. 2 (2015): 193–223.

Leong, Elaine. *Recipes and Everyday Knowledge: Medicine, Science, and the Household in Early Modern England.* University of Chicago Press, 2019.

SELECTED BIBLIOGRAPHY

Leong, Elaine, and Alisha Rankin, eds. *Secrets and Knowledge in Medicine and Science, 1500–1800*. Ashgate, 2011.

Leong, Elaine, and Alisha Rankin. "Testing Drugs and Trying Cures: Experiment and Medicine in Medieval and Early Modern Europe." *Bulletin of the History of Medicine* 91, no. 2 (2017): 157–82.

Lev, Efraim, and Zohar Amar. *Practical Materia Medica of the Medieval Eastern Mediterranean According to the Cairo Genizah*. Brill, 2007.

Lu, Di. *The Global Circulation of Chinese Materia Medica, 1700–1949: A Microhistory of the Caterpillar Fungus*. Palgrave Macmillan, 2023.

Maehle, Andreas-Holger. *Drugs on Trial: Experimental Pharmacology and Therapeutic Innovation in the Eighteenth Century*. Rodopi, 1999.

Maydom, Katrina. "New World Drugs in England's Early Empire." PhD diss., University of Cambridge, 2019.

McCants, Anne E. "Exotic Goods, Popular Consumption, and the Standard of Living: Thinking About Globalization in the Early Modern World." *Journal of World History* 18, no. 4 (2007): 433–62.

McCants, Anne E. "Poor Consumers as Global Consumers: The Diffusion of Tea and Coffee Drinking in the Eighteenth Century." *Economic History Review* 61, no. S1 (2013): 172–200.

McVaugh, Michael R. "Determining a Drug's Properties: Medieval Experimental Protocols." *Bulletin of the History of Medicine* 91, no. 2 (2017): 183–209.

Miller, James Innes. *The Spice Trade of the Roman Empire*. Clarendon Press, 1969.

Mintz, Sidney W. *Sweetness and Power: The Place of Sugar in Modern History*. Elisabeth Sifton Books, 1985.

Müller-Wille, Staffan. "Nature as a Marketplace: The Political Economy of Linnaean Botany." *History of Political Economy* 35, no. 5 (2003): 154–72.

Mullini, Roberta. *Healing Words: The Printed Handbills of Early Modern London Quacks*. Peter Lang, 2015.

Nappi, Carla. *The Monkey and the Inkpot: Natural History and its Transformation in Early Modern China*. Harvard University Press, 2009.

Newson, Linda A. *Making Medicines in Early Colonial Lima, Peru: Apothecaries, Science and Society*. Brill, 2017.

Norton, Marcy. *Sacred Gifts, Profane Pleasures: A History of Tobacco and Chocolate in the Atlantic World*. Cornell University Press, 2008.

Nutton, Vivian. "Ancient Mediterranean Pharmacology and Cultural Transfer." *European Review* 16, no. 2 (2008): 211–17.

Ogilvie, Brian W. *The Science of Describing: Natural History in Renaissance Europe*. University of Chicago Press, 2008.

Palmer, Richard. "Pharmacy in the Republic of Venice in the Sixteenth Century." In

SELECTED BIBLIOGRAPHY

The Medical Renaissance of the Sixteenth Century, edited by Andrew Wear, R. K. French, and Iain M. Lonie. Cambridge University Press, 1985.

Parsons, Christopher M. *A Not-So-New World: Empire and Environment in French Colonial North America*. University of Pennsylvania Press, 2018.

Pépy, Émilie-Anne. "Les Femmes et les plantes: Accès négocié à la botanique savante et résistance des savoirs vernaculaires (France, XVIIIe siècle)." *Genre & Histoire* 22 (2018), online at: http://journals.openedition.org/genrehistoire/3654, accessed March 24, 2023.

Porter, Roy, and Mikuláš Teich, eds. *Drugs and Narcotics in History*. Cambridge University Press, 1997.

Pugliano, Valentina. "Pharmacy, Testing, and the Language of Truth in Renaissance Italy." *Bulletin of the History of Medicine* 91, no. 2 (2017): 233–73.

Raj, Kapil. *Relocating Modern Science: Circulation and the Construction of Knowledge in South Asia and Europe, 1650–1900*. Palgrave Macmillan, 2007.

Rampling, Jennifer M. *The Experimental Fire: Inventing English Alchemy, 1300–1700*. University of Chicago Press, 2020.

Rankin, Alisha. *The Poison Trials: Antidotes, Wonder Drugs, and the Problem of Proof in Early Modern Europe*. University of Chicago Press, 2021.

Riddle, John M. *Goddesses, Elixirs, and Witches: Plants and Sexuality throughout Human History*. Palgrave Macmillan, 2010.

Riddle, John. *Dioscorides on Pharmacy and Medicine*. University of Texas Press, 1985.

Rivest, Justin. "Testing Drugs and Attesting Cures: Pharmaceutical Monopolies and Military Contracts in Eighteenth-Century France." *Bulletin of the History of Medicine* 91, no. 2 (2017): 362–90.

Romaniello, Matthew P. "True Rhubarb? Trading Eurasian Botanical and Medical Knowledge in the Eighteenth Century." *Journal of Global History* 11 (2016): 3–23.

Romaniello, Matthew P. "Who Should Smoke? Tobacco and the Humoral Body in Early Modern England." *The Social History of Alcohol and Drugs* 27, no. 2 (2013): 156–73.

Schiebinger, Londa. *Plants and Empire: Colonial Bioprospecting in the Atlantic World*. Cambridge, Mass.: Harvard University Press, 2004.

Schiebinger, Londa. *Secret Cures of Slaves: People, Plants, and Medicine in the Eighteenth-Century Atlantic World*. Stanford University Press, 2017.

Schiebinger, Londa, and Claudia Swan, eds. *Colonial Botany: Science, Commerce, and Politics in the Early Modern World*. University of Pennsylvania Press, 2005.

Schivelbusch, Wolfgang. *Tastes of Paradise: A Social History of Spices, Stimulants, and Intoxicants*. Pantheon Books, 1992.

Schmidt, Benjamin. *Inventing Exoticism: Geography, Globalism, and Europe's Early Modern World*. University of Pennsylvania Press, 2015.

SELECTED BIBLIOGRAPHY

Sismondo, Sergio, and Jeremy A. Greene. *The Pharmaceutical Studies Reader*. Wiley-Blackwell, 2015.

Spary, E. C. "Publishing Virtue: Medical Entrepreneurship and Reputation in the Republic of Letters." *Centaurus* 62, no. 3 (2020): 498–521.

Teigen, Philip M. "Taste and Quality in 15th- and 16th-century Galenic Pharmacology." *Pharmacy in History* 29, no. 2 (1987): 60–68.

Van Arsdall, Anne, and Timothy Graham, eds. *Herbs and Healers from the Ancient Mediterranean Through the Medieval West: Essays in Honor of John M. Riddle*. Ashgate, 2012.

Walker, Timothy D. "Acquisition and Circulation of Medical Knowledge Within the Early Modern Portuguese Colonial Empire." In *Science in the Spanish and Portuguese Empires, 1500–1800*, edited by Daniela Bleichmar, Paula DeVos, Kristin Huffine, and Kevin Sheehan, 247–70. Stanford University Press, 2009.

Walker, Timothy D. "The Medicines Trade in the Portuguese Atlantic World: Acquisition and Dissemination of Healing Knowledge from Brazil (c. 1580–1800)." *Social History of Medicine* 26, no. 3 (2013): 403–31.

Wall, Wendy. *Recipes for Thought: Food and Taste in the Early Modern English Kitchen*. University of Pennsylvania Press, 2015.

Wallis, Patrick. "Consumption, Retailing, and Medicine in Early-Modern London." *Economic History Review* 61, no. 1 (2008): 26–53.

Wallis, Patrick. "Exotic Drugs and English Medicine: England's Drug Trade, c. 1550–c. 1800." *Social History of Medicine* 25, no. 1 (2012): 20–46.

Watson, Gilbert. *Theriac and Mithridatium: A Study in Therapeutics*. Wellcome Historical Medical Library, 1966.

Wear, Andrew. *Knowledge and Practice in English Medicine, 1550–1680*. Cambridge University Press, 2000.

Wimmler, Jutta. *The Sun King's Atlantic: Drugs, Demons and Dyestuffs in the Atlantic World, 1640–1730*. Brill, 2017.

Winterbottom, Anna E. "Of the China Root: A Case Study of the Early Modern Circulation of Materia Medica." *Social History of Medicine* 28, no. 1 (2015): 22–44.

Winterbottom, Anna. *Hybrid Knowledge in the Early East India World*. Palgrave Macmillan, 2016.

Withington, Philip. "Where Was the Coffee in Early Modern England?" *Journal of Modern History* 92, no. 1 (2020): 40–75.

Zebroski, Bob. *Brief History of Pharmacy: Humanity's Search for Wellness*. Routledge, 2016.

LIST OF CONTRIBUTORS

SAMIR BOUMEDIENE is a researcher at the Institut d'histoire des représentations et des idées dans les modernités, Centre National de la Recherche Scientifique (Lyon) and has previously held postdoctoral positions at the Max Planck Institute (Berlin) and the University of Cambridge. His PhD, completed at the University of Lorraine and Casa de Velázquez (Madrid), was devoted to the history of medicinal plants in the New World. In revised form, it was published in 2016 under the title *La Colonisation du savoir*. He has subsequently published several articles on the history of drugs, medicine and plants. His current research, begun at the Villa Medici and the Dutch Institute in Rome, focuses on the notion of discovery in the early modern period and on the history of questionnaires.

ALEXANDRA COOK is honorary professor of philosophy at the University of Hong Kong and a fellow of the Linnean Society of London. She was also a fellow of the Institute of Advanced Study, Nantes (2018–2019) and the Historisches Kolleg, Munich (2014). Her books include *Jean-Jacques Rousseau and Botany, the Salutary Science* (Voltaire Foundation, 2012), awarded the John Thackray Medal of the Society for the History of Natural History (London), and a translation of Rousseau's botanical writings in *Collected Writings of Rousseau*, vol. 8 (Dartmouth, 2000). Cook's current project is *Bioprospecting Asia: Laurent Garcin and the Circulation of Scientific Knowledge*.

CONTRIBUTORS

PAULA DE VOS is professor of history at San Diego State University. She has written articles for the *Journal of World History, Colonial Latin American Review, Eighteenth-Century Studies, Journal of Interdisciplinary History, Journal of Ethnopharmacology*, and *Isis*, among others. She also served as coeditor of *Science in the Spanish and Portuguese Empires* (Stanford, 2009) and is author of *Compound Remedies: Galenic Pharmacy from the Ancient Mediterranean to New Spain* (Pittsburgh, 2020), recipient of the 2022 Edward Kremers Award from the American Institute for the History of Pharmacy. She is also coeditor-in-chief of the journal *History of Pharmacy and Pharmaceuticals*.

BARBARA DI GENNARO SPLENDORE is research fellow at the Italian Council for Agricultural Research and Economics in Padua. Her research focuses on the history of early modern science, medicine, and pharmacy. She has been awarded a Marie Skłodowska-Curie Global Fellowship at the Universities of Lausanne and Bologna for a project on the knowledge of mushrooms and history of mycology in early modern Italy. Her first book, *The State Drug: Theriac Pharmacy and Politics in Early Modern Italy* (Harvard University Press, 2025) discusses the age-old intersection of power and pharmacy.

HJALMAR FORS heads the Hagströmer Library at Karolinska Institutet, Stockholm, Sweden. He is a PhD (2003) and docent (2014) in the history of science, with both degrees conferred by Uppsala Universitet. His primary research expertise concerns the history of empirical sciences, that is, alchemy, chemistry, mining knowledge, natural history, medicine, and pharmacy, during the seventeenth and eighteenth centuries. He collaborates with natural scientists carrying out reproductions of scientific experiments and craft practices (experimental history of science). Among his several publications can be mentioned the acclaimed monograph *The Limits of Matter: Chemistry, Mining and Enlightenment*, published by the University of Chicago Press in 2015.

CLARE GRIFFIN is an associate professor at Indiana University Bloomington and the author of *Mixing Medicines: The Global Drug Trade and Early Modern Russia* (McGill-Queen's University Press, 2022). She is currently working on her second monograph, which explores the embodied experiences of soldiers in the context of Russian colonialism from the seventeenth century to the present day. She has published in a range of venues, from the pop culture site *Pajiba* to literary outlets like *Wordgathering: A Journal of Disability Poetry and Literature*, academic blogs including *All of Us: The Disability History*

CONTRIBUTORS

Association Blog, as well as academic journals such as *Social History of Medicine* and *Chronotopos—A Journal of Translation History*.

WOUTER KLEIN earned his PhD in the history of pharmacy from Utrecht University in 2018. His research interests include the trajectories of new therapeutic substances; the dynamics of commodification of exotic natural products; and the global trade in medicinal goods and knowledge in the early modern period. Based on thousands of eighteenth-century newspaper ads for secret remedies and commercial transactions in pharmaceutical substances, his PhD research focuses on what can be termed the "first golden age" of medical advertising, and claims that newspaper readers had their own "pharmaceutical literacy," related to their medical world of largely plant-based medicine.

SEBASTIAN KROUPA is an assistant professor in the Department of Social Studies of Medicine at McGill University. His research focuses on early modern histories of medical and natural knowledge in the Indian and Pacific Oceans. Sebastian has published on topics including Indigenous tattooing in the Philippines, networks of knowledge exchange, and Renaissance conceptions of monstrosity. He has coedited a special issue on "Science and Islands in Indo-Pacific Worlds" (2018). Currently, Sebastian is working on a monograph that uses the Philippines as a case study to examine how knowledge was produced at the frontiers of empires by diverse, transcultural agents and communicated globally.

KATRINA MAYDOM completed her PhD in the history and philosophy of science at the University of Cambridge, funded by the Wellcome Trust. Her thesis investigated the production, trade, and consumption of New World drugs in the early English empire. It won the Kenneth Emsley Prize for History from St Edmund's College. She was also awarded the Burnby Memorial Prize by the British Society for the History of Pharmacy. Her published work has explored the use of American drugs by the London apothecary James Petiver, and the interpretation of early modern apothecary prescriptions.

JUSTIN RIVEST is an assistant professor at Kenyon College in Gambier, Ohio. His work focuses on the history of early pharmaceutical monopolies, particularly in France. He is completing a book on early pharmaceutical monopolists and their role in supplying standardized drugs to large scale consumers such as the French army, navy, overseas trading companies, and missionary societies

CONTRIBUTORS

circa 1670–1750. His work has been published in the *Bulletin of the History of Medicine*, *The Canadian Journal of History*, *Ambix*, *Early Science and Medicine*, and *The New England Journal of Medicine*.

E. C. SPARY is a professor in the history of modern knowledge at the University of Cambridge, UK. Her monographs include *Utopia's Garden* (University of Chicago Press, 2000), on French natural history in the late eighteenth century, *Eating the Enlightenment* (University of Chicago Press, 2012), on the relationship between food and reason in eighteenth-century Paris, and *Feeding France* (Cambridge University Press, 2014), on the beginnings of industrial food production in France. She has published and researched on the history of the sciences and medicine in the long eighteenth century, in France and the wider world. Her current research addresses the introduction of exotic plants to French consumption in the decades around 1700.

INDEX

Abulcasis. *See* al-Zahrāwī
Academies: French, 215–16, 226, 294, 329; Russian, 36, 38. *See also* Royal Society
account books, 81, 193–94, 198, 201
Acosta, José de, 187
Aetius of Amida, 158
Africa, 22, 45–49, 55, 61, 86, 97–98, 153–55, 157, 160, 162–63, 165–67, 169, 211–12, 216, 240, 247, 257, 264, 305
Africans, 47, 55
Ago, Renata, 8
agrimony, 149, 213, 215
Albarelli, 26, 28, 205–6, 220
alchemy, 15, 81, 142, 156–57, 159–60, 170–74, 177–79, 235, 237–38, 255, 311
Aldrovandi, Ulisse, 89–90, 289
aloe, 206, 246
aloe-wood, 221
Alpini, Prospero, 90–92
alum, 81

Amazon river, 196
amber, 192
Americas, 13–14, 33–34, 42–43, 52, 54–56, 59–60, 67–68, 70, 81–83, 108, 111, 131–32, 134–38, 140–41, 150–51, 153–55, 157, 161, 168–69, 183–86, 188–94, 198, 202, 211–12, 215–16, 219, 221, 223, 225, 228, 257–59, 261–62, 264, 273–74, 289, 299, 305; North, 5, 54, 135, 143, 151, 215, 219, 261; South, 52, 123, 168, 212, 221, 259, 262; Spanish, 61, 137, 151, 183, 187
Amerindians, 23, 44, 52, 61, 132, 142, 168, 183, 185–87, 191, 196, 247, 258, 260
Amsterdam, 41, 107, 109–10, 114–20, 122–25, 127, 129, 203, 240, 295
Andromachus the Elder, 20, 86, 91–93, 97, 99, 262
angelica, 132, 137–38, 151, 162, 165
animals, 9, 26, 36, 55, 72, 86, 100, 112, 157, 160–61, 165, 187, 189, 232, 248, 255, 259, 295, 311, 325

INDEX

anise, 54, 66, 149, 213, 221
annatto, 192
antidotes, 20, 26, 81, 85, 92, 96–97, 146, 166, 168, 232–33, 289, 311, 325. *See also* poisons
antimony, 26
antiquity, 93, 173, 177, 248, 258, 282, 288
apothecaries, 10, 16, 23–24, 26–28, 34, 37–38, 41, 47, 50, 52, 54–55, 58, 68, 76, 86–102, 106–7, 109–10, 117, 127, 141, 154, 156, 158, 163, 170–71, 188–89, 193, 197–99, 201–6, 209–13, 215–20, 222–24, 226–28, 230, 233, 240, 244–46, 248, 251, 259, 276, 280, 290, 317, 319–21
Arabian Peninsula, 47–49, 79, 83, 91–92, 100, 154, 160, 162, 164–67, 212
Arabs, 45, 47, 49, 51, 100, 253
aristolochia, 96, 221
Armenians, 11, 53
aromatics, 145, 161, 163–64, 167, 169, 244
arsenic, 237, 241, 246
arthritis, 163
asafoetida, 167, 310
Asia, 13, 33–34, 43–44, 53, 68, 83–84, 92, 97–98, 132, 137–38, 144, 153–54, 157, 160, 162–64, 167, 231, 233–34, 237–38, 240–42, 253, 257, 264, 274, 306, 320, 328; Central, 53, 154; East, 33–34, 43, 53, 157, 160, 162–63, 167, 212, 264; Middle East, 10, 22, 33, 155–56, 162–63, 165–66, 169, 306; Near East, 154, 156, 160, 306; South, 43, 160, 162–63, 167, 169, 233–37, 240, 243–44, 253; Southeast, 68, 92, 132, 137–38, 144, 154, 157, 160, 162–64, 167, 240–41, 274. *See also* East Indies; Levant; Orient
assemblages, 21, 29, 277
Atlantic, 13, 19, 21, 35, 84, 108, 153, 161, 191, 197, 294, 298, 305
auctions, 107–8, 114, 122–27, 295
Averroes. *See* Ibn Rushd

Avicenna. *See* Ibn Sīnā
Aztecs. *See* Mexica

Bacon, Francis, 143–44, 303
Bacon, Roger, 173
Bado, Sebastiano, 194, 196, 198, 201
balm or balsam: drug, 69, 82, 90, 93–95, 102, 169, 188, 191, 225, 290; plant, 69, 162, 215
Barberini family, 93–95
barks, 8, 17, 21, 35, 54–55, 69–70, 76, 85, 106–14, 117–29, 132, 142, 144, 146, 168, 184, 194, 196–202, 215, 221, 224–25, 247–48, 250, 255, 262–63, 267, 293–95, 297–98, 308, 312, 317, 325
Basel, 246, 331
Bassand, Joannes Baptista, 119–20, 129
Batavia, 97, 212, 240, 242. *See also* Java
Bauer, Ralph, 15
Bauhin, Caspar, 224, 233
bay laurel, 64, 162, 165
bdellium, 167–68
benzoin, 205, 309
betel, 9
betony, 149, 162, 165, 215
bezoars, 26, 141, 190–91, 198, 231–36, 252, 276, 325
bioprospecting, 6, 19–20, 57, 71, 76, 83, 87, 89, 97, 102
blood-letting: phlebotomy, 51–52, 62, 140
Boerhaave, Herman, 106, 119–21, 128–29, 240, 242, 297, 329
Bontius, Jacobus, 4, 328
borage, 69, 162, 213, 215
Borromeo, Carlo (Cardinal), 193
botanists, 39, 78, 223, 226, 240, 243, 321
botany, 14, 22, 36, 39, 51–52, 54, 86, 89, 93, 97, 112, 114, 196, 240–41, 243, 280, 288, 299, 335
Boulduc family, 111, 113, 215, 224, 230
Bourguet, Louis, 243
boxes, 70, 124, 190–91, 193, 198, 255, 317

352

INDEX

Boyle, Robert, 145–46, 251
Brazil, Brazilians, 51–52, 55, 96–97, 144, 186
Breen, Benjamin, 29, 247
bugloss, 213, 215
Burman family, 241–43, 328

cabinets. *See* collections
cacao, 82, 85, 123, 192, 226–28, 315
Cádiz, 13, 123
cafés: coffeehouses, 3, 8, 48, 324
calcination, 159, 172–73, 177
Caldera de Heredia, Gaspar, 196–98, 317
Caliphates, 158
camphor, 167, 223
cañafistula, 167
cannabis, 29
capers, 162, 165, 297
cardamom, 162, 164, 225
Caribbean (Antilles), 111–12, 212, 223, 241, 272, 320
cartography, 10, 36, 42, 44–45, 55, 257, 267; Godunov map, 39–42
cascarilla, 21, 69, 106–15, 119, 122–23, 125–29, 184, 191, 197, 294, 312. *See also* cinchona
cassia, 141, 162–63, 222–23, 308. *See also* cinnamon
cataracts, 165
Catesby, Mark, 111–12, 114
Catholicism, 23, 61, 166. *See also* Christianity; Jesuits; religion
Ceccarelli, Ippolito, 88–90
Celtic nard, 162–63
Ceylon. *See* Sri Lanka
chamomile, 65, 69, 162, 165, 215, 363
Champier, Symphorien, 85–86
charity, 23, 27, 61, 100, 188, 199
Charles II (King of England), 135
Charles de l'Écluse. *See* Clusius, Carolus
Charleton, Walter, 140
chaste berry, 70, 162, 165
chemistry, 21, 28, 97, 99, 110, 113, 139–41, 151, 173, 177, 206, 219, 252, 302, 306, 311

chest complaints, 147, 167–68, 304
chicory, 70, 162, 166, 213, 215, 218
Chifflet, Jean–Jacques, 201
chili pepper, 12–13, 81–82, 192
China, 41, 83, 154, 164, 212, 237–38, 260–64, 274. *See also* empires: Chinese
China root, 139–40, 143–45, 149, 213, 225, 261–62, 323
chocolate, 13, 84, 154, 191–94, 198, 203–4, 226–27, 324
Christianity, 10, 20, 49–50, 60–61, 79, 86, 93, 100, 166, 185–87, 247. *See also* Catholicism; Protestantism; religion
chymistry. *See* chemistry
cinchona (Peruvian bark, Jesuits' bark), 8, 13, 17, 21, 54, 69, 76, 81–82, 85, 106–15, 117–23, 125–29, 131, 133, 154, 195–96, 204, 225–26, 247, 262, 274, 293–95, 297–99, 303, 312, 319. *See also* cascarilla
cinnamon, 10, 34, 41, 50, 54, 68, 89, 109, 145, 162–64, 168, 192, 213, 219, 223, 248, 260, 298–99, 308. *See also* cassia
climate, 19, 23, 35, 44, 59, 61, 72, 80, 82–84, 143, 146–47, 164, 166, 269, 333
cloves, 53, 132, 136–37, 151, 167, 213, 221
Clusius, Carolus (Charles de l'Écluse), 9, 154, 228
Cobo, Bernabé, 187
cocoa. *See* cacao
coffee, 203–4, 222, 225–27, 324
collecting, 51, 223, 240–41, 244, 332
collections, 39, 54, 106, 154, 157, 161, 189, 206, 210, 239–41, 256, 259, 325
colleges, 74, 93–94, 96, 140, 185, 187–89, 192–93, 312; Jesuit, 73–74, 96, 185, 187–92, 193, 312; medical, 93–94, 96, 102, 140, 189–91
Collins, Samuel, 47–49, 51–52, 54–55
collyria, 174, 177–78
colonialism/colonies, 12–13, 19, 35, 44, 57–63, 68, 71–76, 79, 81–84, 89, 97,

INDEX

101–2, 132, 134, 136–38, 143, 146, 150–51, 257–58, 266–69, 273
Columbian Exchange, 84, 153, 157, 161, 168
commerce. *See* trade
commodities/commodification, 6, 10, 12, 14–15, 17–18, 20, 22–25, 35, 39, 41, 44, 48, 53–54, 56, 58, 76, 80, 82, 84–85, 101, 106–7, 114, 122, 129, 132, 134, 138, 142, 151–53, 163, 183–84, 189, 191, 193–94, 197–200, 218, 220–22, 224–26, 228, 231, 266, 335
confections, 174–75, 178, 219
constitutions, 19, 59–60, 72, 79–80, 148, 247. *See also* healing: Galenic
consumption, 3, 7, 9–12, 14–18, 20–21, 23, 25, 28–29, 44, 48–49, 57–60, 73, 79, 82–85, 87, 125, 142, 145, 150, 152, 203–5, 215, 225, 227, 259–60, 265–66
contraceptives, 167
contrayerva, 119, 168–69, 188
conversion (religious), 23, 78–79, 185–88
Cook, Harold J., 28
Cooper, Alix, 53–54, 89
copahu, 225
copaiba, 188
copal, 169, 311
copper, 249
coral, 223
correspondence, 4–5, 16, 26, 37, 88, 94–95, 119–20, 129, 185, 189–94, 198–99, 201, 223, 240, 243, 285, 298, 322
costus, 86, 162, 164
courts, 3–4, 6–7, 10, 16, 24, 37, 45, 47–48, 88, 95, 100, 119, 194, 226–27
Crawford, Matthew J., 17
cream of tartar, 81, 120
Croll, Oswald, 142
crotón, 108, 110–12, 312
Cromwell, Oliver, 135
cubebs, 68–69, 97, 167–68
cuahmochit (Mexican acacia, quamochil), 82–83, 287

Culpeper, Nicholas, 15, 53, 302–3
cultural transfer, 9, 29, 82, 88, 128, 156, 184, 236, 238. *See also* domestication; naturalization
cumin, 225
cure, 9, 15–16, 19, 22, 58–59, 72, 75, 83, 85, 120, 138, 140, 143, 146, 151, 166, 185–86, 196, 201, 224, 228, 244, 247, 253, 263, 267, 329
curiosity, 9, 12, 16, 18, 189, 192, 194, 228, 255, 258, 269
curiosities, 17, 22, 52, 154, 184, 189, 191–94, 198, 201–2, 228, 256, 258–59, 325. *See also* collections; specimens
Curth, Louise Hill, 17, 107
cuscuta, 223

Da Orta, García, 233, 325
Dal Pozzo, Cassiano, 95
Davies, Surekha, 44
De Monchy, Salomon, 121
De Renou, Jean, 139
De Vivo, Filippo, 95
decoction, 74, 120, 139, 142–44, 146, 149, 159, 170
Delamare, Nicolas, 205, 324
Delft, 34, 41, 115, 200, 296, 363
diet, 4, 59, 140, 142–45, 151, 200
dietetics. *See* regimen
Dioscorides, Pedanius, 74, 157–58, 160–61, 163–64, 309
discovery. *See* invention
diseases, 51–52, 59, 61, 82, 97, 115, 119–21, 128, 132–33, 138–52, 165, 174, 185, 188, 201, 234, 240, 247, 258, 264, 296–97, 301, 304. *See also specific diseases*
distillation, 74, 81, 97, 145–46, 159, 171–73, 177–78, 215, 217–19, 224, 228, 287
Dodoens, Rembrandt, 154
domestication: naturalization (of drugs or species), 8–9, 12–13, 15, 19–20, 22, 28–29, 44, 82–83, 85–87, 90, 95–97, 102, 112, 152, 155, 159–60, 162–63,

354

INDEX

166, 169, 178, 308. *See also* cultural transfer

drugs, 4, 20, 54, 100, 128, 290; adoption, 7, 14–16, 21, 23, 29, 51, 76, 79–81, 86, 105, 132, 157, 161, 168, 194, 274, 309; advertising, 17, 20–21, 99, 106–8, 115–19, 121, 123, 125–29, 244, 293–94, 297; American, New World, 9, 13, 54–55, 67, 82, 131, 133, 136, 138, 140–42, 144, 150, 168, 184–85, 188–89, 202, 221; in antiquity, 22, 160, 311; appeal of, 234; Brahmin, 244; buying and selling, 10, 16–18, 21–24, 27, 34–35, 57, 68, 79, 84, 89, 107, 114, 117–18, 122–23, 160, 164, 189, 193–94, 203, 205–6, 209–10, 222–24, 226–28, 234, 238, 246–47, 251–52, 259, 295, 321; chemical, chymical, alchemical, 81, 87, 98, 141–42, 159, 174, 205–6, 217–19, 224, 241, 246, 249, 329; Chinese, 263; classification, 87; compound, 92,157, 173–74, 177, 205, 216, 226, 250, 322; collections, 206; as commodities, 13, 18, 23–24, 76, 82, 106, 222–24, 247, 310; consumption, 10, 57, 59, 73, 84, 210; curiosities, 184, 189, 191–93, 223, 225; and discovery, 196; display of, 26, 28, 205; Eastern or Oriental, 9, 13; European, 19–20, 34–35, 50, 57, 68, 76, 83–84, 97, 102, 132, 138, 186; exotic or foreign, 9, 12, 14, 28, 35, 37, 51, 54–55, 58, 60, 84–85, 88, 98, 101–2, 114, 141, 151, 162, 213, 215, 228, 250, 273, 308; foods and, 59, 203, 206; Galenic, 19, 57–60, 67–68, 72, 74–76, 79–81, 85, 92, 102, 141, 154, 157, 159–61, 163, 165, 168, 174, 205; gifts, 184, 189, 193; global trade, 11–19, 34–35, 50, 60, 79–80, 84, 89, 102, 153–54, 193–94, 206, 219, 231, 250, 320; identification, 22, 87, 91; importation, 6, 8, 12, 14, 28, 54, 57–58, 68, 72, 75–76, 84, 89, 96–97, 101–2, 123, 137, 144, 151, 153, 155, 157, 184, 225, 233, 263; indigenous, 75–76, 78, 102, 186–87, 202, 253, 265, 324; inventories, 67–69, 76; Levantine, 11, 22, 273; knowledge, 17, 23, 27, 86–88, 91, 131, 189, 223, 252–53, 332; manufacture of, 24, 62, 89, 99–100, 218–19, 250–53, 261, 332; mobility, 15, 18, 29, 63–67, 71, 75, 105–6, 108, 231, 252; new, 6–7, 9, 12, 15, 21, 24–25, 28, 76, 83, 86–87, 89–90, 96–98, 105, 131, 133, 141, 186, 189, 204–5, 224–27, 266; non-European, 4, 7, 10, 14, 20; opposition to, 4, 6, 9, 14, 25–28, 102; Paracelsian, 68, 81, 224; Philippine, 74; Portuguese, 20; preservation, 63, 68, 145, 157, 165, 215, 217, 220, 222, 225, 246, 250, 298; in print, 133–38, 150–52, 299; processing, 51, 99, 142, 145, 168, 170, 192, 205–6, 218–19, 251–52; properties, 3–4, 6–8, 10, 12–14, 22–23, 25, 51, 53, 60, 75, 78–79, 90, 99, 108, 113, 131, 138–44, 149, 151, 126, 151, 157, 159, 163–69, 173, 184, 186, 197–200, 218, 232–34, 241, 243, 247–49, 261–66, 299, 311, 325, 332; regulation, 14, 37, 62, 93, 205–6, 217, 227, 324; secret, 17, 106–8, 117–19, 127–28, 193, 233–34, 237, 242, 251–52, 259, 265; Siddha, 248, 252; simple, 51, 65, 67, 75, 89–91, 102, 151, 157, 159–78, 194, 203, 205–6, 210–13, 215–16, 218, 220–25, 228, 243, 259, 265, 299, 321–22; slave, 236; substitution, 14, 22, 58, 74–76, 80, 82, 87, 96–97, 125, 132, 221, 233, 266, 289, 299; supply, 8, 11, 18, 58, 63–66, 71, 73, 84, 128, 217, 225, 228; testing, assaying, 9, 16, 60, 80, 91–92, 113–14, 160, 184, 192, 194, 199–200, 215–17, 243, 246, 249; value, price, 24, 100, 141, 219–22, 246, 250, 253; writings on, 18, 20, 23, 28, 57, 81, 88–93, 120–21, 133, 158, 160–64, 170, 188, 194, 259

INDEX

Du Chesne, Joseph, 99
Du Laurens, André, 143
Dutch Republic. *See* Netherlands

Earle, Rebecca, 44, 58–59
East. *See* Orient
East Indies, 10, 14, 79, 212, 238, 240, 256, 258, 261–62, 264, 328
Easterby-Smith, Sarah, 29
Ecuador, 197
Egypt, 28, 86, 90–93, 100–101, 160, 212, 305
electuaries, 65, 142, 145, 175, 177–78, 205, 219
élites, 6, 10, 37, 47, 88, 193, 197, 258–59, 324, 333
elixirs, 70, 173, 188, 206, 218–19
emetics, 26, 79–80, 119–20, 311
empires, 23, 53, 56; Chinese, 41; classical, 10, 27, 86, 92, 154, 160, 164; colonial, 12, 17, 267; Dutch, 88, 97, 99, 101; English, 132, 135; European, 13, 15–17, 19; French, 132, 204; Islamic, 157, 161, 168, 307–8; Mughal, 10, 51, 274; Ottoman, 10, 13, 19, 48–49, 88, 100–102, 212, 274; Portuguese, 17, 88, 96–97, 99, 101, 131, 212; Russian, 19, 34, 39–45, 48–50, 55–56; Safavid, 10, 274; Spanish, 17, 22, 44, 57, 60–61, 63, 67–68, 71–73, 81, 90, 108, 125, 131, 156, 158, 161, 168, 188, 212, 262. *See also* colonies
enemas, 120, 174, 177–78
England, 3–4, 19, 86, 93, 132–36, 140, 148, 150–51, 203, 225, 243, 258, 294, 300, 324
English, 53, 61, 100, 111, 114, 118, 132–33, 135, 225, 273, 285, 317, 319
environments. *See* climate
epidemics, 21–22, 95, 106–9, 111, 113–17, 119–22, 125–29, 185, 258, 296–97
essences, 99, 173, 177–78, 205
ethnicity, race, 19, 43–45, 47, 49–50, 59, 61–63, 73, 83

euphorbia, 112, 225
Eurocentrism, 10, 13, 27, 268
evacuation, 73–74, 139, 143, 146. *See also* purgatives
evaporation, 173, 177
Evelyn, John, 144
exchange, 12, 15, 41, 84, 160–65, 189–92, 194, 198, 202, 231, 234–38, 251–53, 256–57, 269, 307–9. *See also* drugs: buying and selling; gifts
exoticism, 5–6, 8–29, 33–41, 43–45, 47, 50–56, 58, 84–92, 102, 114, 141, 146, 150, 152–53, 155, 157, 159, 162–63, 165, 168, 170, 178–79, 187, 189, 192, 194, 204, 215, 228, 234, 250, 256–60, 264–66, 268, 272–73, 276, 288, 302, 308
extraction: drug processing technique, 145, 171–72, 177, 218; resource, 19, 25, 76, 79, 87, 184, 197, 265–68
extracts, 27, 177–78, 222

Fabri, Honoré de (pseud. Antimus Conygius), 201
Fagon, Guy-Crescent, 113, 295
febrifuges, 21, 81, 108, 117, 125–26, 154, 195–96, 199, 202, 247, 295, 312
fennel, 66, 150, 162, 166–67
fenugreek, 65, 162–63
fevers, 8, 21, 106–8, 113–22, 121–29, 133, 145, 147, 149, 166, 168, 184, 188, 194, 196–98, 200–201, 246–47, 249, 262–63, 296–98, 330
Florence, 94, 227, 317
Florida, 35, 54, 132, 135
flowers, 65, 69, 132, 165, 219, 224, 255, 272, 321
foreignness, 6, 9–10, 12, 14–15, 22, 24, 34–35, 37–39, 44, 50–51, 53–55, 58, 60, 68, 71–72, 75–76, 78, 80, 83, 85–92, 95, 98, 101–2, 141, 154–57, 159, 163, 165, 168–69, 178, 213, 234, 256, 260, 269, 302, 308. *See also* domestication; drugs: exotic; exoticism; indigeneity; "otherness"

INDEX

formularies, 156–59, 173–75, 177, 311.
  See also pharmacopeias; recipes
France, 3–4, 6–8, 11–13, 19, 26–28, 33, 46, 85–86, 93–95, 109, 114, 144, 197, 203–29, 273, 275, 294, 299, 320, 324.
  See also empire: French
frankincense, 10, 162, 166
Frederick II, Holy Roman Emperor, 246
French, 7, 11–12, 100, 109–10, 113–14, 139, 143, 225–27, 273
friendship, 190, 193, 199, 238, 242–44.
  See also gifts; patronage
Fuchs, Leonhart, 154

galbanon, 223
Galen, Claudius, 85, 133, 141, 154, 158, 174, 233
Gannibal, Avram Petrovich, 45–47
Garcin, Laurent, 233–53, 325, 327–32
gardens, 11–12, 89, 215, 217, 224, 228; botanical, 17, 22, 86, 154; Garden of Eden, 10, 273
gargles, 174, 177–78
Geneva, 99, 241–42, 244, 331
Geoffroy, Étienne François, 113, 226
Gerard, John, 142, 154
German Lands, 38, 86, 93, 113–14, 259, 262, 301–2
Gesner, Konrad, 154
gifts, 23, 100–101, 184, 189–94, 198–99, 232–36, 238, 253
ginger, 9, 162, 164, 167, 213, 320
ginseng, 8, 13, 204, 267, 299, 319
globalization, 18, 27, 29, 60, 84, 160, 204
Gómez, Pablo F., 15
Greece, 91–92, 160, 305–6
Greiff, Frederic, 99, 292
grocers, 10, 23–24, 203–6, 209–13, 215–20, 223–28, 230–32, 259, 319–21
gout, 5, 138, 147, 149, 163, 167, 301
guaiacum, 21, 24, 51–52, 54, 70, 81–82, 88, 135–37, 139–44, 149, 151, 154, 168–69, 188, 199, 205, 213, 221, 300, 322
Guatemala, 188

guilds: corporations 23–24, 38, 123, 203, 205–6, 209, 219, 222–24, 226, 228–29, 319–20, 324
Gujarat, India, 241
gum arabic, 162, 167, 213, 310, 325
gums, 65–66, 69, 157, 162–63, 165–69, 175, 177, 213, 224, 309–10, 325
Gunn, Geoffrey, 237

Hakluyt, Richard, 142–43
Hartman, John, 142
Harvey, Gideon, 138, 146, 301
healers: medical practitioners, 11–12, 21, 23, 27, 55, 96, 185–86, 206, 223, 237, 267
healing, 10, 21, 27, 51–52, 61, 73, 78, 83, 185–86, 320; Ayurvedic 97, 161, 237–38, 325; "Brahmin," 24, 233–34, 236–38, 244, 247, 251, 253, 325; Chinese, 4, 41, 51, 163–64, 233, 237–38, 253, 263–64, 267, 308, 335; Galenic, 22, 25–27, 44, 57–61, 68, 71–73, 75–76, 79–84, 87, 92, 96, 132, 137, 139–41, 145–46, 151, 154, 163, 201, 247, 262, 264, 266, 276, 301, 306–7; Persian 233; Siddha 24, 236–37, 249, 251–52, 325; Indigenous, 82, 156, 168, 185–87, 234–36, 244, 253, 287; Filipino, 72–76, 78, 80, 83, 283, 285, 287; Islamic, 68, 81, 154, 157, 161, 168, 178, 305–6, 309; magical 61–62, 72, 78, 186, 283; Mesoamerican, 13, 82; Native American, 52, 132; Siddha Tamil 24, 234, 237–38, 246, 248–50, 252, 325–26. *See also* pharmacy
health, 3, 6, 9, 11, 16, 27, 51, 53, 61, 72, 75, 94–95, 143–44, 146, 174, 184–86, 193, 206, 209, 237, 252
hellebore, 70, 162, 165
Helmontianism, 140
hemp-agrimony, 162
Heng, Geraldine, 44–45, 47, 50
herbalism, 24, 54, 91, 215, 217, 238, 321
herbals, 39, 52–53, 133, 154, 158, 224, 308

INDEX

herbaria. *See* collections
hermodactyl, 66, 162–63
Hernández, Francisco, 82, 154
Hippocratic Corpus, 3, 133, 158, 160, 174, 307
Holy Land, 10
hospitals, 58, 60–64, 67–69, 71–75, 81–83, 97, 100, 154, 194–95, 198–99, 209, 317
humanism, 4, 9, 26, 91, 99
humoral medicine. *See* healing: Galenic
hyssop, 162, 165–66, 215

Ibn Rushd (Averroes), 158
Ibn Sahl, 158, 175
Ibn Sīnā (Avicenna), 158, 175
igasud, 78–80
Ignatius of Loyola, 78, 189
incense, 10, 162–63, 165–66, 168–69, 174, 178, 225, 263
India, 6, 22, 51–52, 83, 86, 91–92, 154, 161, 164, 166–68, 175, 212, 237–38, 241, 247, 258, 263, 325. *See also* Asia: South; East Indies; empires: Mughal
Indian Ocean, 164
Indians (*Indios*). *See* Amerindians; indigenous: peoples
indigenous agency, 33–34, 78, 236, 253; bodies, 59–60, 72, 80; cultures, 183, 186, 285; drugs, 14, 19, 53, 76, 78, 82, 84, 86, 96–97, 168, 215, 217–18, 234, 248, 253; languages, 74–75, 78, 186; knowledge, 61, 78–79, 81, 96, 156, 185–86, 234–35, 244, 253, 287; peoples 59–61, 72, 78–80, 132, 183; territory, 61. *See also* healing
indigeneity, 11, 20, 27, 85–86, 89, 102, 215, 273. *See also* exoticism
Indonesia, 39, 163–64, 168, 240–41, 248. *See also* Asia: Southeast
infusions, 64, 146, 170, 173, 175, 178, 188, 200
intoxicants/intoxication, 8, 11. *See also* psychedelics
invention: discovery, 6, 29, 57, 75, 78, 89, 92, 108, 117, 194, 196–98, 226, 233, 236, 241, 244, 251
inventories, 24, 67–69, 76, 81, 186, 190, 193, 203–6, 210, 213, 219, 221, 223–26, 228, 230, 318, 320, 322
ipecacuanha, 168–69, 186, 199, 204, 223, 225–26, 274, 297, 313, 319
Iran. *See* Persia; empires: Safavid
iron, 249–50, 252
Islam: cities, 154; cultures, 48–49, 178, 305, 309; knowledge, 68, 306; expansion, 154, 161; religion, 11, 49. *See also* empires: Islamic; Muslims; pharmacy
Islamophobia, 48–49
Istanbul, 100–101
Italy, 85–103, 154, 156, 158, 183, 198, 201, 223, 306
Ivernois, Jean-Antoine de, 243–44

jalap, 9, 13, 51–52, 81–82, 154, 169, 205
Java, 168, 241. *See also* Batavia
Jesuits, Society of Jesus, 8, 18, 20, 23–24, 57–61, 72–79, 96–97, 102, 183–202, 225, 232–33, 259, 264, 312, 315, 318, 334
Jesuits' bark. *See* cinchona
Jews, 11, 50, 91. *See also* religion: Judaism
juleps, 178

Kamel, Georg Joseph, 58–60, 72–83, 282
Kircher, Athanasius, 187, 189
Kivelson, Valerie, 36, 43, 56

La Condamine, Charles Marie de, 113–14
La Mettrie, Julien Offray de, 120
laboratories/workshops 6, 73, 98, 171, 216–19. *See also* pharmacies
lambatives, 175, 177
laurel, Alexandrian, 249. *See also* bay
Lausanne, 244, 246
lavender, 34, 64, 70, 162
Le Fèvre, Nicaise, 145–46
lead, 249

INDEX

leaves, 6, 65, 70, 83, 132, 142, 165, 224, 297, 323, 325
Le Fèvre, Nicaise, 145–46
Leiden, 115–16, 121, 129, 240, 296
Lémery, Nicolas, 109–10, 156, 228
Leopold Wilhelm of Austria, Archduke, 201
letters. *See* correspondence
Levant, 10–11, 13, 100, 154, 164, 166, 211–12, 216, 220, 273
licorice, 86, 89, 162, 166, 175, 213, 218, 222
lily, 70, 249
liniments, 165, 174, 178
Linnaeus, Carolus, 112, 243; 266–67. *See also* names
L'Obel, Matthias, 224
Loja, Peru, 108, 125, 184, 194, 196–97, 199–200
London, 48, 73, 145–46, 154
Louis XIV (King of France), 5, 109, 113, 197, 239, 327
Lugo, Juan de, Cardinal, 190–91, 194–95, 199, 314, 317
luxury, 10–11, 101, 226

mace, 34, 50, 162, 164, 309
Madrid, 156, 158, 188, 190, 192–93, 198–200, 306, 316
magic. *See* healing: magical
Malabar Coast, India, 12, 164, 233, 241, 244
mallow, 162, 166
Manfredi, Antonio, 93–95
Manila, 12, 19–20, 57–64, 67–75, 81–84, maps. *See* cartography
marshmallow, 162, 166, 213, 215
mastic, 65–66, 132, 137–38, 151, 162, 165–66
materia medica. *See* drugs
Maynwaring, Everard, 141
Mazarin, Jules, 5
McCants, Anne, 203
mechoacan, 66, 81–82, 135–36, 141, 151, 154, 169, 223

medical faculties, 4–7, 26–28, 201, 205, 246
medical marketplace, 6, 9, 12, 14, 18, 23, 26, 93, 96, 98, 100–101, 107–8, 116–17, 123, 125–27, 129, 133, 137–38, 145, 150–51, 184, 188, 192, 197–202, 205–6, 217–18, 234, 238, 265
medical practitioners. *See* healers
medicine. *See* healing
Mediterranean, 10, 20, 22, 26–27, 86, 100, 133, 155–57, 160–62, 164–67, 169, 178–79, 212, 225, 248, 253, 282, 306
merchants, 18, 24, 86, 89, 194, 255–56; Armenian, 11, 53; Asiatic, 13, 53; Batavian, 242; Chinese, 13; Dutch, 109, 129, 225, 294, 328; English, 11, 48, 225; European, 11–12, 87, 299; French, 11, 27, 100, 205 6, 210, 213, 215, 219–30, 320; Genoan, 11; Jewish, 11; Indian, 53; Ottoman, 13; Persian, 263; Portuguese, 11, 232, 273; Spanish, 190, 199, 202, 225; Swiss, 244; Venetian, 11
mercury, 141, 219, 237, 241, 246, 255, 326
Mercury (classical deity), 255–56, 265
Messina, 88, 92, 94
Mesue (Yūḥannā Ibn Māsawaih) or Pseudo-Mesue, 68, 158, 170, 175, 177, 307
Mexica, Aztecs, 52, 82–83
Mexican acacia. *See* cuahmochit
Middle Ages, 10, 13, 15, 28, 44–45, 50, 86, 88, 93, 133, 154, 156–58, 161, 163, 165, 167–68, 170–71, 173–75, 177–79, 193, 209, 259, 305–6, 309, 311, 322, 334
Milan, 88, 193
missionaries, 8, 12, 18, 23–24, 57, 60–62, 76, 78, 87–88, 183, 185–94, 202, 250, 318
minerals, 26, 81, 157, 160–61, 173, 244, 248–49, 255, 311, 325
mithridate, 96, 98–99, 193

INDEX

Monardes, Nicolás, 81, 132–33, 142–43, 148–50, 154, 168–69, 233
Morel, Pierre, 144
Moscow, Muscovy, 18, 34–45, 48–50, 53–56
musk, 192, 233
Muslims, 45, 48–49, 55, 287, 305. *See also* Islam
Musschenbroek, Pieter van, 242–43
myrrh, 10, 65, 162, 166–67, 206

names/nomenclature (of drugs), 21, 54, 81, 253, 262, 264, 300, 320; indigenous, 83, 253; Linnean, 261, 320
Naples, 94, 183
Nappi, Carla, 8
Native Americans. *See* Amerindians
Netherlands, 6, 21, 38, 52, 97, 99, 106–7, 114, 119, 201, 238–39, 242–43, 247, 262, 327. *See also* empires; merchants: Dutch
networks, 11–12, 18, 20, 22–24, 60, 72, 76, 84, 86, 89, 94, 118, 151, 153, 190, 200, 202, 212, 215, 227, 244
Neuchâtel, 24, 233, 242–46, 329
newspapers/journals, 16, 21, 107–8, 117, 123–24, 128, 242–44, 246, 260, 294–95, 298
Nieremberg, Juan Eusebio, 187, 314
Norton, Marcy, 13
Nuremberg, 94, 99
nutmeg, 9, 34, 48, 50, 53, 132, 136–37, 151, 162, 164, 168, 213, 221, 309

Occidentalism, 33–34, 39
Ogilvie, Brian, 154
oils, 51, 64, 142, 145, 151, 157, 165, 174–75, 177–78, 205–6, 218–19, 224
ointments, 64, 69, 157, 165, 174–75, 178. *See also* unguents
opium, 9, 86, 89, 98–99, 106, 119, 165, 223, 293, 297
opobalsam, 85, 90, 92–96, 102
opopanax, 162, 166
Oribasius, 158

Orientalism, 10–11, 13, 33–35, 39, 49–50, 55, 178–79, 256–57, 262, 264, 267–69, 273–74, 333, 336. *See also* exoticism; Occidentalism
"otherness," 11, 33–38, 49–50, 52–53, 55, 267
Ovalle, Alonso de, 187, 193, 198
Oviedo, Gonzalo Fernández, 187

Padua, 47, 94–95
Palacios y Bayá, Félix, 22, 156–59, 161, 163, 177, 306
Panutio, Vincenzo, 93–94
Paracelsus, 15, 141, 301–2
Paracelsianism, 26
Paraguay, 193
pareira brava, 225–26
Paris, 4–7, 18, 23–24, 27, 45, 201, 203–12, 214–20, 223, 226–28, 230, 243, 276
Patin, Guy, 4–9, 15, 25–28, 201, 276
patronage, 5, 29, 63, 93, 97, 190, 192, 201, 225, 242, 324
Paul of Aegina, 158, 174
Paulli, Simon, 7
pepper, 82, 162, 164, 213, 222, 308. *See also* chili pepper; cubebs
perfumes, 163, 165, 174, 178
Persia, Persians, 11, 48, 52, 100, 164, 166–67, 212, 233, 241. *See also* empires: Safavid
Peru, 58, 108–9, 111, 125, 183, 187, 190–91, 193–94, 196–97, 262, 316
Peruvian balm, 169, 188
Peruvian bark. *See* cinchona
pessaries, 174, 177
Peter the Great (Tsar of Russia), 36–37, 45, 54
pharmacies/apothecary shops, 10, 12, 26–28, 58, 62–63, 67–68, 71, 73–74, 76, 88, 90, 94, 97, 99, 106, 117, 154, 188–89, 191, 193, 198, 205–6, 209–10, 223–24, 227–28, 239, 263, 265, 312, 320–21. *See also* drugs: buying and selling

INDEX

pharmacopeias: Chinese, 267; classical, 262; Dutch, 97, 126, 129; English, 133; European, 60, 71, 88, 156, 159, 169, 175; French, 3, 8, 16, 221; Galenic, 188; German, 38; Islamic, 161, 167–68; Italian, 88; Latin, 38, 68; Roman, 87–89, 91–92, 96; Russian, 54; Spanish, 60, 67–68, 81–82, 156; Sri Lankan, 97; Strasbourg, 99; pharmacy, European, 58, 87, 154, 177, 179; Galenic, 10, 19–21, 38, 85–87, 92, 96–99, 101–2, 154–75, 177–79, 188, 205, 266, 282, 289, 299, 305–6, 311; Islamic 81, 168, 305; polypharmacy 26, 28, 157, 174; Spanish, 156, 178; Western, 22, 174, 178

Philippines, 19, 57–84, 188, 212

physicians, 4–9, 26–27, 38, 47–48, 51–52, 60, 62–63, 73, 79, 85–99, 102, 106–7, 109, 113, 117–21, 127–28, 132, 138–44, 149–51, 154, 158, 186, 188, 194, 196–201, 217–18, 223, 226, 228, 232, 234, 243–46, 248, 250, 252–53, 255–56, 263, 294–95, 318, 325, 328–29. *See also* healers

pills, 65, 70, 142, 175, 177–78, 218; Maduran, 24, 231, 233–35, 242–44, 246–53, 330

plague/pestilence, 37, 85

plants, 4, 6, 9–10, 14, 17, 22–24, 39, 48, 51–54, 58, 72–75, 78–80, 82–83, 86–87, 89, 91–92, 96–97, 110, 112–13, 132, 138–40, 145–46, 152, 154, 157, 160–62, 166–69, 178, 183, 185–87, 189, 197, 204–5, 210–25, 228, 238, 240–42, 244, 248–50, 253, 255, 259–61, 263, 299, 301, 308, 311–12, 320, 323, 328, 332; cultivation 12, 17, 22, 96, 132, 138, 154, 160, 166–67, 215, 308; transplantation 12, 162–63, 306

plasters, 65, 157, 165, 174–75, 178

polychrest salt, 119–20, 219

polypody fern, 66

pomegranate, 69, 162, 297

Pomet, Pierre, 12–13, 109–10, 113, 210, 212, 215, 219, 223–24, 226, 320

popes/Papacy, 93–95, 101–2, 191, 315

Portugal/Portuguese, 19, 61, 89–90, 131, 226, 232–33, 239, 264, 273, 275. *See also* drugs; empires; merchants: Portuguese

powders, 65–66, 80, 142, 145, 175, 177–78, 196, 199, 226, 249

pox, 81–82, 137, 139, 141–42, 145, 147–50, 185, 188, 221. *See also* venereal disease

prescriptions, 48, 50, 54–55, 62, 79, 129, 137, 140–41, 145, 151–52, 163, 166, 218, 242, 247

preserves, 65, 69, 175, 177–78, 220

Prieto, Andrés, 187

print, 15–17, 21, 68, 78, 93–95, 107, 110, 114, 131–35, 146–47, 149–50, 152, 158, 161, 187–88, 193, 206, 210, 224, 233, 259, 299–300

Protestantism, 18, 201, 238–39, 242, 250–52, 327, 329

Pugliano, Valentina, 223, 299

purgatives, 27, 65, 79–82, 120–21, 136, 139, 143, 154, 162–63, 166, 168–69, 200, 222, 249, 263–64, 311

purging nut/Indian physick nut, 225

quamochil. *See* cuahmochit
quinine, 8, 262, 272

race. *See* ethnicity

Raleigh, Walter, 144, 146, 303

recipes/recipe books, 16, 20, 23, 26, 54–55, 86, 88, 90–93, 97, 99, 133, 144–46, 152, 159, 175, 177, 184, 188, 192, 194, 199–202, 210–13, 220, 232, 246, 249, 292, 297, 303, 311

regimen, 173, 243

religion, 13, 43, 49–50, 59, 61–63, 71, 75, 78–79, 94, 105, 166, 169, 183–87, 189–90, 192–93, 198–99, 212, 233–34, 244, 247, 299, 325, 329; Christianity, 10, 2045, 49–50, 60–61,

INDEX

79, 86, 89, 93, 100, 166, 185–87, 247; Franciscan, 187, 190; Jews, 11, 50, 91; Oratorian, 210–13; Orthodox, 45, 50, 166; Pietism, 250–52; shamans, 23, 45. *See also* Catholicism; Jesuits; Jews; missionaries; Protestantism

remedies. *See* drugs

Remezov, Semyon, 40, 42–43, 56

Republic of Letters, 236, 242–43, 251

resins, 65, 81–82, 93, 145, 163, 165–67, 169, 175, 191, 311

al-Rāzī (Rhazes), 158

rheumatism, 163, 252

rhubarb, 19, 41, 70–71, 106, 141, 213, 263–64, 267, 297, 320, 335

Rivière, Lazare, 139

Roëll, Willem, 120

Romaniello, Matthew P., 44

Rome, 18, 57, 76, 88–96, 101–2, 183, 185, 188, 190–95, 197, 199–200, 315, 317

roots, 24, 66, 69–70, 82, 85, 119, 132, 136, 138–39, 142–45, 149, 154, 164–69, 186, 213, 224–25, 249, 255, 261–64, 272, 297, 313, 321, 323, 325, 335

roses, 64–66, 68–69, 162, 165–66, 213, 218

rosemary, 34, 66, 162, 165

Rotterdam, 109, 115, 294, 296

Rousseau, François, 218, 223, 230

Royal Society, London, 48, 144, 243–44, 251, 329

rue, 162–63

Rumpf, Isaac Augustijn, 233–36, 326

Rumpf, Rumphius, Georg Eberhard, 248, 326, 328

Russia, Russians, 16, 19, 34–56, 192. *See also* empires: Russian

sage, 13, 69, 162–63, 213

Sahagún, Bernardino de, 187

Said, Edward, 12–13, 33–36, 39, 43, 49, 256–57, 267–68. *See also* Orientalism

Salerno, 158

salts, 81, 118–20, 125–26, 142, 151, 177–78, 205, 219, 255

Salvador i Riera, Joan, 239–40, 243

sandalwood, 71, 167–68, 297

sarsaparilla, 54, 66, 70, 81–82, 88, 139–40, 142–45, 149, 154, 168–69, 213, 225, 261–62, 299, 323

sassafras, 21, 34, 50–51, 54, 56, 81–82, 131–52, 169, 225, 299–300

scammony, 65, 162–63, 206

Schmidt, Benjamin, 5, 33–36, 39–40, 42–43, 52, 55

Seba, Albertus, 110, 114, 129, 295

seeds, 66, 69–70, 78, 149, 164–65, 224, 255, 308, 316, 321, 325

Séguier, Pierre, 3, 5

Sennert, Daniel, 138–39, 143, 302

Serapion the Younger, 60, 79–80, 287

Seville, 13, 57, 158, 190–93, 196–200, 316

shops, 6–7, 10, 12, 26–28, 58, 68, 76, 88, 90, 94, 96–97, 99, 106, 117–18, 154, 188–89, 198, 205–6, 209–10, 215, 218–20, 223–24, 227–28, 240, 259, 320–21. *See also* drugs: buying and selling; pharmacies

Siberia, 39–43, 45, 53, 55

Sibille, Mathurin, 215, 217

Siena, Kevin, 24

silk, 11, 260

Silk Roads, 11, 19, 22, 160, 164

snakebite, 165, 167–68

soldiers, 12, 62, 73, 97, 183, 186, 194

Spain, 16, 19, 61, 68, 71, 92–93, 122–23, 125, 154, 156, 158, 175, 178, 183, 191, 196–98, 200, 226–27, 239, 275, 298, 306. *See also* empires: Spanish

specimens. *See* collections; drugs

spermaceti, 34, 50

Spice Islands (Moluccas, Maluku), 34, 39, 57, 164, 167

spices, 10–11, 15, 34, 39, 53–57, 101, 163–64, 166–68, 206, 220, 222, 247, 255, 259–60, 273, 308

Spon, Charles, 4–5

Sri Lanka, 20, 24, 97, 233–34, 241, 248, 326

St. Ignatius bean, 20, 57–58, 60, 71, 77–78, 83, 287
stavesacre, 225
sublimation, 159, 172–73, 177
Subrahmanyam, Sanjay, 18, 43
sudorifics, 136–39, 141, 151, 165, 168
sugar, 13, 48, 65, 120, 157, 175, 177–78, 192, 206, 213, 218, 220, 222
Sumatra, 168
suppositories, 174, 177–78
Syria, 86, 90, 93, 212
syrups, 64, 69–70, 74, 145, 157, 175, 177–78, 206, 218–19, 224

tacamahaca, 66, 81–82, 169, 191, 311
Talbor, Robert, 202
tamarind, 83, 162–63, 167, 225
tanglat, 73–75, 81, 285
tea, 3–9, 12, 29, 84, 120, 203–4, 226–27, 247, 260, 263, 267, 271–72
temperaments. *See* constitutions
texts, 4, 10, 18, 20–22, 29, 37–39, 43–44, 47–49, 52–55, 74, 88, 94, 96, 108–9, 131–38, 142–43, 145–52, 154, 156, 158, 161, 164–65, 170, 174–75, 184, 187–89, 192, 196, 200–202, 234, 242, 256, 258–59, 299–300, 304, 307, 311. *See also* drugs: advertising; drugs: writings on; print
theriac, 20, 26, 28, 65, 85–102, 193, 205, 262, 276, 288–90, 292; *Theriaca cœlestis*, 20, 98–99, 101; *magna*, 86, 96, 98–99, 292; *Triaga brasilica*, 20, 96–97, 292
tinctures, 177–78, 205
tobacco, 9, 13, 29, 84, 109, 123, 138, 151, 154, 168–69, 192, 218, 220, 274, 315
Tobolsk, Siberia, 40–41, 43, 56
Toledo, 158, 314
toothache, 85, 168–69, 304
Tournefort, Joseph Pitton de, 224, 226, 243
toxins. *See* poisons; antidotes
trade: commerce, 6–7, 10–26, 38, 41, 44, 48, 53, 58, 60, 71, 76, 84, 86, 88–89, 92–93, 102, 106, 110, 114, 117–18, 122–23, 125–26, 128–29, 131–32, 136, 141, 150–54, 160–61, 163–64, 179, 184, 189, 191–94, 199, 202–6, 209, 212, 217, 219, 221, 224–28, 231, 234, 237–38, 240, 242, 244, 246, 250, 252–53, 256–68, 273, 294–95, 298, 322, 335–36; monopolies, 8, 11, 24, 219, 221, 226–27, 233, 261, 324. *See also* drugs; merchants
trading companies, 22–24, 251; Dutch, 4, 23–24, 97, 221, 233, 238–40, 242, 244, 273, 326, 328; English, 318
tragacanth gum, 66, 162, 167, 310
translation and language, 18, 26, 35–39, 48–50, 52–55, 60, 73, 75–76, 80, 82, 85–86, 88–92, 95, 108, 125–26, 132–34, 138–40, 142–45, 148, 154, 156–58, 161, 173, 175, 179, 186, 190, 201, 212, 236–37, 250, 252, 256, 262, 288, 302–3, 306, 311, 313, 320, 325, 333
trituration, 159, 170, 173, 248–49
Tronchin, Theodore, 121, 126
turbith, 225
Turks, 47–49, 100–102
turpentine, 64, 162, 165

unguents, 64, 69, 206, 219. *See also* ointments
United Provinces. *See* Netherlands
universities, 4–7, 12, 22, 26–28, 38, 158, 201, 205, 242, 246

Valentini, Michael Bernhard, 254–56, 259, 261
vander Linden, Johannes Antonides, 5–6
van Helmont, Jan Baptista. *See* Helmontianism
van Raat, Hendrik, 109–11, 294
van Ranouw, Willem, 109–11, 114, 129, 294
van Rheede tot Drakenstein, Hendrik Adriaan, 244, 328
van Swieten, Gerard, 121

INDEX

vanilla, 183, 191–92, 198, 223, 226–28, 324
Vaud, 246
Vega, Juan de, 196–99, 316–17
venereal disease: pox 144, 148, 150, 154, 168–69, 304
venom. *See* poisons; antidotes
Venice, 11, 20, 88–90, 93–94, 100–102, 263, 273, 290
Venner, Tobias, 138
Vienna, 57, 119, 192
vipers, 26, 89, 206
Virginia, 132, 134–35, 144, 146
Viva, Francesco, 183–84, 197–98
VOC (Vereenigde Oostindische Compagnie). *See* trading companies (Dutch)
vomic nut, 60, 79–80, 225

Wallis, Patrick, 17, 154
wars, 16, 19, 94–95, 225, 239–40, 260, 265, 272, 291. *See also* soldiers
Warsaw, 192
waterlily, 162, 165, 218
waters, 177–78, 205–6, 218–19, 224
Wear, Andrew, 141
Winterbottom, Anna, 24
workshops. *See* laboratories; pharmacies

xylobalsam, 70, 91

al-Zahrāwī (Abulcasis), 158, 311
zedoary, 68–69, 167–68